AF556175

ENCYCLOPAEDIA OF STEM CELLS - 3

PLANT STEM CELLS

By

Dr. Amita Sarkar

Dept. of Zoology

Agra College

Agra (U.P.)

(India)

DISCOVERY PUBLISHING HOUSE PVT. LTD.

NEW DELHI-110 002

First Published-2009

ISBN 978-81-8356-407-6

Published by:

DISCOVERY PUBLISHING HOUSE PVT. LTD.

4831/24, Ansari Road, Prahlad Street,
Darya Ganj, New Delhi-110002 (India)
Phone: 23279245 • Fax: 91-11-23253475
E-mail: dphbooks@rediffmail.com
dphtemp@indiatimes.com
Website: www.discoverypublishinghouse.com

Printed at:

Sachin Printers, Delhi

Preface

The present title *Plant Stem Cells* is the amazing advancement of biotechnology. It provides the various fundamental aspects of stem cell technologies to be understood adequately. It has been compilated for graduate and undergraduate students, research scholars, teachers, practising biochemical engineers, biotechnologists, applied and industrial microbiologists, cell biologists and scientists involved in bioprocessing research and development. The basic concepts have been clearly explained and their functions are adequately highlighted. The presentation of the text is simple and systematic. The selection of chapters and topics is according to the specified syllabi of several Indian Universities and are so structured as to enable the student to move easily from the fundamental to the complex. It is our earnest hope that this title will be of great value to all our students.

To make the work more comprehensive and informative, the author has consulted many authoritative books, research journals, abstracts, monographs etc. He is grateful to all those great scholars whose work are cited or substantially reproduced.

There can be no claim to originality except in the manner of treatment and much of the information has been obtained from the books and scientific journals available in the different libraries.

The author expresses his thanks to his friends and colleagues whose continue inspirations have initiated him to bring out this book.

The author expresses his gratitude to Mr. Wasan and staff of M/s Discovery Publishing Pvt. Ltd. for their whole hearted co-operation in the publication of this book.

1

INTRODUCTION

Most methods of plant transformation applied to GM crops require that a whole plant is regenerated from isolated plant cells or tissue which have been generically transformed. This regeneration is conducted *in vitro* so that the environment and growth medium can be manipulated to ensure a high frequency of regeneration. In addition to a high frequency of regeneration, the regenerable cells must be accessible to gene transfer by whatever technique is chosen. The primary aim is therefore to produce, as easily and as quickly as possible, a large number of regenerable cells that are accessible to gene transfer. The subsequent regeneration step is often the most difficult step in plant transformation studies. However, it is important to remember that a high frequency of regeneration does not necessarily correlate with high transformation efficiency. This chapter will consider some basic issues concerned with plant tissue culture *in vitro*, particularly as applied to plant transformation. It will also look at the basic culture types used for plant transformation and cover some of the techniques that can be used to regenerate whole transformed plants from transformed cells or tissue.

PLANT TISSUE CULTURE

Practically any plant transformation experiment relies at some point on tissue culture. There are some exceptions to this generalisation, but the ability to regenerate plants from isolated cells or tissues *in vitro* underpins most plant transformation systems.

Plasticity and Totipotency

Two concepts, plasticity and totipotency, are central to understanding plant cell culture and regeneration.

Plants, due to their sessile nature and long life span, have developed a greater ability to endure extreme conditions and predation than have animals Many of the processes involved in plant growth and development adapt to environmental conditions. This plasticity allows plants to alter their metabolism, growth and development to best suit their environment. Particularly important aspects of this adaptation, as far as plant tissue culture and regeneration are concerned, are the abilities to initiate cell division from almost any tissue of the plant and to regenerate lost organs or undergo different developmental pathways in response to particular stimuli. When plant cells and tissues are cultured *in vitro* they generally exhibit a very high degree of plasticity which allows one type of tissue or organ to be initiated from another type.In this way, whole plants can be subsequently regenerated.

This regeneration of whole organisms depends upon the concept that plant cells can, given the correct stimuli, express the total genetic potential the parent plant. This maintenance of genetic potential is called '*totipotency*' Plant cell culture and regeneration do, in fact, provide the most compelling evidence for totipotency.

In practical terms though, identifying the culture conditions and stimuli required to manifest this totipotency can be extremely difficult and it is still a largely empirical process.

The Culture Environment

When cultured *in vitro*, all the needs, both chemical and physical, of the plant cells have to met by the culture vessel, the growth medium and the external environment (light, temperature, etc.). The growth medium has supply all the essential mineral ions required for growth and development. In many cases (as the biosynthetic capability of cells cultured *in vitro* may not replicate that of the parent plant), it must also supply additional organic suplements such as amino acids and vitamins. Many plant cell cultures as they are not photosynthetic, also require the addition of a fixed carbon source in the form of a sugar (most often sucrose). One other vital component that must also be supplied is water, the principal biological solvent. Physical factors, such as temperature, pH, the gaseous environment, light osmotic pressure, also have to be maintained within acceptable limits.

Plant Cell Culture Media

Culture media used for the *in vitro* cultivation of plant cells are composed three basic components:

1. Essential elements, or mineral ions, supplied as a complex mixture;

Table 1.1 Some of the elements important for plant nutrition and their physiological function.

Element	*Function*
Nitrogen	Component of proteins, nucleic acids and some coenzymes Element required in greatest amount
Potassium	Regulates osmotic potential, principal inorganic cation
Calcium	Cell wall synthesis, membrane function, cell signalling
Magnesium	Enzyme cofactor, component of chlorophyll
Phosphorus	Component of nucleic acids, energy transfer, component of intermediates in respiration and photosynthesis
Sulphur	Component of some amino acids (methionine, cysteine) and some cofactors
Chlorine	Required for photosynthesis
Iron	Electron transfer as a component of cytochromes
Manganese	Enzyme cofactor
Cobalt	Component of some vitamins
Copper	Enzyme cofactor, electron-transfer reactions
Zinc	Enzyme cofactor, chlorophyll biosynthesis
Molybdenum	Enzyme cofactor, component of nitrate reductase

2. An organic supplement supplying vitamins and/or amino acids; and
3. A source of fixed carbon; usually supplied as the sugar sucrose.

For practical purposes, the essential elements are further divided into the following categories:

1. Macroelements (or macronutrients);
2. Microelements (or micronutrients); and
3. An iron source.

Complete, plant cell culture medium is usually made by combining several different components, as outlined in Table 1.2.

Media Components

It is useful to briefly consider some of the individual components of the stock solutions.

Macroelements

As is implied by the name, the stock solution supplies those elements required in large amounts for plant growth and development. Nitrogen, phosphorus, potassium, magnesium, calcium and sulphur (and

Table 1.2 Composition of a typical plant culture medium.

Essential element	*Concentration in stock solution (mg/l)*	*Concentration in medium(mg/l)*
Macroelements		
NH_4NO_3	33000	1650
KNO_3	38000	1900
$CaCl_2.2H_2O$	8800	440
$MgSO_4.7H_2O$	7400	370
KH_2PO_4	3400	170
Microelements		
KI	166	0.83
H_3BO_3	1240	6.2
$MnSO_4.4H_2O$	4460	22.3
$ZnSO_4.7H_2O$	1720	8.6
$Na_2MoO_4.2H_2O$	50	0.25
$CuSO_4.5H_2O$	5	0.025
$CoCl_2.6H_2O$	5	0.025
Iron source		
$FeSO_4$ $7H_2O$	5560	27.8
$Na_2EDTA.2H_2O$	7460	37.3
Organic supplement		
Myoinositol	20000	100
Nicotinic acid	100	0.5
Pyridoxine-HCl	100	0.5
Thiamine-HCl	100	0.5
Glycine	400	2
Carbon source		
Sucrose	Added as solid	30000

carbon, which is added separately) are usually regarded as macro-elements. These elements usually comprise at least 0.1 % of the dry weight of plants.

Nitrogen is most commonly supplied as a mixture of nitrate ions (from the KNO_3) and ammonium ions (from the NH_4NO_3). Theoretically, there is an, advantage in supplying nitrogen in the form of ammonium ions, as nitrogen must be in the reduced form to be incorporated into macromolecules. Nitrates ions therefore need to be reduced before incorporation. However, at high centrations, ammonium ions can be toxic to plant cell cultures and uptake of ammonium ions from the medium causes acidification of the medium order to use ammonium ions as the sole nitrogen source, the medium needs to be buffered. High concentrations of ammonium ions can also cause culture problems by increasing the frequency of vitrification (the culture appears pale and 'glassy' and is usually unsuitable for further culture). Using a mixture of nitrate and ammonium ions has the advantage of weakly buffering the medium as the uptake of nitrate ions causes OH^- ions to be excreted.

Phosphorus is usually supplied as the phosphate ion of ammonium, sodium or potassium salts. High concentrations of phosphate can lead to the precipitation of medium elements as insoluble phosphates.

Microelements

These elements are required in trace amounts for plant growth and development, and have many and diverse roles. Manganese, iodine, copper, cobalt, boron, molybdenum, iron and zinc usually comprise the microelements, although other elements such as nickel and aluminium are frequently found in some formulations. Iron is usually added as iron sulphate, although iron citrate can also be used. Ethylenediaminetetraacetic acid (EDTA) is usually used in conjunction with the iron sulphate. The EDTA complexes with the iron so as to allow the slow and continuous release of iron into the medium. Uncomplexed iron can precipitate out of the medium as ferric oxide.

Organic supplements

Only two vitamins, thiamine (vitamin B_1) and myoinositol (considered a B vitamin) are considered essential for the culture of plant cells *in vitro*. However, other vitamins are often added to plant cell culture media for historical reasons.

Amino acids are also commonly included in the organic supplement. The most frequently used is glycine (arginine, asparagine, aspartic acid, alanine, glutamic acid, glutamine and proline are also used), but in many cases its inclusion is not essential. Amino acids provide a source of reduced nitrogen and, like ammonium ions, uptake causes acidification of the medium. Casein hydrolysate can be used as a relatively cheap source of a mix of amino acids.

Carbon source

Sucrose is cheap, easily available, readily assimilated and relatively stable and is therefore the most commonly used carbon source. Other carbohydrates (such as glucose, maltose, galactose and sorbitol) can also be used, and in specialised circumstances may prove superior to sucrose.

Gelling agents

Media for plant cell culture *in vitro* can be used in either liquid or 'solid' forms, depending on the type of culture being grown. For any culture types that require the plant cells or tissues to be grown on the surface of the medium, it must be solidified (more correctly termed '*gelled*'). Agar, produced from seaweed, is the most common type of gelling agent, and is ideal for routine applications. However, because it is a natural product, the agar quality can vary from supplier to supplier and from batch to batch. For more demanding applications, a range of purer gelling agents are available. Purified agar or agarose can be as can a variety of gellan gums.

Plant Growth Regulators

We have already briefly considered the concepts of plasticity and totipotency, The essential point as far as plant cell culture is concerned is that, due to this plasticity and totipotency, specific media manipulations can be used to direct the development of plant cells in culture. Plant growth regulators are the critical media components in determining the developmental pathway of the plant cells. The plant growth regulators used most commonly are plant hormones or their synthetic analogues.

Classes of plant growth regulators

There are five main classes of plant growth regulator used in plant cell culture namely:

1. Auxins;
2. Cytokinins;
3. Gibberellins;
4. Abscisic acid;
5. Ethylene.

Each class of plant growth regulator will be briefly looked at.

Auxins

Auxins promote both cell division and cell growth The most important naturally occurring auxin is IAA (indole-3-acetic acid), but its use in plant culture media is limited because it is unstable to both heat and light. Occasionally, amino acid conjugates of IAA (such as indole-acetyl-L-alanine and indole-acetyl-L-glycine), which are more stable, are used to partially the problems associated with the use of IAA. It is more common, though to use stable chemical analogues of IAA as a source of auxin in plant cell media. 2,4-Dichlorophenoxyacetic acid (2,4-D) is the most commonly auxin and is extremely effective in most circumstances. Other auxins available, and some may be more effective or 'potent' than 2,4-D in some instances.

Table 1.3. Commonly used auxins, their abbreviation and chemical name

Abbreviation/name	*Chemical name*
2,4-D	2,4-dichlorophenoxyacetic acid
2,4,5-T	2,4,5-trichlorophenoxyacetic acid
Dicamba	2-methoxy-3,6-dichlorobenzoic acid
IAA	Indole-3-acetic acid
IBA	Indole-3-butyric acid
MCPA	2-methyl-4-chlorophenoxyacetic acid
NAA	1-naphthylacetic acid
NOA	2-naphthyloxyacetic acid
Picloram	4-amino-2,5,6-trichloropicolinic acid

Cytokinins

Cytokinins promote cell division. Naturally occurring cytokinins are a large group of structurally related (they are purine derivatives) compounds. Of the naturally occurring cytokinins, two have some use in plant tissue culture media. These are zeatin and 2iP (2-isopentyl adenine). Their use is not widespread as they are expensive (particularly zeatin) and relatively unstable. The synthetic analogues, kinetin and BAP (benzylaminopurine), are therefore used more frequently. Non-purine-based chemicals, such as substituted phenylureas, are also used as cytokinins in plant cell culture media. These substituted phenylureas can also substitute for auxin in some culture systems.

Table 1.4. Commonly used cytokinins, their abbreviation

Abbreviation/name	*Chemical name*
BAP	6-benzylaminopurine
2iP (IPA)	[N^6 -(2-isopentyl) adenine]
Kinetin	6-furfurylaminopurine
Thidiazuron	1-phenyl-3-(1,2,3-thiadiazol-5-yl) urea
Zeatin	4-hydroxy-3-methyl-trans-2-butenylaminopurine

Gibberellins

There are numerous, naturally occurring, structurally related compounds termed '*gibberellins*'. They are involved in regulating cell elongation, and are agronomically important in determining plant height and fruit-set. Only a few of the gibberellins are used in plant tissue culture media, GA_3 being the common.

Abscisic acid

Abscisic acid (ABA) inhibits cell division. It is most commonly used in plant tissue culture to promote distinct developmental pathways such as somatic embryogenesis.

Ethylene

Ethylene is a gaseous, naturally occurring, plant growth regulator most commonly associated with controlling fruit ripening in climacteric fruits, and its use in plant tissue culture is not widespread. It does, though, present a particular problem for plant tissue culture. Some plant cell cultures produce ethylene which, if it builds up sufficiently, can inhibit the growth and development the culture. The type of culture vessel used and its means of closure affect the gaseous exchange between the culture vessel and the outside atmosphere thus the levels of ethylene present in the culture.

Plant growth regulators and tissue culture

Generalisations about plant growth regulators and their use in plant cell culture media have been developed from initial observations made in the 1950s There is, however, some considerable difficulty in predicting the effects of plant growth regulators: this is because of the great differences in culture response between species, cultivars and even plants of the same cultivar grown under different conditions.

However, some principles do hold true and have become the paradigm to which most plant tissue culture regimes are based.

Auxins and cytokinins are the most widely used plant growth regulators in plant tissue culture and are usually used together, the

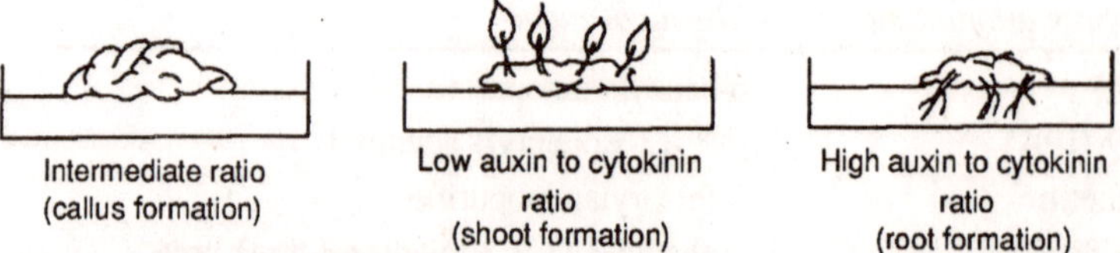

Fig. 1.1. The effect of different ratios of auxin to cytokinin on the growth and morphogenesis of callus.

ratio of the auxin to the cytokinin determining the type of culture established or regenerated. A high auxin to cytokinin ratio generally favours root formation, whereas a high cytokinin to auxin ratio favours shoot formation. An intermediate ratio favours callus production.

Culture Types

Cultures are generally initiated from sterile pieces of a whole plant. These pieces are termed '*explants*', and may consist of pieces of organs, such as leaves or may be specific cell types, such as pollen or endosperm. Many features of the explant are known to affect the efficiency of culture initiation. Generally, younger, more rapidly growing tissuc (or tissue at an early stage of development) is most effective. Several different culture types most commonly used in plant transformation studies will now be examined in more detail.

Callus

Explants, when cultured on the appropriate medium, usually with both an auxin and a cytokinin, can give rise to an unorganised, growing and dividing mass of cells..It is thought that any plant tissue can be used as an explant, if the correct conditions are found. In culture, this proliferation can be maintained more or less indefinitely, provided that the callus is subcultured on to fresh medium periodically. During callus formation there is some degree of dedifferentiation (i.e, the changes that occur during development and specialisation some extent, reversed), both in morphology (callus is usually composed of unspecialised parenchyma cells) and metabolism. One major consequence of this dedifferentiation is that most plant cultures lose the ability to photosynthesise. This has important consequences for the culture of callus the metabolic profile will probably not match that of the donor plant. This necessitates the addition of other components—such as vitamins and most importantly, a carbon source—to the culture medium, in addition to the usual mineral nutrients.

Callus culture is often performed in the dark (the lack of photosynthetic capability no drawback) as light can encourage differentiation of the callus

During long-term culture, the culture may lose the requirement for auxin and/or cytokinin, This process, known as '*habituation*', is common in callus cultures from some plant species (such as sugar beet).

Callus cultures are extremely important in plant biotechnology. Manipulation of the auxin to cytokinin ratio in the medium can lead to the development of shoots, roots or somatic embryos from which whole plants can subsequently be produced.

Callus cultures can also be used to initiate cell suspensions, which are used in a variety of ways in plant transformation studies.

Cell-suspension Cultures

Callus cultures, broadly speaking, fall into one of two categories: compact or friable. In compact callus the cells are densely aggregated, whereas in friable callus the cells are only loosely associated with each other and the callus becomes soft and breaks apart easily. Friable callus provides the inoculum to form cell-suspension cultures. Explants from some plant species or particular cell types tend not to form friable callus, making cell-suspension initiation a difficult task. The friability of callus can sometimes be improved by manipulating the medium components or by repeated subculturing. The friability of the callus can also sometimes be improved by culturing it on 'semi-solid' medium (medium with a low concentration of gelling agent).

When friable callus is placed into a liquid medium (usually the same composition as the solid medium used for the callus culture) and then agitated single cells and/or small clumps of cells are released into the medium. Under the correct conditions, these released cells continue to grow and divide, eventually producing a cell-suspension culture. A relatively large inoculum should be used when initiating cell suspensions so that the released cell numbers build up quickly. The inoculum should not be too large though, as toxic products leased from damaged or stressed cells can build up to lethal levels. Large cell clumps can be removed during subculture of the cell suspension.

Cell suspensions can be maintained relatively simply as batch cultures in conical flasks.They are continually cultured by repeated subculturing into fresh medium. This results in dilution of the suspension and the initiation another batch growth cycle. The degree of dilution during subculture should be determined empirically for each culture. Too great a degree of dilution result in a greatly extended lag period or, in extreme cases, death of the transferred cells.

After subculture, the cells divide and the biomass of the culture increase in a characteristic fashion, until nutrients in the medium are exhausted a toxic by-products build up to inhibitory levels—this is called the '*stationary phase*'. If cells are left in the stationary phase for too long, they will die and the culture will be lost. Therefore, cells should be transferred as they enter the stationary phase. It is therefore important that the batch growth-cycle parameters are determined for each cell-suspension culture.

Protoplasts

Protoplasts are plant cells with the cell wall removed. Protoplasts are most commonly isolated from either leaf mesophyll cells or cell suspensions, although other sources can be used to advantage. Two general approaches removing the cell wall (a difficult task without damaging the protoplast) can be taken—mechanical or enzymatic isolation.

Mechanical isolation, although possible, often results in low yields, poor quality and poor performance in culture due to substances released from damaged cells.

Enzymatic isolation is usually carried out in a simple salt solution with a high osmoticurn, plus the cell wall degrading enzymes. It is usual to use a mix of both celluse and pectinase enzymes, which must be of high quality and purity.

Protoplasts are fragile and easily damaged, and therefore must be cultured carefully. Liquid medium is not agitated and a high osmotic potential is maintained, at least in the initial stages. The liquid medium must be shallow enough to allow aeration in the absence of agitation. Protoplasts can be plated out on to solid medium and callus produced. Whole plants can be regenerated by organogenesis or somatic embryogenesis from this callus. Protoplasts are ideal targets for transformation by a variety of means.

Root Cultures

Root cultures can be established *in vitro* from explants of the root tip of either primary or lateral roots and can be cultured on fairly simple media. The growth of roots *in vitro* is potentially unlimited, as roots are indeterminate organs. AlthoughI the establishment of root cultures was one of the first achievements of modern plant tissue culture, they are not widely used in plant transformation studies.

Shoot Tip and Meristem Culture

The tips of shoots (which contain the shoot apical meristem) can be cultured *in vitro* clumps of shoots from either axillary or adventitious buds. This method can be used for clonal propagation.

Shoot meristem cultures are potential alternatives to the more commonly used methods for cereal regeneration as they are less genotype-dependent and more efficient (seedlings can be used as donor matreial).

Embryo Culture

Embryosbe used as explants to generate callus cultures or somatic embryos. Both immature and mature embryos can be used as explants. Immature, embryo-derived embryogenic callus is the most popular method of monocot plant regeneration.

Microspore Culture

Haploid tissue can be cultured *in vitro* by using pollen or anthers as an explant. Pollen contains the male gametophyte, which is termed the 'microspore'. Both callus and embryos can be produced from pollen. Two main approaches can be taken to produce *in vitro* cultures from haploid tissue.

The first method depends on using the anther as the explant. Anthers (somatic tissue that suurrounds and contains the pollen) can be cultured on solid medium (agar should not be used to solidify the medium as it contains inhibitory substances). Pollen-derived embryos are subsequently produced via dehiscence of the mature anthers. The dehiscence of the anther depends both on its isolation at the correct stage and on the correct culture conditions. In some species, the reliance on natural dehiscence can be circumvented by cutting the wall of the anther, although this does, of course, take a considerable amount of time. Anthers can also be cultured in liquid medium, and pollen released from the anthers can be induced to form embryos, although, efficiency of plant regeneration is often very low. Immature pollen can also be extracted from developing anthers and cultured directly, although this is a very time-consuming process.

Both methods have advantages and disadvantages. Some beneficial to the culture are observed when anthers are used as the explant material There is, however, the danger that some of the embryos produced from culture will originate from the somatic anther tissue rather than the haploid microspore cells. If isolated pollen is used there is no danger of mixed embryo formation, but the efficiency is Iow and the process is time-consuming.

In microspore culture, the condition of the donor plant is of critical importance, as is the timing of isolation. Pretreatments, such as a cold treatment often found to increase the efficiency. These pretreatments can be applied before culture, or, in some species, after placing the anthers in culture.

Plant species can be divided into two groups, depending on whether quire the addition of plant growth regulators to the medium for pollen/ anther culture; those that do also often require organic supplements, e.g. amino acids. Many of the cereals (rice, wheat, barley and maize) require medium supplemented with plant growth regulators for pollen/ anther culture.

Regeneration from microspore explants can be obtained by direct embryogenesis, or via a callus stage and subsequent embryogenesis.

Haploid tissue cultures can also be initiated from the female gametophyte (the ovule). In some cases, this is a more efficient method than using pollen or anthers.

The ploidy of the plants obtained from haploid cultures may not be haploid. This can be a consequence of chromosome doubling during the culture period. Chromosome doubling the (which often has to be induced by treatment with chemicals such as colchicine) may be an advantage, as in many cases haploid plants are not the desired outcome of regeneration from haploid Such plants are often referred to as '*di-haploids*', because they contain two copies of the same haploid genome.

Plant Regeneration

Having looked at the main types of plant culture that can be established *in vitro,* we can now look at how whole plants can be regenerated from these cultures.

In broad terms, two methods of plant regeneration are widely used in plant transformation studies, i.e. somatic embryogenesis and organogenesis.

Somatictic Embryogenesis

In somatic (asexual) embryogenesis, embryo-like structures, which can develop into whole plants in a way analogous to zygotic embryos, are formed from somatiic tissues. These somatic embryos can be produced either directly or indirectly. In direct somatic embryogenesis, the embryo is formed directly from a cell or small group of cells without the production of an intervening callus. Though common from some tissues (usually reproductive tissue such as the nucellus, styles or pollen), direct somatic embryogenesis is generally rare in comparison with indirect somatic embryogenesis.

In indirect somatic embryogenesis, callus is first produced from the explant. Embryos can then be produced from the callus tissue or from a cell suspension produced from that callus. Somatic embryogenesis from carrot is the classical exapmle of indirect somatic embryogenesis and is explained in more detail.

Somatic embryogenesis usually proceeds in two distinct stages. In the initial stage(embryo initiation), a high concentration of 2,4-D is

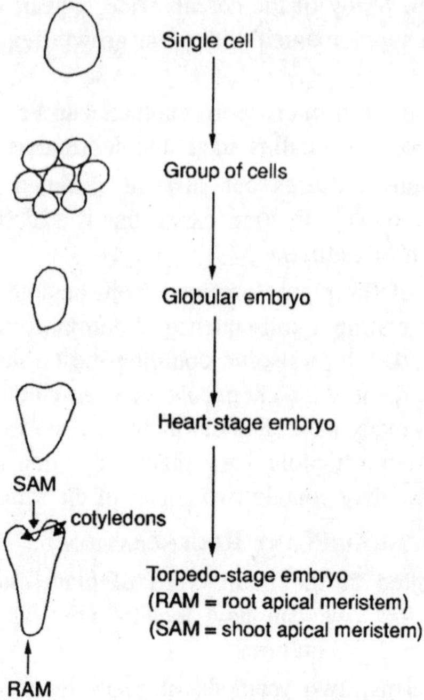

Fig. 1.2. A schematic representation of the sequential stages of somatic embryo development.

used. In the second stage (embryo production) embryos are produced in a medium with no or very of 2,4-D.

In many systems it has been found that somatic embryogenesis is improved by supplying a source of reduced nitrogen, such as specific amino acids or casien hydrolysate.

EXAMPLES

Cereal Regeneration via Somatic Embryogenesis from Immature or Mature Embryos

The prinipal method adopted for the tissue culture and regeneration of a wide range of cereal species is somatic embryogenesis, using cultures initiated from immature zygotic embryos. Embryogenic callus is normally initiated by placing the immature embryo on to a medium containing 2,4-D. Shoot regeneration is initiated by placing the embryogenic callus on a medium with BAP (with or without 2,4-D). These shoots can be subsequently rooted. The medium used for induction of embryogenic callus is usually a modified MS (for Triticeae) or N6 (for rice and maize). Maltose is often used as the carbon source in preference to sucrose and additional organic supplements (such as specific amino acids, yeast extract hydrolysate) are common.

The isolation and culture of immature embryos is, however, a labour-intensive and relatively expensive procedure. An additional problem is the small target size if the immature embryos are to used for biolistic transformation. Alternatives are therefore being sought One alternative is to use mature embryos (or seeds) as the explant to initiate embryogenic callus. This approach has been successfuly applied to several cereal species such as rice and oats. The culture techniques and media used for culture establishment and regeneration from mature embryo/seed-derived culture are fundamentally the same as those used for cultures initiated from immature embroys.

Importancc of Genotype

The major influence on tissue-culture response appears to be genetic, with culture requirement varying between species and cultivars. Model genotypes that responded well to culture *in vitro* were initially used in plant transformation studies. However, most of the genotypes used were not elite, commercial cultivars. The commercial cultivars tended to respond poorly to culture in vitro. One of the main aims is therefore to identify the components that make up a widely applicable, optional culture regime.

Many factors have been investigated for ability to improve the culture response from elite cultivars, including media components (such as alternative carbon sources, macro-and microelement concentrations and composition), media preparation method and donor plant condition and growth conditions.

Organogenesis

Somatic embryogenesis relies on plant regeneration through a process analogous to zygotic embryo germination. Organogenesis relies on the production of organs, either directly from an explant or from a callus culture. There are three methods of plant regeneration via, organogenesis.

The first two methods depend on adventitious organs arising either from a callus culture or directly from an explant. Alternatively, axillary bud formation and growth can also be used to regenerate whole plants from some types of tissue culture.

Organogenesis relies on the inherent plasticity of plant tissues, and is regulated by altering the components of the medium. In particular, it is the auxin to cytokinin ratio of the medium that determines which developmental path the regenerating tissue will take. It is usual to induce shoot formation by increasing the cytokinin to auxin ratio of the culture medium. These shoots can then be rooted relatively simply.

Intergration of Plant Tissue Culture into Plant Transformation Protocols

Various methods of plant regeneration are available to the plant biotechnologist. Some plant species may be amenable to regeneration by a variety of methods, but some may only be regenerated by one method. The various methods that can be used to transform plants will be considered, but it is worthwhile briefly considering the interaction of plant regeneration methodology and transformation methodology here. Not all plant tissue is suited to every plant transformation method, and not all plant species can be regenerated by every method. There is therefore a need to find both a suitable plant tissue culture/ regeneration regime and a compatible plant transformation methodology.

2

BASIC TECHNIQUES

Plant tissue culture is the science of growing plant cells, tissues or organs isolated from the mother plant, on artificial media. It includes techniques and methods used to research into many botanical disciplines and has several practical objectives. Before beginning to propagate plants by tissue culture methods, it is necessary to have a clear understanding of the ways in which plant material can be grown and manipulated in '*test tubes*'. This chapter therefore describes the techniques that have been developed for the isolation and *in vitro* culture of plant material, and shows where further information can be obtained. Both organized and unorganized growth are possible *in vitro*.

HISTORY

Organized Growth

Organized growth contributes towards the creation or maintenance of a defined structure. It occurs when plant organs such as the growing points of shoots or roots (*apical meristems*), leaf initials, young flower buds or small fruits, are transferred to culture and continue to grow with their structure preserved. Growth that is coherently organized also occurs when organs are induced. This may occur *in vitro* either directly upon an organ or upon a piece of tissue placed in culture (an explant), or during the culture of previously unorganized tissues. The process of *de novo* organ formation is called *organogenesis* or *morphogenesis* (the development of form).

Unorganized Growth

The growth of higher plants depends on the organized allocation of functions to organs which in consequence become differentiated, that is to say, modified and specialized to enable them undertake their essential roles. Unorganized growth is seldom found in nature, but occurs fairly frequently when pieces of whole plants are cultured *in vitro*. The cell aggregates, which are then formed, typically lack any recognizable structure and contain only a limited number of the many kinds of specialized and differentiated cells found in an intact plant. A differentiated cell is one that has developed a specialized form (*morphology*) and/or function (*physiology*). A differentiated tissue (e.g. xylem or epidermis) is an aggregation of differentiated cells. So far, the formation of differentiated cell types can only be controlled to a limited extent in culture. It is not possible, for example, to maintain and multiply a culture composed entirely of epidermal cells. By contrast, unorganized tissues can be increased in volume by subculture and can be maintained on semi-solid or liquid media for long periods. They can often also be used to commence cell suspension cultures. Differentiation is also used botanically to describe the formation of distinct organs through morphogenesis.

Tissue Culture

Cultures of Organized Structures

Organ culture is used as a general term for those types of culture in which an organized form of growth can be continuously maintained. It includes the aseptic isolation from whole plants of such definite structures as leaf primordia, immature flowers and fruits, and their growth *in vitro*. For the purposes of plant propagation, the most important kinds of organ culture are:

1. Meristem cultures, in which are grown very small excised shoot apices, each consisting of the apical meristematic dome with or without one or two leaf primordia. The shoot apex is typically grown to give one single shoot.
2. Shoot tip, or shoot cultures, started from excised shoot tips, or buds, larger than the shoot apices employed to establish meristem cultures, having several leaf primordia. These shoot apices are usually cultured in such a way that each produces multiple shoots.
3. Node cultures of separate lateral buds, each carried on a small piece of stem tissue; stem pieces carrying either single or multiple nodes may be cultured. Each bud is grown to provide a single shoot.
4. Isolated root cultures. The growth of roots, unconnected to shoots: a branched root system may be obtained.

5. Embryo cultures, where fertilized or unfertilized zygotic (seed) embryos are dissected out of developing seeds or fruits and cultured *in vitro* until they have grown into seedlings. Embryo culture is quite distinct from somatic embryogenesis.

Cultures of Unorganized Tissues

'*Tissue culture*' is commonly used as a collective term to describe all kinds of *in vitro* plant cultures although strictly it should refer only to cultures of unorganized aggregates of cells. In practice the following kinds of cultures are most generally recognized:

1. *Callus (or tissue) cultures*. The growth and maintenance of largely unorganized cell masses, which arise from the uncoordinated and disorganized growth of small plant organs, pieces of plant tissue, or previously cultured cells.
2. *Suspension (or cell) cultures*. Populations of plant cells and small cell clumps, dispersed in an agitated, that is aerated, liquid medium.
3. *Protoplast cultures*. The culture of plant cells that have been isolated without a cell wall.
4. *Anther cultures*. The culture of complete anthers containing immature pollen microspores. The objective is usually to obtain haploid plants by the formation of somatic embryos directly from the pollen, or sometimes by organogenesis *via* callus. Pollen cultures are those initiated from pollen that has been removed from anthers.

Using Tissue Cultures for Plant Propagation

The objective of plant propagation via tissue culture, termed *micropropagation*, is to propagate plants true-to-type, that is, as clones. Plants obtained from tissue culture are called *microplants* and can be derived from tissue cultures in three ways:

1. From pre-existing shoot buds or primordial buds (*meristems*) which are encouraged to grow and proliferate;
2. Following shoot morphogenesis when new shoots are induced to form in unorganized tissues or directly upon explanted tissues of the mother plant;
3. Through the formation of somatic embryos which resemble the seed embryos of intact plants, and which can grow into seedlings in the same way. This process is called *somatic embryogenesis*.

To obtain plants by the first two of these methods, it is necessary to treat shoots of an adequate size as miniature cuttings and induce them to produce roots. The derivation of new plants from cells, which would not normally have taken part in the process of regeneration, shows that living, differentiated plant cells may express totipotency, i.e. they each retain a latent capacity to produce a whole plant. *Totipotency* is a special characteristic of cells in young tissues and meristems. It can be exhibited by some differentiated cells, e.g. cambial cells and leaf palisade cells but not those which have developed into terminally differentiated structures (e.g. sieve tubes or tracheids).

Theoretically, plant cells, organs, or plants, can all be cloned, i.e., produced in large numbers as a population where all the individuals have the same genetic constitution as the parent. Present tissue culture techniques do not permit this in every case and irregularities do sometimes occur, resulting in '*somaclonal variants*'. Nevertheless, as will be described in the chapters, which follow, a very large measure of success can be achieved and cultures of various kinds can be used to propagate plants.

Initiating Tissue Cultures

Explants

Tissue cultures are started from pieces of whole plants. The small organs or pieces of tissue that are used are called *explants*. The part of the plant (the stock plant or mother plant) from which explants are obtained, depends on:

1. The kind of culture to be initiated;
2. The purpose of the proposed culture;
3. The plant species to be used.

Explants can therefore be of many different kinds. The correct choice of explant material can have an important effect on the success of tissue culture. Plants growing in the external environment are invariably contaminated with microorganisms and pests. These contaminants are mainly confined to the outer surfaces of the plant, although, some microbes and viruses may be systemic within the tissues. Because they are started from small explants and must be grown on nutritive media that are also favourable for the growth of micro-organisms, plant tissue cultures must usually be established and maintained in aseptic conditions. Most kinds of microbial organism, and in particular bacteria and fungi, compete adversely with plant material growing *in vitro*. Therefore, as far as possible, explants must be free from microbial contaminants when they are first placed on a nutrient medium. This usually involves growing stock plants in ways that will minimize infection, treating the plant material with disinfecting chemicals to kill superficial microbes, and sterilizing the tools used for dissection and the vessels and media in which cultures are grown. Some kinds of plants can, however, be micropropagated in non-sterile environments.

Isolation and incubation

The work of isolating and transferring cultured plant material is usually performed in special rooms or inside hoods or cabinets from which microorganisms can be excluded. Cabinets used for isolation can be placed in a draught-free part of a general laboratory, but are much better situated in a special inoculation or transfer room reserved for the purpose. Cultures, once initiated, are placed in incubators or growth rooms where lighting, temperature and humidity can be controlled. The rate of growth of a culture will depend on the temperature (and sometimes the lighting) regime adopted.

Cultural environment

Plant cultures are commenced by placing one or more explants into a pre-sterilized container of sterile nutrient medium. Some explants may fail to grow, or may die, due to microbial contamination: to ensure the survival of an adequate number, it therefore is usual to initiate several cultures at the same time, each being started from an identical organ or piece of tissue. Explants taken from stock plants at different times of the year may not give reproducible results in tissue culture. This may be due to variation in the level of external contaminants or because of seasonal changes in endogenous (internal) growth regulator levels in the stock plant.

Media

Plant material will only grow *in vitro* when provided with specialized media. A medium usually consists of a solution of salts supplying the major and minor elements necessary for the growth of whole plants, together with:

1. Various vitamins (optional);
2. Various amino acids (optional);
3. An energy source (usually sucrose).

Growth and development of plant cultures usually also depends on the addition of plant growth regulators to the medium. Plant growth regulators are compounds, which, at very low concentration, are capable of modifying growth or plant morphogenesis. Many workers define a medium as a completed mixture of nutrients and growth regulators.

This is a rather inflexible method, as growth regulators frequently need to be altered according to the variety of plant, or at different stages of culture, whilst the basic medium can stay unchanged. It is therefore recommended that nutritional and regulatory components should be listed separately. Plant material can be cultured either in a liquid medium or on a medium that has been partially solidified with a gelling agent. The method employed will depend on the type of culture and its objective.

Solidified media

Media which have had a gelling agent added to them, so that they have become semi-solid, are widely used for explant establishment; they are also employed for much routine culture of callus or plant organs (including micropropagation), and for the long-term maintenance of cultures. Agar is the most common solidifying agent, but a gellan gum is also widely used.

Cultures grown on solid media are kept static. They require only simple containers of glass or plastic, which occupy little space. Only the lower surface of the explant, organ or tissue is in contact with the medium. This means that as growth proceeds there may be gradients in nutrients, growth factors and the waste products of metabolism, between the medium and the tissues. Gaseous diffusion into and out of the cells at the base of the organ or tissue may also be restricted by the surrounding medium.

Liquid media

Liquid media are essential for suspension cultures, and are preferred for critical experiments on the nutrition, growth and cell differentiation in callus tissues. They are also used in some micro-propagation work. Very small organs (e.g. anthers) are often floated on the top of liquid medium and plant cells or protoplasts can be cultured in very shallow layers of static liquid, providing there is sufficient gaseous diffusion. Larger organs such as shoots (e.g. proliferating shoots of shoot cultures) can also often be grown satisfactorily in a shallow layer of non-agitated liquid where part of the organ protrudes above the surface. However, some method of support is necessary for small organs or small pieces of tissue, which would otherwise sink below the surface of a static liquid medium, or they will die for lack of aeration.

Many tissues and organs, small and large, also grow well unsupported in a liquid medium, providing it is aerated by shaking or moving. Some kind of agitation is essential for suspension cultures to prevent cells and cell aggregates settling to the bottom of the flask. Other purposes served by agitation include: the provision of increased aeration, the reduction of plant polarity, the uniform distribution of nutrients and the dilution of toxic explant exudates.

There are several alternative techniques. Plant cell suspensions can be cultured very satisfactorily when totally immersed in a liquid culture medium, providing it is shaken (by a rotary or reciprocating shaking machine) or stirred (e.g. by a magnetic stirrer) to ensure adequate aeration. This method may also be used for culturing organs of some plants (e.g. proliferating shoot cultures), but the fragmentation, which occurs, can be disadvantageous.

Periodic immersion may be achieved by growing cultured material in tubes or flasks of liquid medium which are rotated slowly. Steward and Shantz (1956) devised so-called '*nipple flasks*' for this purpose which had several side-arms. They were fixed to a wooden wheel, which was rotated so that tissue in the arms of each flask was alternately bathed in medium and drained or exposed to the air. This technique ensured that callus tissue for which they were used was well aerated. The medium usually became turbid as cells dissociated from the callus and started a cell suspension. Flasks of this sort are seldom used to-day because of their cost. A similar alternating exposure can be achieved by placing calluses in vessels, which are rotated slowly. An alternative to the costly rotating systems to achieve periodic immersion of the cultures, is the increasingly popular. temporary immersion system in which static vessels are periodically or temporarily flooded with culture medium. Medium is pumped from a reservoir container into the culture vessel for experimentally determined time intervals repeated over a 24 hour cycle. This system prevents anoxia and has the advantage that the medium can easily be changed in the reservoir.

Liquid medium in flasks or column bio-reactors (fermentors) can be circulated and at the same time aerated, by the introduction of sterile air. Shearing forces within air-lift reactors are much less than in mechanically-stirred vessels so that plant cell suspensions suffer less damage. Bioreactors are used in the pharmaceutical industry to produce high value plant secondary products and to carry out substrate conversions.

Rather than immersing callus or organ cultures, liquid medium may be slowly dripped onto the growing tissues or applied as a mist and afterwards the liquid drained or pumped away for recirculation. A particular advantage of this technique is the ability to grow cultures in a constant and non-depleted medium; nutrients can be varied frequently and rapidly and their availability controlled by altering either concentration or flow rate. Toxic metabolites, which in a closed container might accumulate and inhibit growth, can be removed continuously. As complicated apparatus is needed, the method has not been widely used.

Problems of Establishment

Phenolic oxidation

Some plants, particularly tropical species, contain high concentrations of phenolic substances that are oxidized when cells are wounded or senescent. Isolated tissue then becomes brown or black and fails to grow.

Minimum inoculation density

Certain essential substances can pass out of plant cells by diffusion. Substances known to be released into the medium by this means include alkaloids, amino acids, enzymes, growth substances and vitamins. The loss is of no consequence when there is a large cluster of cells growing in close proximity or where the ratio of plant material to medium is high. However, when cells are inoculated onto an ordinary growth medium at a low population density, the concentration of essential substances in the cells and in the medium can become inadequate for the survival of the culture. For successful culture initiation, there is thus a minimum size of explant or quantity of separated cells or protoplasts per unit culture volume. Inoculation density also affects the initial rate of growth *in vitro*. Large explants generally survive more frequently and grow more rapidly at the outset than very small ones. In practice, minimum inoculation density varies according to the genotype of plant being cultured and the cultural conditions. For commencing suspension cultures it is commonly about $1\text{-}1.5 \times 10^4$ cells/ml.

The minimum cell density phenomenon is sometimes called a '*feeder effect*' because deficiencies can often be made up by the presence of other cells growing nearby. Suspension cultures can be started from a low density of inoculum by '*conditioning*' a freshly prepared medium - i.e. allowing products to diffuse into it from a medium in which another culture is growing actively, or adding a quantity of filter-sterilized medium which has previously supported another culture. The use of conditioned media can reduce the critical initial cell density by a factor of about 10.

It is possible to overcome the deficiencies of plant cells at low starting densities by adding small amounts of known chemicals to a medium. For example, Kao and Michayluk (1975) have shown that *Vicia hajastana* cells or protoplasts can be cultured from very small initial inocula or even from individual cells: a standard culture medium was supplemented with growth regulators, several organic acids, additional sugars (apart from sucrose and glucose), and in particular, casein hydrolysate (casamino acids) and coconut milk. There is often a maximum as well as a minimum plating or inoculation density for plant cells or protoplasts. In a few cases the effective range has been found to be quite narrow.

Patterns of Growth and Differentiation

A typical unorganized plant callus, initiated from a new explant or a piece of a previously-established culture, has three stages of development, namely:

1. The induction of cell division;
2. A period of active cell division during which differentiated cells lose any specialized features they may have acquired and become dedifferentiated;
3. A period when cell division slows down or ceases and when, within the callus, there is increasing cellular differentiation.

These phases are similarly reproduced by cell suspensions grown in a finite volume of medium (a batch culture), where according to a variety of different parameters that can be used to measure growth (e.g. cell number, cell dry weight, total DNA content) an S-shaped growth curve is generally obtained.

The phases are:

(a) a lag phase;

(b) a period of exponential and then linear growth;

(c) a period when the rate of growth declines;

(d) a stationary phase when growth comes to a halt.

Some differentiation of cells may occur in cell cultures during the period of slowed and stationary growth, but generally it is less marked and less complete than that which occurs in callus cultures. Cultures cannot be maintained in stationary phase for long periods. Cells begin to die and, as their contents enter the nutrient medium, death of the whole culture accelerates. Somewhat similar patterns of growth also occur in cultures of organized structures. These also cease growth and become moribund as the components of the medium become exhausted.

Subculturing

Once a particular kind of organized or unorganized growth has been started *in vitro,* it will usually continue if callus cultures, suspension cultures, or cultures of indeterminate organs are divided to provide new explants for culture initiation on fresh medium. Subculturing often becomes imperative when the density of cells, tissue or organs becomes excessive; to increase the volume of a culture; or to increase the number of organs (e.g. shoots or somatic embryos) for micropropagation. The period from the initiation of a culture or a subculture to the time of its transfer is sometimes called a *passage*. The first passage is that in which the original explant or inoculum is introduced.

Suspensions regularly subcultured at the end of the period of exponential growth can often be propagated over many passages. However, many cultures reach a peak of cell aggregation at this time and aggregation often becomes progressively more pronounced in subsequent passages. Subculture is therefore more conveniently carried out during the stationary phase when cell aggregation is least pronounced. Rapid rates of plant propagation depend on the ability to subculture shoots from proliferating shoot or node cultures, from cultures giving direct shoot regeneration, or callus or suspensions capable of reliable shoot or embryo regeneration.

A further reason for transfer, or subculture, is that the growth of plant material in a closed vessel eventually leads to the accumulation of toxic metabolites and the exhaustion of the medium, or to its drying out. Thus, even to maintain the culture, all or part of it must be transferred onto fresh medium.

Callus subcultures are usually initiated by moving a fragment of the initial callus (an inoculum) to fresh medium in another vessel. Shoot cultures are subcultured by segmenting individual shoots or shoot clusters. The interval between subcultures depends on the rate at which a culture has grown: at 25°C, subculturing is typically required every 4-6 weeks. In the early stages of callus growth it may be convenient to transfer the whole piece of tissue to fresh medium, but a more established culture will need to be divided and only small selected portions used as inocula. Regrowth depends on the transfer of healthy tissues.

Decontamination procedures are theoretically no longer necessary during subculturing, although sterile transfer procedures must still be used. However, when using shoot or node cultures for micropropagation, some laboratories do re-sterilize plant material at this stage as a precaution against the spread of contaminants. Cultures which are obviously infected with microorganisms should not be used for subculturing and should be autoclaved before disposal.

Subculturing Hazards

Several kinds of callus may arise from the initial explant, each with different morphogenic potential. Strains of callus tissue capable of giving rise to somatic embryos and others without this capability can, for instance, arise simultaneously from the culture of grass and cereal seed embryos. Careful selection of the correct strain is therefore necessary if cultures capable of producing somatic embryos are ultimately required. Timing of the transfer may also be important, because if left alone for some while, non-embryogenic callus may grow from the original explant at the expense of the competent tissue, which will then be obscured or lost.

Although subculturing can often be continued over many months without adverse effects becoming apparent, cultures of most unorganized cells and of some organized structures can accumulate cells that are genetically changed. This may cause the characteristics of the culture to be altered and may mean that some of the plants regenerated from the culture will not be the same as the parent plant. Cultures may also inexplicably decline in vigour after a number of passages, so that further subculture becomes impossible.

Types of Tissue Culture

Organ Cultures

Differentiated plant organs can usually be grown in culture without loss of integrity. They can be of two types:

1. Determinate organs which are destined to have only a defined size and shape (e.g. leaves, flowers and fruits);
2. Indeterminate organs, where growth is potentially unlimited (apical meristems of roots and non-flowering shoots).

In the past, it has been thought that the meristematic cells within root or shoot apices were not committed to a particular kind of development. It is now accepted that, like the primordia of determinate organs such as leaves, apical meristems also become inherently programmed (or determined) into either root or shoot pathways. The eventual pattern of development of both indeterminate and determinate organs is often established at a very early stage. For example, the meristematic protrusions in a shoot apex become programmed to develop as either lateral buds or leaves after only a few cell divisions have taken place.

Culture of determinate organs

An organ arises from a group of meristematic cells. In an indeterminate organ, such cells are theoretically able to continue in the same pattern of growth indefinitely. The situation is different in the primordium of a determinate organ. Here, as meristematic cells receive instructions on how to differentiate, their capacity for further division becomes limited.

If the primordium of a determinate organ is excised and transferred to culture, it will sometimes continue to grow to maturity. The organ obtained *in vitro* may be smaller than that which would have developed on the original plant *in vivo,* but otherwise is likely to be normal. The growth of determinate organs cannot be extended by subculture as growth ceases when they have reached their maximum size.

Organs of limited growth potential, which have been cultured, include leaves; fruits; stamens; ovaries and ovules (which develop and grow into embryos) and flower buds of several dicotyledonous plant species.

Until recently, a completely normal development was obtained in only a few cases. This was probably due to the use of media of sub-optimum composition. By experimenting with media constituents, Berghoef and Bruinsma (1979a) obtained normal growth of *Begonia franconis* buds and were thus able to study the effect of plant growth substances and nutritional factors on flower development and sexual expression. Similarly, by culturing dormant buds of *Salix*, Angrish and Nanda (1982a,b) could study the effect of bud position and the progressive influence of a resting period on the determination of meristems to become catkins and fertile flowers. In several species, flowers have been pollinated *in vitro* and have then given rise to mature fruits.

Plants cannot be propagated by culturing meristems already committed to produce determinate organs, but providing development has not proceeded too far, flower meristems can often be induced to revert to vegetative meristems *in vitro.* In some plants the production of vegetative shoots from the flower meristems on a large inflorescence can provide a convenient method of micropropagation.

Culture of indeterminate organs

Meristem and shoot culture

The growing points of shoots can be cultured in such a way that they continue uninterrupted and organized growth. As these shoot initials ultimately give rise to small organized shoots which can then be rooted, their culture has great practical significance for plant propagation. Two important uses have emerged.

Meristem culture

Culture of the extreme tip of the shoot, is used as a technique to free plants from virus infections. Explants are dissected from either apical or lateral buds. They comprise a very small stem apex (0.2-1.0 mm in length) consisting of just the apical meristem and one or two leaf primordia.

Shoot culture or shoot tip culture

Culture of larger stem apices or lateral buds (ranging from 5 or 10 mm in length to undissected buds) is used as a very successful method of propagating plants.

If successful, meristem culture, shoot culture and node culture can ultimately result in the growth of small shoots. With appropriate treatments, these original shoots can either be rooted to produce small plants or '*plantlets*', or their axillary buds can be induced to grow to form a cluster of shoots. Plants are propagated by dividing and reculturing the shoot clusters, or by growing individual shoots for subdivision. At a chosen stage, individual shoots or shoot clusters are rooted. Tissue cultured shoots are removed from aseptic conditions at or just before the rooting stage, and rooted plantlets are hardened off and grown normally.

Embryo culture

Zygotic or seed embryos are often used advantageously as explants in plant tissue culture, for example, to initiate callus cultures. In embryo culture however, embryos are dissected from seeds, individually isolated and '*germinated*' *in vitro* to provide one plant per explant. Isolated embryo culture can assist in the rapid production of seedlings from seeds that have a protracted dormancy period, and it enables seedlings to be produced when the genotype (e.g. that resulting from some interspecific crosses) conveys a low embryo or seed viability.

During the course of evolution, natural incompatibility systems have developed which limit the types of possible sexual crosses. Two kinds of infertility occur:

1. *Pre-zygotic* incompatibility, preventing pollen germination and/or pollen tube growth so that a zygote is never formed;
2. *Post-zygotic* incompatibility, in which a zygote is produced but not accepted by the endosperm. The embryo, not receiving sufficient nutrition, disintegrates or aborts.

Pre-zygotic incompatibility can sometimes be overcome in the laboratory using a technique developed by Kanta *et al.* (1962) called *in vitro* pollination (or *in vitro* fertilization).

Embryo culture has been used successfully in a large number of plant genera to overcome *post-zygotic* incompatibility which otherwise hampers the production of desirable hybrid seedlings. For example, in trying to transfer insect resistance from a wild *Solanum* species into the aubergine, Sharma *et al.* (1980a) obtained a few hybrid plants (*Solanum melongena* x *S. khasianum*) by embryo culture. Embryo culture in these circumstances is more aptly termed embryo rescue. Success rates are usually quite low and the new hybrids, particularly if they arise from remote crosses, are sometimes sterile. However, this does not matter if the plants can afterwards be propagated asexually. Hybrids between incompatible varieties of tree and soft fruits have been obtained by culturing fairly mature embryos.

Fruits or seeds are surface sterilized before embryo removal. Providing aseptic techniques are strictly adhered to during excision and transfer to a culture medium, the embryo itself needs no further sterilization. To ease the dissection of the embryo, hard seeds are soaked in water to soften them, but if softening takes more than a few hours it is advisable to re-sterilize the seed afterwards. A dissecting microscope may be necessary to excise the embryos from small seeds as it is particularly important that the embryo should not be damaged.

Culture of immature embryos (pro-embryos) a few days after pollination frequently results in a greater proportion of seedlings being obtained than if more mature embryos are used as explants, because incompatibility mechanisms have less time to take effect. Unfortunately dissection of very small embryos requires much skill and cannot be done rapidly: it also frequently results in damage which prevents growth *in vitro*. In soybean, Hu and Sussex (1986) obtained the best *in vitro* growth of immature embryos if they were isolated with their suspensors intact. Excised embryos usually develop into seedlings precociously (i.e. before they have reached the size they would have attained in a normal seed).

As an alternative to embryo culture, in some plants it has been possible to excise and culture pollinated ovaries and immature ovules. Ovule culture, sometimes called '*in ovulo embryo culture*', can be more successful than the culture of young embryos. Pro-embryos generally require a complex medium for growth, but embryos contained within the ovule require less complicated media. They are also easily removed from the plant and relatively insensitive to the physical conditions of culture.

Because seedlings, which resulted from ovule culture of a *Nicotiana* interspecific cross all died after they had developed some true leaves, Iwai *et al.* (1985) used leaves of the immature seedlings as explants for the initiation of callus cultures. Most shoots regenerated from the callus also died at an early stage, but one gave rise to a plant, which was discovered later to be a sterile hybrid. Plants were also regenerated from callus of a *Pelargonium* hybrid by Kato and Tokumasu (1983). The callus in this case arose directly from globular or heart-shaped zygotic embryos which were not able to grow into seedlings.

The seeds of orchids have neither functional storage organs, nor a true seed coat, so dissection of the embryo would not be possible. In fact, for commercial purposes, orchid seeds are now almost always germinated *in vitro,* and growth is often facilitated by taking immature seeds from green pods.

Many media have been especially developed for embryo culture and some were the forerunners of the media now used for general tissue culture. Commonly, mature embryos require only inorganic salts supplemented with sucrose, whereas immature embryos have an additional requirement for vitamins, amino acids, growth regulators and sometimes coconut milk or some other endosperm extract. Raghavan (1977b) encouraged the incorporation of mannitol to replace the high osmotic pressure exerted on proembryos by ovular sap. Seedlings obtained from embryos grown *in vitro* are planted out and hardened off in the same manner as other plantlets raised by tissue culture.

Although embryo culture is especially useful for plant breeders, it does not lead to the rapid and large scale rates of propagation characteristic of other micropropagation techniques, and so it is not considered further in this book.

Isolated root culture

Root cultures can be established from root tips taken from primary or lateral roots of many plants. Suitable explants are small sections of roots bearing a primary or lateral root meristem. These explants may be obtained, for example, from surface sterilized seeds germinated in aseptic conditions. If the small root meristems continue normal growth on a suitable medium, they produce a root system consisting only of primary and lateral roots. No organized shoot buds will be formed.

The discovery that roots could be grown apart from shoot tissue was one of the first significant developments of modern tissue culture science. Root culture initially attracted a great deal of attention from research workers and the roots of many different species of plants were cultured successfully.

Plants fall generally into three categories with regard to the ease with which their roots can be cultured. There are some species such as clover, *Datura*, tomato and *Citrus*, where isolated roots can be grown for long periods of time, some seemingly, indefinitely providing regular subcultures are made. In many woody species, roots have not been grown at all successfully in isolated cultures. In other species such as pea, flax and wheat, roots can be cultured for long periods but ultimately growth declines or insufficient lateral roots are produced to provide explants for subculture.

The inability to maintain isolated root cultures is due to an induced meristematic dormancy or '*senescence*', related to the length of time that the roots have been growing *in vitro*. Transferring dormant meristems to fresh medium does not promote regrowth, possibly due to the accumulation of naturally-occurring auxinic growth substances at the root apex. The addition of so-called *anti-auxin*, or *cytokinin* growth regulators can often prolong active growth of root cultures, whereas placing auxins or gibberellic acid in the growth medium, causes it to cease more rapidly. Cultures, which cannot be maintained by transferring root apices, can sometimes be continued if newly-initiated lateral root meristems are used as secondary explants instead.

Isolated plant roots can usually be cultured on relatively simple media such as White (1954) containing 2% sucrose. Liquid media are preferable, as growth in or on a solid medium is slower. This is presumably because salts are less readily available to the roots from a solidified medium and oxygen availability may be restricted. Although roots will accept a mixed nitrate/ammonium source, they will not usually grow on ammonium nitrogen alone. Species, and even varieties or strains, of plants, are found to differ in their requirement for growth regulators, particularly for auxins, in the root culture medium.

Isolated root cultures have been employed for a number of different research purposes. They have been particularly valuable in the study of nematode infections and provide a method by which these parasites can be cultured in aseptic conditions. Root cultures may also be used to grow beneficial mycorrhizal fungi, and to study the process of root nodulation with nitrogen-fixing *Rhizobium* bacteria in leguminous plants. For the latter purpose, various special adaptations of standard techniques have been adopted to allow roots to become established in a nitrate-free medium.

Unlike some other cultured tissues, root cultures exhibit a high degree of genetic stability. It has therefore been suggested that root cultures could afford one means of storing the germ-plasm of certain species. For suitable species, root cultures can provide a convenient source of explant material for the micropropagation of plants, but they will only be useful in micropropagation if shoots can be regenerated from roots. There are however, several ways in which this can be done, although they are likely to be effective in only a small number of plant genera which have a natural tendency to produce suckers, or new shoots from whole or severed roots:

1. From direct adventitious shoots;
2. From shoots or embryos originating indirectly on root callus;
3. By conversion of the apical root meristem to a shoot meristem.

Adventitious shoots form readily on the severed roots of some plant species, and root cuttings are employed by horticulturists to increase plants *in vivo*. Shoot regeneration from roots has not been widely used as a method of micropropagation, even though direct shoot regeneration from roots has been observed *in vitro* on many plants. Sections of fleshy roots used as primary explants are especially likely to form new shoots. Adventitious shoots always develop at the proximal end of a root section while, as a rule, new roots are produced from the distal end. Isolated root cultures would be useful in micropropagation if shoots could be induced to form directly upon them.

Unfortunately plants seem to have a high degree of genetic specificity in their capacity to produce shoots directly on isolated root cultures. Shoot induction often occurs after the addition of a cytokinin to the medium. Seeliger (1956) obtained shoot buds on cultured roots of *Robinia pseudoacacia* and Torrey (1958), shoot buds on root cultures of *Convolvulus*. Direct shoot formation was induced in three species of *Nicotiana* and on *Solanum melongena* by Zelcer *et al.* (1983) but in *N. tabacum* and *N. petunoides* shoots were only obtained after callus formed on the roots. The most optimistic report we have seen comes from Mudge *et al.* (1986), who thought that the shoot formation, which they could induce in raspberry root cultures would provide a convenient and labour-saving method of multiplying this plant *in vitro*.

Plants may also be regenerated from root-derived callus of some species e.g. tomato; *Isatis tinctoria*; *Atropa belladonna*. Embryogenesis, leading to the formation of protocorm-like bodies, occurs in the callus derived from the root tips of certain orchids e.g. *Catasetum trulla* x *Catasetum*; *Epidendrum obrienianum*; *Oncidium varicosum*.

Changing the determined nature of a root meristem, so that it is induced to produce a shoot instead of a root, is a very rare event but has been noted to occur *in vitro* in the orchid *Vanilla planifolia*. The quiescent centre of cultured root tip meristems was changed into a shoot meristem so that cultured root tips grew to produce plantlets or multiple shoots maintained that newly initiated root initials, arising from single nodes of *Nasturtium officinale*, could be made to develop into shoot meristems by placing a crystal of kinetin on each explant which was then transferred to a medium containing 0.05% glucose.

Culture of Unorganized Cells

Callus cultures

Callus is a coherent and amorphous tissue, formed when plant cells multiply in a disorganized way. It is often induced in or upon parts of an intact plant by wounding, by the presence of insects or micro-organisms, or as a result of stress. Callus can be initiated *in vitro* by placing small pieces of the whole plant (explants) onto a growth-supporting medium under sterile conditions. Under the stimulus of endogenous growth regulators or growth regulating chemicals added to the medium, the metabolism of cells, which were in a quiescent state, is changed, and they begin active division. During this process, cell differentiation and specialization, which may have been occurring in the intact plant, are reversed, and the explant gives rise to new tissue, which is composed of meristematic and unspecialized cell types.

During dedifferentiation, storage products typically found in resting cells tend to disappear. New meristems are formed in the tissue and these give rise to undifferentiated parenchymatous cells without any of the structural order that was characteristic of the organ or tissue from which they were derived. Although callus remains unorganized, as growth proceeds, some kinds of specialized cells may again be formed. Such differentiation can appear to take place at random, but may be associated with centres of morphogenesis, which can give rise to organs such as roots, shoots and embryos. The *de novo* production of plants from unorganized cultures is often referred to as plant regeneration.

Although most experiments have been conducted with the tissues of higher plants, callus cultures can be established from gymnosperms, ferns, mosses and thallophytes. Many parts of a whole plant may have an ultimate potential to proliferate *in vitro,* but it is frequently found that callus cultures are more easily established from some organs than others. Young meristematic tissues are most suitable, but meristematic areas in older parts of a plant, such as the cambium, can give rise to callus. The choice of tissues from which cultures can be started is greatest in dicotyledonous species. A difference in the capacity of tissue to give rise to callus is particularly apparent in monocotyledons. In most cereals, for example, callus growth can only be obtained from organs such as zygotic embryos, germinating seeds, seed endosperm or the seedling mesocotyl, and very young leaves or leaf sheaths, but so far never from mature leaf tissue. In sugar cane, callus cultures can only be started from young leaves or leaf bases, not from semi-mature or mature leaf blades.

Even closely associated tissues within one organ may have different potentials for callus origination. Thus when embryos of *Hordeum distichum* at an early stage of differentiation are removed from developing seeds and placed in culture, callus proliferation originates from meristematic mesocotyl cells rather than from the closely adjacent cells of the scutellum and coleorhiza.

The callus formed on an original explant is called '*primary callus*'. Secondary callus cultures are initiated from pieces of tissue dissected from primary callus. Subculture can then often be continued over many years, but the longer callus is maintained, the greater is the risk that the cells thereof will suffer genetic change.

Callus tissue is not of one single kind. Strains of callus differing in appearance, colour, degree of compaction and morphogenetic potential commonly arise from a single explant. Sometimes the type of callus obtained, its degree of cellular differentiation and its capacity to regenerate new plants, depend upon the origin and age of the tissue chosen as an explant. Loosely packed or '*friable*' callus is usually selected for initiating suspension cultures.

Some of the differences between one strain of callus tissue and another can depend on which genetic programme is functioning within the cells (epigenetic differences). Variability is more likely when callus is derived from an explant composed of more than one kind of cell. For this reason there is often merit in selecting small explants from only morphologically uniform tissue, bearing in mind that a minimum size of explant is normally required to obtain callus formation.

The genetic make up of cells is very commonly altered in unorganized callus and suspension cultures. Therefore another reason for cell strains having different characteristics, is that they have become composed of populations of cells with slightly different genotypes.

Cell suspension cultures

Unorganized plant cells can be grown as callus in aggregated tissue masses, or they can be freely dispersed in agitated liquid media. Techniques are similar to those used for the large-scale culture of bacteria. Cell or suspension cultures, as they are called, are usually started by placing an inoculum of friable callus in a liquid medium. Under agitation, single cells break off and, by division, form cell chains and clumps which fracture again to give individual cells and other small cell groups. It is not always necessary to have a previous callus phase before initiating suspension cultures. For example, leaf sections of *Chenopodium rubrum* floated on Murashige and Skoog (1962) medium in the light, show rapid growth and cell division in the mesophyll, and after 4 days on a rotary shaker they can be disintegrated completely to release a great number of cells into suspension.

Because the walls of plant cells have a natural tendency to adhere, it is not possible to obtain suspensions that consist only of dispersed single cells. Some progress has been made in selecting cell lines with increased cell separation, but cultures of completely isolated cells have yet to be obtained. The proportion and size of small cell aggregates varies according to plant variety and the medium in which the culture is grown. As cells tend to divide more frequently in aggregates than in isolation, the size of cell clusters increases during the phase of rapid cell division. Because agitation causes single cells, and small groups of cells, to be detached, the size of cell clusters decreases in batch cultures as they approach a stationary growth phase.

The degree of cell dispersion in suspension cultures is particularly influenced by the concentration of growth regulators in the culture medium. Auxinic growth regulators increase the specific activity of enzymes, which bring about the dissolution of the middle lamella of plant cell walls. Thus by using a relatively high concentration of an auxin and a low concentration of a cytokinin growth regulator in the culture medium, it is usually possible to increase cell dispersion.

However, the use of high auxin levels to obtain maximum cell dispersion will ensure that the cultured cells remain undifferentiated. This may be a disadvantage if a suspension is being used to produce secondary metabolites. Well-dispersed suspension cultures consist of thin-walled undifferentiated cells, but these are never uniform in size and shape. Cells with more differentiated structure, possessing, for example, thicker walls and even tracheid-like elements, usually only occur in large cell aggregates.

Many different methods of suspension culture have been developed. They fall into two main types: batch cultures in which cells are nurtured in a fixed volume of medium until growth ceases, and continuous cultures in which cell growth is maintained by continuous replenishment of sterile nutrient media. All techniques utilize some method of agitating the culture medium to ensure necessary cell dispersion and an adequate gas exchange.

Batch cultures

Batch cultures are initiated by inoculating cells into a fixed volume of nutrient medium. As growth proceeds, the amount of cell material increases until nutrients in the medium are depleted or there is the accumulation of an inhibitory metabolite. Batch cultures have a number of disadvantages that restrict their suitability for extended studies of growth and metabolism, or for the industrial production of plant cells, but they are nevertheless widely used for many laboratory investigations. Small cultures are frequently agitated on orbital shakers onto which are fixed suitable containers, which range in volume from 100 ml (Erlenmeyer conical flasks) to 1000 ml (spherical flasks); the quantity of medium being approximately the same as the flask volume. The shakers are usually operated at speeds from 30-180 rpm with an orbital motion of about 3 cm. Alternatively, stirred systems can be used.

Continuous cultures

Using batch cultures, it is difficult to obtain a steady rate of production of new cells having constant size and composition. Attempts to do so necessitate frequent sub-culturing, at intervals equivalent to the doubling time of the cell population. Satisfactorily balanced growth can only be produced in continuous culture, a method, which is especially important when plant cells are to be used for the large-scale production of a primary or secondary metabolite. Continuous culture techniques require fairly complicated apparatus. Agitation of larger cultures in bio-reactors is usually achieved by stirring with a turbine and/or by passing sterile air (or a controlled gaseous mixture) into the culture from below and releasing it through plugged vents. Mechanically stirred reactors damage plant cells by shearing. This is minimized in air-lift reactors.

Use of suspension cultures in plant propagation

The growth of plant cells is more rapid in suspension than in callus culture and is also more readily controlled because the culture medium can be easily amended or changed. Organs can be induced to develop in cell suspensions: root and shoot initiation usually commences in cell aggregates. Somatic embryos may arise from single cells. Cells from suspensions can also be plated onto solid media where single cells and/or cell aggregates grow into callus colonies from which plants can often be regenerated. For these reasons suspension cultures might be expected to provide a means of very rapid plant multiplication. There are two methods:

1. Plants may be obtained from somatic embryos formed in suspensions. Once embryos have been produced, they are normally grown into plantlets on solid media, although other methods are potentially available;
2. Cells from suspensions are plated onto solid media where single cells and/or cell aggregates grow into callus colonies from which plants can often be regenerated.

In practice neither of these techniques has been sufficiently reliable for use in plant propagation.

Immobilized cell cultures

Plant cells can be captured and immobilized by being cultured in a gel which is afterwards solidified. This technique has only limited application to plant micropropagation, but is now employed quite widely when plant cells are grown for the production of their secondary products or for the bio-transformation of chemical compounds.

Cultures of Single Cell Origin

Single cell clones

Cultures can be initiated from single plant cells, but only when special techniques are employed. Frequently these comprise passing suspension-cultured cells through a filter which removes coarse cell aggregates and allows only single cells and very small cell clusters to pass through. Small groups of cells are then assumed to have originated from single cells. The suspension obtained is usually plated onto (or incorporated into) a solidified medium in Petri dishes at a sufficient density to permit cell growth, but with the cells sufficiently dispersed so that, when growth commences, individual callus colonies can be recognized under a binocular microscope and transferred separately to fresh medium. Cell lines originating from single cells in this way are sometimes called single cell clones or cell strains.

Each cell clone has a minimum effective initial cell density (or minimum inoculation density) below which it cannot be cultured. The minimum density varies according to the medium and growth regulators in which the cells are placed; it is frequently about 10-15 cells/ml on standard media. Widely dispersed cells or protoplasts will not grow because they lose essential growth factors into the surrounding medium. The minimum inoculation density can therefore be lowered by adding to a standard medium either a filtered extract of a medium in which a culture has been previously grown (the medium is then said to be conditioned), or special organic additives (when it is said to be supplemented).

Cells or protoplasts plated at a density which is insufficient for spontaneous cell division may also be nurtured into initial growth by being '*nursed*' by tissue growing nearby. One way of doing this is to place an inoculum onto a filter paper disc (a raft) or some other inert porous material, which is then put in contact with an established callus culture of a similar species of plant, the cells of which are called *nurse cells*, and the tissue a feeder layer. An alternative technique is to divide a Petri dish into compartments. Nurse tissues cultured in some segments assist the growth of cells or protoplasts plated in the other areas.

Another method of producing cell colonies which are very likely to have had a single cell origin, has been described by Bellincampi *et al.* (1985). A filtered cell suspension with a high proportion of single cells, is cultured at high density in a medium which contains only 0.2% agar. At this concentration the agar does not solidify the medium, but keeps apart the cell colonies growing from individual cells, preventing them from aggregating. When clusters of approximately 10-15 cells have been formed, they can be plated at a dilution of 50 (20% plating efficiency) to 200 plating units/ml (60% plating efficiency) on a medium gelled with 1% agar where they grow as separate callus colonies. Plating efficiency is the percentage of plating units (cell aggregates in this case) which give rise to callus colonies.

The establishment of single cell clones is one way to separate genetically different cell lines from a mixed cell population. By artificially increasing the genetic variation between cells in a culture, and then applying a specific selection pressure, resistant cell lines have been obtained (e.g. those resistant to certain drugs, herbicides or high levels of salt), and in some instances plants with similar resistances have then been regenerated from the resulting cells or callus.

Separated cells

Single cells can be separated directly from intact plants. They are often more easily isolated and less liable to damage than protoplasts, because the cell wall remains intact. Consequently, single cells can be used in robust operations, such as direct physiological studies. It has been said that, for this purpose, they are more representative of differentiated tissues than cells derived from tissue cultures; but the disruption caused by separation may induce atypical responses.

Mechanical separation

In some plant species, disrupting the tissue mechanically can separate intact cells of certain organs. Viable mesophyll cells, for example, can be obtained easily from *Asparagus cladodes* and from leaves of *Macleaya cordata*. These cells can be grown either in suspension or solid culture and induced into morphogenesis, including somatic embryo formation simply placed pieces of the young cotyledons of sweet potato in water inside an abrasive tube in which a vortex was created. After removing debris, a cell suspension could be obtained from which cells grew and formed callus when plated on nutrient agar.

However, the capacity to isolate separated cells directly from higher plants appears to be limited. The type of tissue used seems to be important both to permit cell separation and to obtain subsequent growth. Cells separated from the leaves, instead of from the cotyledons, of sweet potato had no capacity for growth, and it was not possible to even separate cells by mechanical means from several other plants.

Enzymatic separation

Cell separation can be assisted by treating plant tissue with enzyme preparations such as crude pectinase or polygalacturonase, which loosen the attachment between individual cells in a tissue. Zaitlin first used this technique in 1959 to separate viable cells from tobacco leaves. Methods of isolation have been described by Takebe *et al.* (1968); Servaites and Ogren (1977) and Dow and Callow (1979). Cells isolated in this way can be suspended in culture medium and remain metabolically active.

Separated cells from leaf tissue of tobacco pre-infected with Tobacco Mosaic Virus have been used to study the formation of viral RNA's in the infected cells, and for studies on the interaction between leaf tissue cells and elicitor chemicals produced by fungal pathogens. Button and Botha (1975) produced a suspension of single cells of *Citrus* by macerating callus with 2-3% Macerase enzyme: the degree of dispersion of cells from suspension cultures can also be improved by enzyme addition.

Protoplasts

A protoplast is the living part of a plant cell, consisting of the cytoplasm and nucleus with the cell wall removed. Protoplasts can be isolated from whole plant organs or tissue cultures. If they are then placed in a suitable nutrient medium, they can be induced to re-form a cell wall and divide. A small cluster of cells eventually arises from each cell and, providing the protoplasts were originally plated at a relatively low density, can be recognized as one of many discrete '*callus colonies*'. Plants can often be regenerated from such callus. Protoplast culture therefore provides one route whereby plants can be multiplied, but it is not yet used for routine micropropagation work, although the number of species in which plant regeneration has been achieved is steadily increasing.

At present isolated protoplasts are used chiefly in research into plant virus infections, and for modifying the genetic information of the cell by inserting selected DNA fragments. Protoplasts may also be fused together to create plant cell hybrids. Genetically modified cells will be only of general practical value if whole plants having the new genetic constitution can be regenerated from them. The ability to recover plants from protoplast cultures is therefore of vital importance to the success of such genetic engineering projects in plant science.

Methods of protoplast preparation

There are several different methods by which protoplasts may be isolated:

(a) by mechanically cutting or breaking open the cell wall;

(b) by digesting away the cell wall with enzymes;

(c) by a combination of mechanical and enzymatic separation.

For successful isolation it has been found essential to cause the protoplast to contract away from the cell wall, to which, when the cell is turgid, it is tightly and pressed. Contraction is achieved by plasmolyzing cells with solutions of salts such as potassium chloride and magnesium sulphate, or with sugars or sugar alcohols (particularly mannitol). These osmotica must be of sufficient concentration to cause shrinkage of the protoplasm, but of insufficient strength to cause cellular damage.

In the past, protoplasts have been mechanically isolated from pieces of sectioned plant material, but only very small numbers were obtained intact and undamaged. This method has therefore been almost completely replaced by enzymatic isolation techniques. Commercially available preparations used for protoplast isolation are often mixtures of enzymes from a fungal or bacterial source, and have pectinase, cellulase and/or hemicellulase activity: they derive part of their effectiveness from being of mixed composition. Protoplasts are usually isolated using a combination of several different commercial products. Plasmolysis helps to protect the protoplast when the cell wall is ruptured during mechanical separation and also appears to make the cell more resistant to the toxic effects of the enzymes used for cell wall digestion. It also severs the plasmodesmata linking adjacent cells and so prevents the amalgamation of protoplasms when the cell walls are digested away.

Tissue from an entire plant to be used for protoplast separation, is first surface sterilized. Some further preparation to allow the penetration of osmotic solutions and the cell wall degrading enzymes, is often advantageous. For instance, when protoplasts are to be separated from leaf mesophyll, the epidermis of the leaf is first peeled away, or the leaf is cut in strips and the tissue segments are then plasmolyzed. The next step is to incubate the tissue with pectinase and cellulase enzymes for up to 18 hours in the same osmoticum, during which time the cell walls are degraded. Agitation of the incubated medium after this interval causes protoplasts to be released. They are washed and separated in solutions of suitable osmotic potential before being transferred to a culture medium.

Less severe and prolonged enzymatic cell digestion is required if plant tissue is first treated to mild mechanical homogenization before cellulase treatment. Another technique calls for the sequential use of enzymes; firstly pectinase to separate the cells, and then, when separation is complete, cellulase to digest the cell walls. The yield of viable protoplasts can sometimes be increased by pre-treatment of the chosen tissue with growth substances before separation is attempted. Protoplasts are also commonly isolated by enzymatic treatment of organs or tissues that have been cultured *in vitro*. Cells from suspension cultures, which have been subcultured frequently, and are dividing rapidly, are one suitable source.

The successful isolation of viable protoplasts capable of cell division and growth, can depend on the manner in which the mother plant was grown. For example, Durand (1979) found that consistently successful protoplast isolation from haploid *Nicotiana sylvestris* plants depended on having reproducible batches of young plants *in vitro*. The composition of the medium on which these plants were cultured had a striking effect on protoplast yield and on their ability to divide. A low salt medium devoid of vitamins was particularly disadvantageous. The light intensity under which the plants were grown was also critical.

Protoplast culture

Isolated plant protoplasts are very fragile and particularly liable to either physical or chemical damage. Thus if they are suspended in a liquid medium, it must not be agitated, and the high osmotic potential of the medium in which isolation was carried out must be temporarily maintained. As growth depends on adequate aeration, protoplasts are usually cultured in very shallow containers of liquid or solid media; fairly high plating densities (5×10^4 to 10^5 protoplasts/ml) may be necessary, possibly because endogenous chemicals are liable to leak away from such unprotected cells. To promote growth, it may also be beneficial to add to the medium supplementary chemicals and growth factors not normally required for the culture of intact cells.

The capability of plant protoplasts to divide appears to be closely related to their ability to form a cell wall. The type of wall that is produced initially can be controlled to some extent by the nature of the culture medium. A non-rigid wall can be produced on tobacco mesophyll protoplasts, for example, by culture in a medium containing a relatively high concentration of salts; but although such cells will divide 2-3 times, further cell division does not occur unless a rigid wall is induced to be formed by a change in the culture medium.

Under favourable circumstances formation of a cell wall seems to occur as soon as protoplasts are removed from hydrolysing enzyme preparations, and the first signs of cellulose deposition can be detected after only about 16 hours in culture medium. Once wall formation is initiated, the concentration of osmoticum is reduced to favour cell growth. This is readily accomplished in a liquid medium, but where protoplasts have been plated onto a solidified medium it will be necessary to transfer the cells on blocks of agar, to another substrate.

When it has formed a cell wall, the regenerated plant cell generally increases in size and may divide in 3-5 days. If further cell divisions occur, each protoplast gives rise to a small group of intact cells and then a small callus colony. Green chloroplasts in cells derived from leaf mesophyll protoplasts, lose their integrity and disappear as callus formation proceeds. Protoplasts may originate from cells of the intact plant, which are not all of the same genetic composition. If such cells are grown in liquid medium, they may stick together and form common cell walls. Colonies of mixed callus will result which could give rise to genetically different plants or plant chimeras.

To avoid cell aggregation, protoplasts should be freely dispersed and cultured at as low a density as possible. This may mean that, as in the culture of intact cells at low density, nurse tissue, or a conditioned or specially supplemented medium, must be employed. A method of the latter kind was devised by Raveh *et al.* (1973). A fabric support has been used to suspend protoplasts in a liquid medium so that media changes can be made readily.

An entire plant was first regenerated from callus originated from an isolated protoplast in 1971. Since then plants have been produced from the protoplasts of a wide range of species, using indirect shoot morphogenesis or indirect embryogenesis. The direct formation of somatic embryos from cultured protoplasts is also possible.

Protoplast fusion

Although fusion of plant protoplasts was observed many years ago, it has become especially significant since methods have been developed for protoplast isolation and subsequent regeneration into intact plants. Isolated protoplasts do not normally fuse together because they carry a superficial negative charge causing them to repel one another. Various techniques have been discovered to induce fusion to take place. Two of the most successful techniques are the addition of *polyethylene glycol* (PEG) in the presence of a high concentration of calcium ions and a pH between 8-10, and the application of short pulses of direct electrical current (electro-fusion). By mixing protoplasts from plants of two different species or genera, fusions may be accomplished:

(a) between protoplasts of the same plant where fusion of the nuclei of two cells would give rise to a homokaryon (synkaryon);

(b) between protoplasts of the same plant species (intravarietal or intraspecific fusion);

(c) between protoplasts of different plant species or genera (interspecific or intergeneric fusion).

Fusions of types (b) and (c) above can result in the formation of genetic hybrids (heterokaryocytes), which formally could only be obtained rarely through sexual crossings. By separating the fused hybrid cells from the mixed protoplast population before culture, or by devising a method whereby the cells arising from fused cells may be recognized once they have commenced growth, it has been possible to regenerate new somatic hybrid (as opposed to sexually hybrid) plants. Some novel interspecific and intergeneric hybrid plants have been obtained by this means. A fusion of the cytoplasm of one kind of plant with the nucleus of another is also possible. Such cybrid plants can be useful in plant breeding programmes for the transfer of cytoplasmic genes.

Cytodifferentiation

In an intact plant there are many kinds of cells all having different forms and functions. Meristematic cells, and soft thin-walled parenchymatous tissue, are said to be undifferentiated, while specialized cells are said to be differentiated. The cells of callus and suspension cultures are mainly undifferentiated, and it is not yet possible to induce them to become of just one differentiated type. This is partly because culture systems are usually designed to promote cell growth: differentiation frequently occurs as cells cease to divide actively and become quiescent. Furthermore, the formation of differentiated cells appears to be correlated with organ development, therefore the prior expression of genes governing organogenesis may often be required. The *in vitro* environment can also be very different to that in the whole plant where each cell is governed by the restraint and influence of other surrounding cells. In suspension cultures, for example, cells are largely deprived of directional signals, influences from neighbouring differentiated tissues, and correlative messages that may normally pass between adjacent cells by way of interconnecting strands of protoplasm (plasmodesmata).

The differentiated state is also difficult to preserve when cells are isolated from a plant. Askani and Beiderbeck (1988) tried to keep mesophyll cells in a differentiated state. The character of palisade parenchyma cells with regard to size, cell form, colour and size, and distribution of chloroplasts could be preserved for 168 h, but after this the chloroplasts became light green, their distribution was no longer homogeneous and some of the cells began to divide. Differentiated cells are most effectively produced *in vitro* within organs such as shoots and roots; even here there may not be the full range of cell types found in intact plants *in vivo*.

Differentiated Cells in Callus and Cell Cultures

Three types of differentiated cells are commonly found in callus and cell cultures; these are vessels and tracheids (the cells from which the water-conducting vascular xylem is constructed), and cells containing chloroplasts (organelles carrying the green photosynthetic pigment, chlorophyll). Phloem sieve tubes may be present but are difficult to distinguish from undifferentiated cells.

Tracheid formation

Callus cultures are more likely to contain tracheids than any other kind of differentiated cell. The proportion formed depends on the species from which the culture originated and especially upon the kind of sugar and growth regulators added to the medium. Tracheid formation may represent or be associated with an early stage in the development of shoot meristems. Nodules containing xylem elements in callus of *Pelargonium* have, for example, been observed to develop into shoots when moved to an auxin-free medium.

The rapid cell division initiated when tissue is transferred to a nutrient medium usually occurs in meristems formed around the periphery of the explant. Cell differentiation does not take place in callus cultures during this phase but begins when peripheral meristematic activity is replaced or supplemented by the formation of centres of cell division deeper in the tissue. These internal centres generally take the form of meristematic nodules that may produce further expanded and undifferentiated cells (so contributing to callus growth) or cells that differentiate into xylem or phloem elements. Nodules can form primitive vascular bundles, with the xylem occurring centrally and the phloem peripherally, separated from the xylem by a meristematic region.

Chloroplast differentiation

The formation and maintenance of green chloroplasts in cultured plant cells represents another form of cellular differentiation which is easy to monitor, and which has been studied fairly extensively. When chloroplast-containing cells from an intact plant are transferred to a nutrient medium they begin to dedifferentiate. This process continues in the event of cell division and results in a loss of structure of the membranes containing chlorophyll (thylakoids) and the stacks (grana) into which they are arranged, and the accumulation of lipid-containing globules. The chloroplasts eventually change shape and degenerate.

Callus cells frequently do not contain chloroplasts but only plastids containing starch grains in which a slightly-developed lamellar system may be apparent. All the same, many calluses have been discovered that do turn green on continued exposure to light and are composed of a majority of chloroplast-containing cells.

Chloroplast formation can also be connected with the capacity of callus to undergo morphogenesis. Green spots sometimes appear on some calluses and it is from these areas that new shoots arise. By subculturing areas with green spots, a highly morphogenic tissue can sometimes be obtained. The formation of chloroplasts and their continued integrity is also favoured by cell aggregation. When green callus tissue is used to initiate suspension cultures, the number of chloroplasts and their degree of differentiation are reduced. Nevertheless, there can be some increase in chlorophyll content during the stationary phase of batch cultures.

The level of chlorophyll so far obtained in tissue cultures is well below that found in mesophyll cells of whole plants of the same species, and the rate of chlorophyll formation on exposure of cultured cells to the light is extremely slow compared to the response of etiolated organized tissues. The greening of cultures also tends to be unpredictable and even within individual cells, a range in the degree of chloroplast development is often found.

In the carbon dioxide concentrations found in culture vessels, green callus tissue is normally photomixotrophic (i.e. the chloroplasts are able to fix part of the carbon that the cells require) and growth is still partly dependent on the incorporation of sucrose into the medium. However, green photoautotrophic callus cultures have been obtained from several different kinds of plants. When grown at high carbon dioxide concentrations (1-5%), without a carbon source in the medium, they are capable of increasing in dry weight by photosynthetic carbon assimilation alone.

Photoautotrophic cell suspensions have also been obtained. They too normally require high carbon dioxide levels, but cell lines of some species have been isolated capable of growing in ambient CO_2 concentration. Why cultured cells do not freely develop fully functional chloroplasts is not fully known.

MORPHOGENESIS

Nature and Induction

New organs such as shoots and roots can be induced to form on cultured plant tissues. Such freshly originated organs are said to be adventive or adventitious. The creation of new form and organization, where previously it was lacking, is termed morphogenesis or organogenesis. Tissues or organs that have the capacity for morphogenesis/organogenesis are said to be *morphogenic* (*morphogenetic*) or *organogenic* (*organogenetic*). So far it has been possible to obtain the *de novo* (adventitious) formation of:

1. Shoots (*caulogenesis*) and roots (*rhizogenesis*) separately. The formation of leaves adventitiously *in vitro* usually denotes the presence of a shoot meristem. Sometimes leaves appear without apparent shoot formation: opinions are divided on whether such leaves can have arisen *de novo,* or whether a shoot meristem must have been present first of all and subsequently failed to develop.
2. Embryos that are structurally similar to the embryos found in true seeds. Such embryos often develop a region equivalent to the suspensor of zygotic embryos and, unlike shoot or root buds, come to have both a shoot and a root pole. To distinguish them from zygotic or seed embryos, embryos produced from cells or tissues of the plant body are called *somatic embryos* (or *embryoids*) and the process leading to their inception is termed *embryogenesis*. The word '*embryoid*' has been especially used when it has been unclear whether the embryo-like structures seen in cultures were truly the somatic equivalent of zygotic embryos. Somatic embryogenesis is now such a widely observed and documented event that *somatic embryo* has become the preferred term.
3. Flowers, flower initials or perianth parts. The formation of flowers or floral parts is rare, occurring only under special circumstances and is not relevant to plant propagation.

Haploid Plants

Anther and Pollen Culture

In 1953 Tulecke discovered that haploid tissue (i.e. tissue composed of cells having half the chromosome number that is characteristic of a species), could be produced by the culture of *Ginkgo* pollen. Little notice was taken of his work until Guha and Maheshwari (1964, 1967) managed to regenerate haploid plants from pollen of *Datura innoxia* by culturing intact anthers. Since then a great deal of research has been devoted to the subject.

The basis of pollen and anther culture is that on an appropriate medium the pollen microspores of some plant species can be induced to give rise to vegetative cells, instead of pollen grains. This change from a normal sexual gametophytic pattern of development into a vegetative (*sporophytic*) pattern, appears to be initiated in an early phase of the cell cycle when transcription of genes concerned with gametophytic development is blocked and genes concerned with sporophytic development are activated.

The result is that in place of pollen with the capacity to produce gametes and a pollen tube, microspores are produced capable of forming haploid pro-embryos (somatic embryos formed directly from the microspores), or callus tissue. The formation of plants from pollen microspores in this way is sometimes called *androgenesis*. Haploid plants are more readily regenerated by culturing microspores within anthers than by culturing isolated pollen. The presence of the anther wall provides a stimulus to sporophytic development. The nature of the stimulus is not known but it may be nutritional and/or hormonal. Embryogenesis has only been induced from isolated pollen of a very small number of plants.

The number of plants species from which anther culture has resulted in haploid plants is relatively few. It comprised about 70 species in 29 genera up to 1975 and 121 species or hybrids in 20 families by 1981-1982 and by now, very many more. The early stages of embryogenesis or callus formation without plant regeneration have been obtained in several other kinds of plants. Fifty-eight per cent of the reports of embryogenesis or plant regeneration in Maheshwari *et al.* (1982) was attributable to species within the family Solanaceae. Species in which haploid plants can be regenerated reliably and at high frequency remain. a comparatively small part of the total. They again mainly comprise Solanaceous species such as *Datura*, *Nicotiana*, *Hyoscyamus*, *Solanum* and some brassicas.

Gynogenesis

Another theoretical source of haploid plants in angiosperms is the female egg nucleus or ovum; this is contained within the nucellus of an ovule in a specialized cell (the *megaspore* or *embryo sac*). The ovum cannot be separated readily from other associated nuclei in the megaspore and so haploid plants can normally be produced from it, only by stimulating the development of unfertilized ovules into seedlings. In some species (e.g. *Gerbera jamesonii*; maize; sugar beet; onion), some haploid plants can be obtained by culturing unpollinated ovules, ovaries or flower buds. In some other plants, larger numbers of haploids are obtained if ovaries are pollinated by a distantly-related species (or genus) or with pollen which has been irradiated with X- or γ-rays.

Successful pollination results in stimulation of endosperm growth by fusion of one of the generative nuclei of the pollen tube with the central fusion nucleus of the megaspore, but fusion of the other generative nucleus with the egg cell does not occur and the egg cell is induced to grow into a seedling without being fertilized (*gynogenesis*). An alternative technique, which has resulted in haploid *Petunia* seedlings is to treat ovaries with γ-rays and then pollinate them with normal pollen. Gynogenesis has so far been employed much less frequently than androgenesis for the production of haploids.

Haploid cells and haploid plants produced by androgenesis or gynogenesis have many uses in plant breeding and genetics. Most recent research on anther culture has concentrated on trying to improve the efficiency of plantlet regeneration in economically important species. Haploid plants of cereals are particularly valuable in breeding programmes, but in the Gramineae, the frequency and reliability of recovery through anther culture is still too low for routine use.

3

Genetic Engineering of Cells

At present, various bioengineering methods as applied to plant breeding are steadily gaining in importance, including such techniques as microclonal propagation of valuable elite plants, embryo and meristem cultures, another culture, cell culture breeding based on somaclonal variability, somatic hybridization of protoplasts, and others. A lot of hope is pinned to gene engineering. The Soviet government gives a great deal of attention to the development of bioengineering in our country, and 22 laboratories and teams dedicated to this research have been established in breeding centres of the Soviet Union.

Method of Sterile Tissue and Cell Cultures

Isolated tissue and organ cultures have always fascinated investigators ever since their concept emerged at the turn of the century, when G. Haberlandt (1902) was the first to propose cultivation of plant cells as a tool for proving their totipotency—that is, ability to generate a whole organism from its part, as is found in the fertilized egg cell. He wrote that cultivation of isolated cells in nutrient media would at least provide the experimenter with a possibility to approach many problems from an entirely new angle. The first cells cultivated by scientists in artificial media were those of animal rather than plant organisms. The first report about this breakthrough appeared in 1907. As regards plant cells, they turned out to be more difficult to control, and botanists were able to tame them only by the early thirties.

Currently, these methods are not only used in research but have also gained wide practical application in plant breeding, vegetable gardening, fruit growing, and especially, floriculture, more specifically in producing virus-free planting stock. A method developed way back in the fifties by G. Morel and C. Martin from the National Institute of Agronomic Research in France permits one to indefinitely propagate a single plant from pieces of its stem, a bud, and so on. One rose bush may yield anywhere from 200 to 400 thousand offspring per year. What makes this method so interesting is that it allows mass propagation of species impossible to propagate traditionally by cuttings or those whose propagation by seed is undesirable (e.g., hybrid varieties of asparagus and other crops). The method was applied at first to such ornamentals and vegetable crops as dahlias, carnations, orchids, strawberry, potatoes, and eventually to fruit trees. At present, it is used for vegetative propagation and even planting, on test plots, of oil and coconut palms, coffee, and sugarcane.

Companies specializing in mass production of such material sprang up in many countries. The enterprise VEB PAC Jungpflanzen in Dresden, GDR, for example, fully meets the country's requirements in virus-free stock plants of chrysanthemum, carnation, freesia, garden geranium, hydrangea, and strawberry.

Over the past three decades, the isolated tissue and organ culture method has become the most important tool of Soviet plant breeders, especially to overcome the difficulties arising in distant hybridization.

The basic method of tissue culture were developed for relatively few convenient model objects of no particular economic importance, such as *Nicotiana tabacum*, *Daucus carota*, *Petunia*, and some wild plants. In fundamental research at the international level these plants still play an important role, although the current trend is to involve other species, especially cultivated ones. The right choice of the object species is decisive in the tissue culture method because the cultivation conditions elaborated for one species most often cannot be extrapolated to others with equal success. As a rule, even for different varieties of the same species the basic procedure must be modified.

Some plant families are especially hard to experiment with in tissue culture. Since our current level of knowledge about all pertinent aspects of economically important grain cereals is limited, only a few of them lend themselves to breeding experiments based on this method. The best results so far have been obtained on rice. In contrast, if we take *Solanaceae*, particularly tobacco, the procedure has been expanded broadly enough to incorporate haploid plants from another culture into standard breeding programs.

The advent of the tissue culture method has led to development of better varieties, faster use of the valuable source material and unique individual forms; in other words, it has substantially enhanced the effectiveness of plant breeding. Application of the tissue culture method imposes certain limitations and requirements, namely: (1) the breeder is not free in selecting the object, and his choice is confined to particular cultivated plants; (2) the published results of laboratory experiments must be accurately and reliably reproducible as well as applicable to large-scale work; (3) the potential of tissue culture must be duly taken into account in planning the procedures for improving a particular crop.

At present, virtually any part of a plant can be passed into sterile culture in spite of the marked differences between individual species, organs, and stages of their development. Different problems with sterilization arise in initiating the primary culture, depending on how heavily the starting material is contaminated.

When the method is used in plant breeding and growing, only such organs and tissues should be cultivated in which organogenesis may be induced to an adequate degree. Some potentialities of the plant tissue culture method as applied to breeding experiments are summarized in Table 3.1.

The method comprises the following steps: selection of a variety (genotype) appropriate to the task at hand and the right organ for initiation of the culture; initiation of sterile culture; creation of the right conditions (nutrient mixture, temperature, lighting, etc.) for stimulating plant development along the desired path in line with a particular plant breeding or growing program; regeneration of viable plants; transplantation of the stock selected for breeding or growing into soil or soilless substrate (hydroponics) so as to make it available to the breeder for his subsequent work.

The areas of application of the tissue culture method can be grouped according to the following objects: expansion of the genetic stock for plant breeding by producing new source material; preservation and propagation of valuable elite plants and strains; and obtaining virus-free material for cultivation of fruit and other crops and keeping it healthy. These three main groups of objectives are closely interrelated and intertwined.

Expansion of Genetic Stock for Plant Breeding

The standard plant breeding procedures based on combinations, recombinations, and selection adversely affect genetic variability or,

Table 3.1. Vitro culture for research and plant breeding purposes.

Cultivated organs	*Objective*
Apical portion of shoots (apical meristem + diferen entiated tissue of the shoot)	To propagate individual plants under conditions (cloning)
	To derive lines in breeding for heterosis
	To produce virus-free material in varieties and selection forms (preservation of maternal plants)
	To conduct physiological studies, particularly into transition from vegetative to reproductive phase
Flower buds, ovaries, ovules	To overcome sexual incompatibility in interspecific and intergeneric crosses by *in vitro* fertilization
	To resort to propagation under sterile conditions with induced polyembryony
Anthers, pollen	To produce haploid tissues and plants
	To rationally produce certain homozygotes
Embryo	To overcome postgamic incompatibility in interspecific and intergeneric crosses and to propagate hybrid plants
	To interrupt seed dormancy
	To resort to propagation under sterile conditions with polyembryonic and somatic embryogenesis
Cells	To conduct genetic and physiological studies
	To perform mutations and selection at the cellular level
	To prepare cell suspensions
	To obtain and isolate protoplasts
	To overcome pro- and postgamic incompatibility by somatic hybridization

	To resort to genetically identical propagation (in exceptional cases)
	To transfer carriers of genetic information (chromosomes, plastids, mitochondria, etc.) and, if possible, DNA
Callus tissue can be induced in cultures of all plant organs and tissues	To utilize enhanced genetic variability in producing new source materials
	To resort to genetically identical propagation in species and varieties characterized by cytological stability through induced regeneration or somatic embryogenesis
	To prepare suspended cultures
	To isolate protoplast cells
	To conduct cytological and physiological studies

in other words, lead to genetic erosion, whereby the genetic resources are improverished and plants become genetically vulnerable. Only through continuous expansion of the genetic resources can plant breeding remain effective in the future. In this respect, tissue culture methods have made an important contribution.

Interspecific and intergeneric hybrids are instrumental in expanding the genetic resources for cultivated plants, true allopolyploids being especially effective. What may gain in importance in future plant breeding is work with cell protoplasts, based not only on successful somatic interspecific hybridization but also on genetic manipulation to pass on genetic information (cell nuclei, chloroplasts, plastids, and, if possible, chromosomes and DNA).

Ways to Overcome Progamic Incompatibility (*in vitro* fertilization)

The progamic incompatibility can be overcome by fertilization *in vitro* (in a tube). Such an approach implies successful cultivation of ovaries and ovules *in vitro* as well as creating appropriate conditions for germination of the pollen on a sterile nutrient medium.

In 1951, J.P. Nitsch developed a method for cultivating tomato and cucumber ovaries and was the first to have achieved *in vitro* formation of viable seeds in cucumbers. Later, investigators decided to conduct the entire process of pollen germination, pollen tube growth, fertilization, and development of fruit from the maternal plant *in vitro*. In 1962, K. Kanta and collaborators were the first to report about *in vitro* pollination of opium poppy (*Papaver somniferum*). Recent publications clearly show that more progress can be achieved in this field with emphasis on cultivated plants.

Ways to Overcome Postgamic Incompatibility (Embryo Culture)

Ever since F.Z. Laibach (1925) was for the first time successful in growing young embryos of the usually sterile interspecific flax hybrid *Linum perenne* × *L. austricum* into healthy sprouts, this method has often been used, especially over the past 25 years, by many workers in difficult intergeneric and interspecific crosses. Today, embryo culture is indispensable in difficult crossing programs.

Worthy of note are the successful crosses *Triticum* × *Agropyrum*, *Hordeum* × *Secale*, *Elymus* × *Triticum*, *Tripsacum* × *Zea*, and others. This method has also led to such hybrids as *Hordeum jubatum* × *Secale cereale*, *Hordeum maritinum* × *Secale cereale*, and the first hybrid *Hordecale*.

The use of embryo culture at the International Maize and Wheat Improvement Centre in Mexico has produced good results in triticale breeding.

Anther Culture

In 1964, S. Guha and S.C. Maheshwari discovered embryo-like structures in an *in vitro* culture of Jimsonweed (*Datura*) anthers and two years later made them grow into a haploid plant. They traced the origin of the latter to a haploid pollen grain. Important contributions to mass production of haploids from pollen have been made by J.P. Nitsch and coworkers. The elaborated method involves anther culture to induce *in vitro* androgenesis in plants.

G.Z. Melchers (1972, 1977) proposed new approaches to mass production of haploids, highly promising for plant breeding, and suggested that the anther culture method should be integrated into standard breeding procedures. Chinese workers have used anther culture to produce a short-stem, early-ripening and high-yielding rice variety. In our country, the method has been instrumental in deriving promising forms of barley, triticale, tobacco, potato, and many other crops. For example, two promising spring barley varieties, Istok and Odessky, have been developed at the All-Union Institute of Plant Breeding and Genetics. The development of Istok took an unusually short period of time—four years instead of the typical 8 to 12 years. Prospective rice and potato varieties are also in the works. In view of the major role to be played by somatic cell genetics in the future, the importance of haploid plants is steadily increasing.

Pollen Culture

When anther culture is used, division of the diploid cells of the anther tissue and the subsequent emergence of diploid embryoids from them can hardly be eliminated. Therefore, much more promising are experiments in which *in vitro* androgenesis proceeds on isolated pollen rather than anthers. The first successful results are obtained on cabbage and tomato pollen by T. Kameya and K. Xinata in 1970. Their achievement opens up new possibilities for mass production of haploids useful in plant breeding.

Cell Culture and Somaclonal Breeding

While resorting to thc tissue culture, method, experimenters use separate portions of living tissues, which are essentially aggregations of cells homogeneous in structure, identical in function and sharing a common origin in the development of a particular organ. Then the question arose whether a single cell could be isolated as a further step. This turned out to be quite possible.

Plant cell and isolated protoplast cultures have yielded spectacular results in recent years. Breeders are especially interested in selection of specific mutants from cell lines. To do this, a reliable technique is necessary to induce morphogenesis and regeneration of karyologically normal, viable plants. The regeneration of plants from individual cells and protoplasts has so far been easy in the case of Solanaceae, the family that includes tomato and tobacco, as well as some other families including such crops as carrots, sweet orange, rape, and asparagus. Plants have also been regenerated from isolated cells of potato and protoplasts of alfalfa.

Preservation and Propagation of Valuable Elite Plants and Strains *in vitro*

An important application of the tissue culture method has to do with propagation of valuable elite plants with a view to obtaining genetically identical clones for various uses in plant breeding and growing. The method permits: preservation and propagation of individual genotypes as starting forms for accomplishing a number of tasks in practical and experimental breeding; rapid and effective propagation of new valuable varieties; preservation and effective propagation of lines for hybrid seed production involving vegetable, ornamental, and other crops; inexpensive propagation of highly productive genotypes of forest and ornamental trees as well as stock for fruit trees; propagation of virus-free material under sterile conditions; and preservation of the existing varieties of vegetatively propagated cultivated plants and major cross-pollinated crops.

Cloning Based on Apical Meristem Culture under Sterile Conditions

In addition to the apical meristem, the stem tips transplanted onto an artificial nutrient medium contain other components as well. This is also true with regard to production of virus-free material. Stem tips are genetically stable in most cases when cultivated *in vitro*. Continuous growth and regeneration of roots are achievable without any difficulty. Incorporation of the *in vitro* method into breeding procedures is essential for preservation of valuable plants.

Regeneration of Viable Plants from Other Organs

Stems, pieces of leaves, flower organs, bulbs, and other parts of plants have been successfully used in cloning under sterile conditions. Vegetable crop breeders have been able to propagate leek from sliced bulbs. Red cabbage has been found to exhibit high regenerative capacity. A strain preservation method has been developed for asparagus, Brussels sprouts, and cauliflower. Using underground buds of asparagus permits us to achieve better root formation. Cauliflower inflorescences have proved to be ideal for propagation under sterile conditions, the multiplication ratio achievable within a single passage may be about 1:800. Positive results have been attained with liliaceous, iridaceous and amaryllidaceous explantates from bulbs, corms, rhizomes, leaves, inflorescences, and ovules, as well as star-of-Bethlehem explantates from the stem, leaves, ovules, calyx lobes, and bulb scales. The resulting plants are identical with the maternal form. Flower buds of freesia have been used to produce young plants through induced adventitious formation. Good results have been obtained with narcissus explantates from the leaf base, pedicel, and ovule.

Cloning under Sterile Conditions as an Important Component of Breeding some Cultivated Plants for Heterosis

The breeding for heterosis not only leads to 20-40 percent higher yields, but also gives more homogeneous hybrids, as compared to the standard varieties. Therefore, F_1 hybrids, also referred to as heterotic, of some vegetable crops become a prerequisite for their commercial cultivation. As regards cauliflower, for instance, its commercial growing calls for single-pass mechanical harvesting, which can be guaranteed only by using F_1 hybrids because even the best standard varieties have to be harvested four or five times. The production of elite lines, just as their vegetative propagation and preservation in genetically identical form, can be ensured only by cloning under conditions of sterility. Strains of some vegetable crops cannot be produced by inbreeding in most cases because of the pronounced depression that follows. Therefore, there is very reason to consider propagation of strains by the sterile culture method for heterotic seed production economically justified. This can be ensured by carefully studying the theoretical aspects of the optimal mode of propagation of a sterile culture and refining the actual cloning procedure so that *in vitro* plant growing becomes attractive and feasible at minimal cost.

Obtaining Virus-free Material and Keeping it Healthy

When we say "virus-free", the implication is that infected plants have to be rid of viruses. In this case we are speaking exclusively of vegetatively propagated plants in which this propagation mode is conducive to further spread of viral diseases. This is why the measures taken to get rid of viruses must always be associated with primary seed production; that is, the planting stock must be kept healthy.

There is evidence that the concentration of viruses decreases toward the growth cone. The growth cone itself is often free of viruses. This finding was instrumental in eliminating viruses from the source material by isolating the meristem under sterile conditions and ensuring it *in vitro* differentiation. It is very difficult to isolate a meristem ranging in size from 0.05 to 0.1 mm, and the chances that plants will differentiate from it are minimal. Therefore, it is isolated together with the first leaf primordia, and we are now essentially dealing with the apical portion of a shoot as large as 0.1 to 1 mm. While this ensures more reliable production of virus-free material, the degree of its differentiation increases. Moreover, this method is more convenient in practical use. Experimenters do not have to concern themselves with selection of the appropriate nutrient medium. They have at their disposal the readily available Murashige-Skoog, Bayes, White, and Heller nutrient media.

To give an idea of the nutrient media suitable for the purpose, Table 3.2. gives the composition of the frequently employed Murashige-Skoog medium.

If the meristem culture is maintained to keep the material healthy on a large, commercial scale, particular attention should be given to the following: varietal purity must be ensured throughout all propagation steps and one must provide for continuous, even growth of a large number of meristems and a high rate of growth of the differentiated plants after their transplantation into pots. Once taken, the meristem

Table 3.2. Murashige-Skoog Medium

Macrosalts	*ppm*	*mM*	*Micronutrients*	*ppm*	*µm*
KNO_3	1900	18.3	H_3BO_3	6.2	100.0
KH_2PO_4	170	1.3	$MnSO_4 \cdot 4H_2O$	22.3	100.0
NH_4NO_3	1650	20.6	$FeSO_4 \cdot 7H_2O$	27.8	100.0
$MgSO_4 \cdot 7H_2O$	370	1.5	Na_2-EDTA$\cdot 2H_2O$	37.3	100.0
$CaCl_2 \cdot 2H_2O$	440	3.0	$CoCl_2 \cdot 6H_2O$	0.025	0.15
			$CuSO_4 \cdot 5H_2O$	0.025	0.1
			$ZnSO_4 \cdot 7H_2O$	8.6	30.0
			$Na_2MoO_4 \cdot 2H_2O$	0.25	1.0
			KJ	0.83	5.0

Ionic composition	*mM*	*Ionic composition*	*µM*
Na^+	0.2	Mg^{++}	100.0
K^+	20.1	Fe^{++}	100.0
NH_4^+	20.6	Co^{++}	0.15
Mg^{++}	1.5	Cu^{++}	0.1
Ca^{++}	3.0	Zn^{++}	30.0
SO_4^{--}	1.7	BO_3^{---}	100.0
NO_3^-	39.4	MoO_4^{--}	1.0
PO_4^{---}	1.3	J^-	5.0
Cl^-	6.0		

Organic Additives to Murashige-Shook Medium

Components	*ppm*	*mM*	*Components*	*ppm*	*µM*
Edamine	1000	—	Glycine	2.0	26.7
Myoinositol	100	0,6	Thiamine-HCl	0.1	0.3
Sucrose	30,000	87.6	Pyridoxin-HCl	0.5	2.5
			Nicotinic acid	0.5	4.1
			Indoleacetic acid	2.0	11.4
pH 5.7-5.8			Kinetin	0.2	0.9

must be kept in a fully controlled indoor environment at 22°C and under illumination at 1000 to 3000 lux for 16 hours.

Application of the tissue culture methods opens up new possibilities for enhancing the effectiveness of breeding procedures and also offers new approaches to developing valuable highly productive crop varieties.

Propagation of individual plants under sterile conditions to obtain genetically identical material is essential as a preliminary step in the propagation of new valuable varieties, breeding stock, and lines or ridding them of viruses, on the one hand, and as a precursor to the prospective methods of haploid-yielding anther or pollen culture, callus culture, and suspension culture of cells or their somatic fusion, on the other.

To sum up, the tissue culture method is especially promising as a means for substantially expanding the genetic resources for plant breeding based on interspecific and intergeneric hybridization, preservation propagation of valuable elite plants, and obtaining virus-free plants of field and fruit crops. The method has been successfully integrated into those of artificial fertilization; embryo, callus, anther, pollen and somatic cell cultures; protoplast culture (including fusion of somatic cells); and apical meristem culture. In each particular case, the criterion of its effectiveness is the degree of control over the differentiation process.

The Potential of Genetic and Gene Engineering for Plant Breeding

The development of new crop varieties of the so-called intensive type, which is to say responding well to improvements in farming practices, is a task of paramount importance. It can be accomplished by resorting to modern genetic methods of breeding. We should like to spend some time discussing the potential of genetic engineering as applied to plant breeding. Although genetic engineering is in the forefront of bioengineering, the two terms are not synonymous. By genetic engineering is meant application of integrated genetic, cytological, or molecular-genetic methods of create organisms with desired properties. When manipulations are at the level of individual genes or their fragments, we are speaking of gene engineering.

Depending on the task at hand, the following levels of application of bio- and genetic engineering methods are involved: (1) molecular, when we are dealing with individual fragments of genes; (2) gene; (3) chromosome; (4) plasmid; (5) cellular; (6) tissue; (7) organism; and (8) population.

Many aspects of genetic engineering in its broad sense have already been discussed. These include development of new varieties based on genetic recombination with accurate calculations using genetic maps of chromosomes, substitution of chromosomes of one species by those of another in distant hybridization, incorporation of valuable genes into individual chromosomes by way of translocation, creation of new specific forms by combining sets of chromosomes belonging to different species through allopolyploidization, and so on. Here, we should like to dwell on the potential of preprogrammed genotype alterations at the level of cells, molecules, and genes, achieved through genetic engineering.

Genetic Engineering at Cellular Level

The advent of some new techniques has broadened the potential of genetic engineering at the cellular level. A decisive role was played by the method of somatic cell hybridization based on union of isolated protoplasts (plant cell contents separated enzymatically from the rigid envelope). Such protoplasts from different plants, including those belonging to different species, fuse easily, whereby the parental genes become combined in the hybrid cell. In this way, fused cells of, for example, barley and carrot, soybean and maize, and other pairs can be obtained. Such fusion, however, makes sense only if it ultimately results in a completely viable organism. Then, the hybrid cell can be used to grow a complete plant. This method enables new plant forms not existing in nature to be synthesized.

Isolation of protoplasts from plant cells was made possible by treatment of mesophyll cells in a hypertonic medium with the enzymes pectinase and cellulase acting on the connective tissues of leaves and cell envelopes. In this culture, plant cell protoplasts form green spheres separated from the surrounding medium only by the cytoplasmic membrane. To ensure fusion of the protoplasts, sodium nitrate, polyethylene glycol, and other substances are used as inductors. When protoplasts from two parents fuse into one, the entire plant can be cloned with a full complement of the parental characters. Fusion of cells belonging to different species gives somatic, or parasexual, hybrids. This technique is highly promising for genetic engineers, especially when it comes to creation of genomes that cannot be obtained generatively because of the complete sexual incompatibility between the parents.

The method of somatic hybridization includes the following consecutive steps: preparation of the appropriate tissue under sterile conditions; preparation of a cell suspension from the tissue, for example, mesophyll of the parental plant leaves; enzymatic dissolution of cell envelopes to obtain bare protoplasts; stimulation of the cells; inducing nuclear fusion; isolation of the fused heterokaryons; initiation of cell division; and regeneration of hybrid plants.

Even a simplified presentation of the process clearly shows how difficult it is to produce somatic hybrids. None the less, in 104 cases hybrid plants were successfully produced by the somatic cell fusion method. The resulting hybrids included 35 intraspecific, 48 interspecific, 9 intergeneric, 12 intertribal, and 16 interfamily ones. These hybrids belonged primarily to the family Solanaceae (genera *Nicotiana*, *Datura*, *Solanum*, and *Petunia*). Positive results have also been obtained with umbellifers, crucifers, and legumes. Of greatest interest from the practical standpoint today is somatic hybridization of different species belonging to the same genus. Most of the successful experiments have involved *Nicotiana* and *Solanum*. In 1978, a potato-tomato hybrid called "pomato" was produced by this technique. This was merely another demonstration of the method's possibilities, but the plant itself was completely worthless from the practical point of view. According to a report published in the United States in 1984, bioengineers had developed a new potato variety capable of defending itself against pests. It was a hybrid between Bolivian potato and an insectivorous plant. With the aid of the hairs forming a solid cover on leaves and stems, the plant easily dispatched attacking insects. The approaching pests were trapped by the hairs exuding a viscous toxic liquid. The hybrid turned out to be highly effective, killing 90 percent of the "*aggressors*" careless enough to contact the plant. Growers of this insectivorous potato did not have to buy insecticides any more. The only catch was that this 'weapon' was equally effective against pests and beneficial insects.

The evolution of this method as a new experimental technique is marked by the following three stages:

1972-1976—lack of confidence and criticism;

1977-1980—excessive optimism;

from 1981 on—more realistic assessment corroborated by intensive fundamental studies.

Protoplastic fusion also opens up other possibilities, more realistic for the foreseeable future. If most of the genetic information in a cell of higher plants is contained in the nucleus, some of it is known to be found in such cytoplasmic organelles as chloroplasts (ensuring photosynthesis) and mitochondria (ensuring cell respiration). For instance, genes responsible for herbicide resistance are located in chloroplasts, while those responsible for male sterility are in mitochondria. The existence of plants with male sterility permits mass production of hybrid seed without emasculation of the maternal forms. The genetic novelty of somatic hybrids became evident in recent years as regards combination of nuclear and extrachromosomal inheritance. They are radically different from sexual hybrids: the former inherit extranuclear genes from both parents, whereas the latter do so only along the maternal line. It is believed that somatic hybridization may result in 12 viable nucleus-cytoplasm combinations, while the sexual process gives only two.

Obtaining different nucleus-cytoplasm combinations opens up quite new possibilities for plant breeding. A good example of practical application of this method is provided by the work with rape carried out at the National Institute of Agronomic Research in France. The objective was to obtain rape plants with male sterility. The first series of experiments based on classical interspecific hybridization of rape and radish had resulted in androsterile plants of rape in which the nucleus belonged to rape and the cytoplasm belonged to radish. The male sterility of these plants was caused by the mitochondria of radish, which was rather fortunate for agronomists. However, the presence of radish chloroplasts in cells resulted in yellowing of the plants and their lacking nectaries which are essential to attract pollinator bees. G Pelletier and coworkers (1983) continued their experiments to obviate these problems. They induced fusion of protoplasts from sterile and normal rape plants and eventually obtained plants with the nucleus and chloroplasts from normal rape, while the only cell components inherited from radish were mitochondria. Such plants retained their male sterility yet did not turn yellow and had nectaries. The latter is believed to stem from recombination in mitochondrial DNA.

Thus, somatic hybridization may lead to many higher plant hybrids that are otherwise impossible to obtain. On the other hand, it becomes quite obvious that practical breeding procedures will be most effective if classical methods, whose potential is yet to be exhausted, are rationally combined with new techniques of genetic engineering.

It should be noted that the development of somatic hybridization methods went hand in hand with elaboration of another approach, namely cloning of protoplasts for practical breeding purposes. In some plant species (tobacco, tomato, carrot, etc.), an individual protoplast may give rise to a whole plant after cell envelope regeneration and callus formation. Over the past seen or eight years, promising practical results have been obtained in this direction. In this case, protoplasts may serve both as good recipients of foreign genetic information and as a source material for somatic hybridization.

Gene Engineering and its Potential for Plant Breeding

Gene engineering was begotten by molecular genetics. This is a new field of genetic engineering, which permits manipulation of individual genes and their components. There is still something incongruous in the word combination "*gene engineering*", although it has already entered into common parlance: on the one hand, the term "*engineering*" is so remote from biology, belonging rather to the realm of technology, while on the other, "*gene*" has to do with hereditary factors. Indeed, gene engineering based on careful analysis of the maternal carrier of genetic information provides a tool for reconstructing heredity. Applications of gene engineering methods have already made it possible to commercially produce a number of biologically active substances of great medical importance for diagnosis and treatment of hereditary diseases. These include, primarily, insulin indispensable for treatment of diabetes. Its production has already been put on a commercial basis in several countries. The method resides in that genes of certain organisms are incorporated into the genomes of others, sometimes belonging to quite remote species, which cannot be done by traditional hybridization techniques. Gene engineering makes it possible, in principle, to integrate genes of practically any organisms, and even chemically synthesized genes, into the genome of a selected cell. Practical results in this field have so far been confined to micro-organisms, bacteria, and phages.

What is it precisely that gene engineering may offer to plant breeders? According to some molecular biologists, the time will come when breeders, hopelessly outdated empiricists, will have to give way to the wizards who, manipulating DNA as if it were the key to the mysteries of Life, will eradicate famine at a snap of their fingers. Unfortunately, this is not to happen. Analysis of the experience gained in this field suggests that applied gene engineering will not work any wonders. At the same time, the potential it holds in some very specific areas if beyond and doubt, especially that incorporation of foreign genes into plant cells had become an accomplished fact in the eighties.

To improve a crop variety, an appropriate gene is introduced into the plant cell with the aid of special vectors, such as *Agrobacterium tumefaciens* and *A. rhizogenes*. Then, the transformed cell is used to regenerate a full-fledged plant with new biological properties in a tissue culture, which will serve as a source of seed for the new variety. This process is illustrating the gene-engineering manipulations with plants, involving *A. tumefaciens*. These bacteria induce a cancerous tumour (A) in plants. They contain the Ti plasmid whose segment (tDNA) is capable of being integrated into the chromosomal DNA of the plant cell. The infection induces synthesis of compounds known as opines which serve as food for the bacteria. It is precisely this infection mechanism that permits incorporation of foreign genes into plants.

The concept boils down to incorporation of tDNA into the site and transformation of the plant cell by such a recombinant plasmid. The cell must develop into a complete plant, with the incorporated gene depriving tDNA of its carcinogenic properties. This method has made it possible to isolate a herbicide-resistance gene which was transferred into tobacco cells in an attempt to regenerate resistant plants from them. A group of American investigators transformed sunflower cells by a gene of phaseolin (reserve protein of beans), whose expression in the regenerated plants was adequate and which was inherited by the progeny. Another group of workers transplanted a gene of one of the photosynthetic enzymes (a small subunit thereof, to be precise) which manifested itself in the progeny.

Using *Agrobacterium rhizogenes* as a vector has some specific aspects to be taken into account. It contains the Ri plasmid which also has tDNA capable of being integrated into the chromosomal DNA present in plant cells. In this case, tDNA causes vigorous root formation known as the hairy root syndrome. This tDNA is functional in the sense that the transformed root cells synthesize opines. This vector offers the advantage of simpler and faster regeneration from roots, as compared to that from cancerous tumour cells.

What are the trends to be followed by gene engineering in plant breeding? There are many problems to be resolved. Take, for instance, the problem of supplying nutrients to plants. Although they are surrounded by atmospheric nitrogen, the latter is not available to them without intermediaries. Unfortunately, there are not many such intermediaries in nature; actually, nitrogen-fixing bacteria are the transferring genes of the *nif* group from the bacteria fixing atmospheric nitrogen to cereal crops, thereby eliminating the need to treat the latter with nitrogen fertilizers. Alas, this idea may never cease to be a dream. Firstly, as many as 17 genes will have to be transferred. And even if we suppose that this can be done and the genes are made to work (in wheat genome, for example), according to specialists, such a plant will have its yield of dry matter reduced by 20 to 30 percent because of the energy spent to fix nitrogen.

Some plants, particularly legumes, enter into symbiotic relations with nitrogen-fixing bacteria, which allows farmers to save on fertilizers necessary for their treatment. However, most of farm crops, including cereals, are not capable of such symbiosis ensuring nitrogen fixation.

According to scientists, one of the ways to create "*self-fertilizing*" crops is to impart them the ability to associate symbiotically with nitrogen-fixing bacteria by gene engineering techniques.

Another promising method involves transformation of bacteria themselves. Among the most realistic projects in the immediate future is transfer of genes from nitrogen-fixing bacteria into different soil microflora not capable of fixing atmospheric nitrogen and not symbiosing with plant roots. The spectrum of nitrogen-fixing soil microorganisms must be expanded. This can be attained by creating new types of nodule bacteria. If the problem of atmospheric nitrogen fixation is solved by gene engineering, modern agriculture would undergo drastic changes and millions of rubles spend in the production of nitrogen fertilizers could be saved, to say nothing of cleaner environment.

The reconstruction of plants based on gene engineering, however, is not confined only to the problem of nitrogen fixation. Some scientists see the potential of gene transfer as an effective and useful way to render plants immune to herbicides and pesticides. This is not an easy task, though, for such genes must first of all be found, isolated, and made functional. Nevertheless, many companies, primarily American, have been displaying a great deal of interest in applying gene engineering to plant breeding and invest a lot of money into this activity. The management of these companies obviously proceed from the assumption that once such a variety is created, it will be widely used as a donor of the herbicide immunity gene. The company developing this variety also produces the herbicide in question. In this situation, it is indeed interested in offering its own proprietary gene for incorporation into somebody else's crop varieties, thereby ensuring a market for its herbicide.

It is also quite realistic to talk about transfer of genes responsible for resistance to pathogens. However, breeders have already been disappointed in these plant organisms, quite tempting but not effective in the long run because of the incessant emergence of new pathogens. A quarter of a century ago (1963), J.E. Van der Plank, a phytopathologist, described a phenomenon that dooms such breeding programs to failure.

The development of new varieties involves traits and characters controlled by many genes at a time, and not all of them can be manipulated by gene engineers. Therefore, most of traits and characters of plants will always remain beyond the reach of gene engineering. Even more serious difficulties can be encountered. If the transferred resistance genes have adversely affected the productivity of plants, this does not mean at all that other genes will behave in a similar manner. The incorporation of the Opaque-2 gene into the genotype of maize is a typical case in this respect. In 1964, it was used at Purdue University (USA) for enriching maize kernels with the amino acid lysine in order to enhance their nutritive value. The gene was transferred using methods of classical genetics. It was soon found that the introduction of the Opaque-2 gene had led to a 15 percent decrease in the yielding capacity of the transformed varieties, while the kernels had become brittle and prone to diseases. To restore the original properties lost in the course of experimentation, workers of the International Maize and Wheat Improvement Centre in Mexico had to spent 25 years of hard work.

It is believed that gene engineering is especially useful in studying the processes of plant development and differentiation, which will help breeders to better plant and conduct their work. Today, molecular biology is already offering a number of interesting ancillary techniques. For example, R.A. Owens and T.O. Diener (1981) used enzymes (probes) inherent in DNA to detect the virus of a most dangerous disease of potato, potato spindle tuber, thereby providing the basis for a very simple diagnostic method.

Workers of the Plant Breeding Institute in Cambridge, Great Britain, were able to isolate in this way rye chromosome fragments from the wheat genome after both species had been crossed.

Applications of gene engineering are not limited to the above examples. The scientists' imagination knows no limits. Among other promising areas of activity of gene engineers is making plants resistant to late spring and early autumn frosts currently inflicting heavy damage on agriculture. Even in the United States, where climatic conditions are much more favourable for farming than in our country, losses due to such frosts exceed a billion dollars each year. As a matter of fact, the "culprits" are bacteria conducive to ice formation in plant cells. It has been established that two species of bacteria, *Pseudomonas syringae* and *Erwina herbicola*, thriving on plant surfaces act as nuclei of ice crystals. In the absence of these bacteria on leaf surfaces, the water in plant tissues does not freeze when the temperature drops several degrees below zero but becomes supercooled. In such cases, plants can tolerate temperatures down to –8°C.

Frost damages plants only if water freezes. At a temperature several degrees below zero, for supercooled water to start freezing a "nucleus" of crystallization is needed. The above-mentioned bacteria serve precisely as such "nuclei". Ice crystals form around them and eventually destroy plant cells and tissues. Since such bacteria occur on many crops, including vegetables, fruit trees and shrubs, industrial crops, and grasses grown all over the world, the damage caused by frost amounts to billions of rubles.

Therefore, attempts are now being made to rid crops of these bacteria by both traditional methods and gene engineering. One of such attempts involves synthesis of bacteriophagous viruses that would selectively kill bacteria belonging to both species on plant surfaces. In laboratory experiments conducted on leguminous crops, the rate of kill achieved by using this method was 90 percent within a few hours.

Another approach to rendering plants frost-resistant is transformation of such bacteria by gene engineering so that they would not nucleate ice crystals. The recently developed strains of these bacteria will be sprayed over crop fields or plantations early in the growing season with a view to replacement of the natural ones. The potential of this method has already been tested in laboratory experiments. Scientists have been successful in removing, from the genotype of *P. syringae*, genes responsible for the ability to initiate crystallization of supercooled water.

4

MICROPROPAGATION

Plants can be propagated through their two developmental life cycles; the *sexual*, or the *asexual*. In the sexual cycle new plants arise after fusion of the parental gametes, and develop from zygotic embryos contained within seeds or fruits. In most cases seedlings will be variable and each one will represent a new combination of genes, brought about during the formation of gametes (meiotic cell division) and their sexual fusion. By contrast, in the vegetative (asexual) cycle the unique characteristics of any individual plant selected for propagation (termed the mother plant, stock plant or ortet) are usually perpetuated because, during normal cell division (*mitosis*), genes are typically copied exactly at each (mitotic) division. In most cases, each new plant (or ramet) produced by this method may be considered to be an extension of the somatic cell line of one (sexually produced or mutant) individual. A group of such asexually reproduced plants (ramets) is termed a clone. In the natural environment sexual and asexual reproduction have their appropriate selective advantages according to the stage of evolution of different kinds of plants. Plants selected and exploited by man also have different propensities for propagation by seed or by vegetative means.

SEED VERSUS SOMA

Propagation Using Seeds

Seeds have several advantages as a means of propagation:

1. They are often produced in large numbers so that the plants regenerated from them are individually inexpensive;
2. Many may usually be stored for long periods without loss of viability;

3. They are easily distributed;
4. Most often plants grown from seed are without most of the pests and diseases which may have afflicted their parents.

For many agricultural and horticultural purposes it is desirable to cultivate clones or populations of plants which are practically identical. However, the seeds of many plants typically produce plants which differ genetically, and to obtain seeds which will give uniform offspring is either very difficult, or impossible in practical terms. Genetically uniform populations of plants can result from seeds in three ways:

1. From inbred (*homozygous*) lines which can be obtained in self-fertile (*autogamous*) species. Examples of autogamous crops are wheat, barley, rice and tobacco.
2. From F1 seeds produced by crossing two homozygous parents. Besides being uniform, F1 plants may also display hybrid vigour. F1 seeds of many flower producing ornamentals and vegetables are now available, but due to high production costs, they are expensive.
3. From apomictic seedlings. In a few genera, plants that are genotypically identical to their parents are produced by apomixis. Seeds are formed without fertilization and their embryos develop by one of several asexual processes that ensure that the new plants are genetically identical to the female parent (i.e. they have been vegetatively reproduced).

Some plants do not produce viable seeds, or do so only after a long juvenile period. Alternatively, to grow plants from seed may not provide a practical method of making new field plantings. In such instances vegetative propagation is the only means of perpetuating and multiplying a unique individual with desirable characteristics.

Vegetative Propagation

Many important crop plants are increased vegetatively and grown as clones. They include cassava, potato, sugar cane and many soft (small) fruits and fruit trees. A very large number of herbaceous and woody ornamental plants are also propagated by these means. Suitable methods for vegetative propagation have been developed over many centuries. These traditional '*macro-propagation*' techniques (or '*macro-methods*') which utilize relatively large pieces of plants, have been refined and improved by modern horticultural research. For instance, methods of applying fine water mist to prevent the desiccation of cuttings, better rooting composts and the control of temperature in the rooting zone, have considerably enhanced the rate at which many plants of horticultural or agricultural interest can be multiplied. Research to improve macropropagation methods continues, but has lost some impetus in recent years with the continued extension of tissue culture for plant multiplication.

Whether it will be most rewarding to propagate a plant by seed, by traditional vegetative techniques, or by tissue culture, will often not only depend on the plant species, but also on the development of proven techniques, relative costs and agronomic objectives. The extent to which tissue culture methods can be used for genetic manipulations and for propagation is changing continuously. Until recently, potato plants have been raised from seed during breeding programmes to select new varieties: tissue culture may have been employed to multiply certain lines and to propagate disease-tested stocks of established cultivars, while macropropagation of field-grown tubers has been used to provide normal planting material. New research into genetic manipulations and methods of propagation using tissue culture techniques, can alter this situation: diversity can be introduced and controlled through genetic engineering while certified stock of new varieties can be produced on a large scale by micropropagation.

The selection of a propagation method for any given plant is constrained by its genetic potential. For example, some plants readily produce adventitious shoot buds on their roots, while others do not; trying to propagate a plant, which does not have this capability, from root cuttings or root explants, will be more problematic both *in vivo* and *in vitro*. Plant tissue culture does overcome some genetically imposed barriers, but a clear effect of genotype is still apparent. It is not yet possible to induce an apple tree to produce tubers!

PROPAGATION *IN VITRO*

Advantages

Methods available for propagating plants *in vitro* are largely an extension of those already developed for conventional propagation. *In vitro* techniques have the following advantages over traditional methods:

1. Cultures are started with very small pieces of plants (explants), and thereafter small shoots or embryos are propagated (hence the term '*micropropagation*' to describe the *in vitro* methods). Only a small amount of space is required to maintain plants or to greatly increase their number. Propagation is ideally carried out in aseptic conditions (avoiding contaminations). The often used term "*axenic*" is not correct, because it means "free from any association with

other living organisms". Once cultures have been started there should be no loss through disease, and the plantlets finally produced should be ideally free from bacteria, fungi and other micro-organisms.

2. Methods are available to free plants from specific virus diseases. Providing these techniques are employed, or virus-tested material is used for initiating cultures, certified virus-tested plants can be produced in large numbers. Terminology such as virus-free and bacteria-free should not be used, as it is impossible to prove that a plant is free of all bacteria or viruses. One can only prove that a plant has been freed from a specific contaminant provided the appropriate diagnostic tools are available.
3. A more flexible adjustment of factors influencing vegetative regeneration is possible such as nutrient and growth regulator levels, light and temperature. The rate of propagation is therefore much greater than in macropropagation and many more plants can be produced in a given time. This may enable newly selected varieties to be made available quickly and widely, and numerous plants to be produced in a short while. The technique is very suitable when high volume production is essential.
4. It may be possible to produce clones of some kinds of plants that are otherwise slow and difficult (or even impossible) to propagate vegetatively.
5. Plants may acquire a new temporary characteristic through micropropagation which makes them more desirable to the grower than conventionally-raised stock. A bushy habit (in ornamental pot plants) and increased runner formation (strawberries) are two examples.
6. Production can be continued all the year round and is more independent of seasonal changes.
7. Vegetatively-reproduced material can often be stored over long periods.
8. Less energy and space are required for propagation purposes and for the maintenance of stock plants (ortets).
9. Plant material needs little attention between subcultures and there is no labour or materials requirement for watering, weeding, spraying etc.; micropropagation is most advantageous when it costs less than traditional methods of multiplication; if this is not the case there must be some other important reason to make it worthwhile.

Disadvantages

The chief disadvantages of *in vitro* methods are that advanced skills are required for their successful operation.

1. A specialized and expensive production facility is needed; fairly specific methods may be necessary to obtain optimum results from each species and variety and, because present methods are labour intensive, the cost of propagules is usually relatively high. Further consequences of using *in vitro* adaptations are although they may be produced in large numbers, the plantlets obtained are initially small and sometimes have undesirable characteristics.
2. In order to survive *in vitro*, explants and cultures have to be grown on a medium containing sucrose or some other carbon source. The plants derived from these cultures are not initially able to produce their own requirement of organic matter by photosynthesis (i.e. they are not autotrophic) and have to undergo a transitional period before they are capable of independent growth. More recently techniques have been proposed which allow the production of photo-autotrophic plants *in vitro*.
3. As they are raised within glass or plastic vessels in a high relative humidity, and are not usually photosynthetically self-sufficient, the young plantlets are more susceptible to water loss in an external environment. They may therefore have to be hardened in an atmosphere of slowly decreasing humidity and increased light. The chances of producing genetically aberrant plants may be increased.

Techniques

The methods that are theoretically available for the propagation of plants *in vitro* and described in the following sections. They are essentially:

1. By the multiplication of shoots from axillary buds:
2. By the formation of adventitious shoots, and/or adventitious somatic embryos, either (a) directly on pieces of tissue or organs (explants) removed from the mother plant; or (b) indirectly from unorganized cells (in suspension cultures) or tissues (in callus cultures) established by the proliferation of cells within explants; on semi-organized callus tissues or propagation bodies (such as protocorms or pseudobulbils) that can be obtained from explants (particularly those from certain specialized whole plant organs).

The techniques that have been developed for micropropagation are described in greater detail in the following sections of this Chapter.

In practice most micropropagated plants are produced at present by method (i), and those of only a few species (which will be instanced later) by method (ii). Shoots and/or plantlets do not always originate in a culture by a single method. For example, in shoot cultures, besides axillary shoots, there are sometimes adventitious shoots formed directly on existing leaves or stems, and/or shoots arising indirectly from callus at the base of the explant. The most suitable and economic method for propagating plants of a particular species could well change with time. There are still severe limitations on the extent to which some methods can be used. Improvements will come from a better understanding of the factors controlling morphogenesis and genetic stability *in vitro*.

Rooting

Somatic embryos have both a root and a shoot meristem. Under ideal conditions they can grow into normal seedlings. The shoots procured from axillary or adventitious meristems are miniature cuttings. Sometimes these small cuttings form roots spontaneously, but usually they have to be assisted to do so. The small rooted shoots produced by micropropagation are often called *plantlets*.

Stages of Micropropagation

Professor Murashige defined three steps or stages (I-III) in the *in vitro* multiplication of plants. These have been widely adopted by both research and commercial tissue culture laboratories because they not only describe procedural steps in the micropropagation process, but also usually represent points at which the cultural environment needs to be changed. Some workers have suggested that the treatment and preparation of stock plants should be regarded as a separately numbered stage or stages. We have adopted the proposal of Debergh and Maene (1981) that such preparative procedures should be called Stage 0. A fourth stage (IV), at which plants are transferred to the external environment, is now also commonly recognized.

Requirements for the completion of each stage of micropropagation vary according to the method being utilized; the progress of cultures will not always fit readily into neat compartments. Furthermore, it is not always necessary to follow each of the prescribed steps. The stages are therefore described here for general guidance but should not be applied too rigidly.

Stage 0: Mother plant selection and preparation

Before micropropagation commences, careful attention should be given to the selection of stock plants. They must be typical of the variety or species, and free from any symptoms of disease. It may be advantageous to treat the chosen plant (or parts of it) in some way to make *in vitro* culture successful. Steps to reduce the contamination level of explants were considered sufficiently important by Debergh and Maene (1981) to constitute a separate essential stage in a commercial micropropagation programme.

Growth, morphogenesis and rates of propagation *in vitro* can be improved by appropriate environmental and chemical pre-treatment of stock plants. Procedures to detect and reduce or eliminate systemic bacterial and virus diseases may also be required. Disease indexing and disease elimination should be a defmite part of all micropropagation work; but these precautions are unfortunately often omitted, sometimes with adverse consequences.

It seems appropriate to include all procedures adopted in plant selection and pre-treatment within 'Stage 0'. The recognized numbering of Murashige's stages is then unaltered.

Stage I: Establishing an aseptic culture

The customary second step in the micropropagation process is to obtain an aseptic culture of the selected plant material. Success at this stage firstly requires that explants should be transferred to the cultural environment, free from obvious microbial contaminants; and that this should be followed by some kind of growth (e.g. growth of a shoot tip, or formation of callus). Usually a batch of explants is transferred to culture at the same time. After a short period of incubation, any container found to have contaminated explants or medium is discarded. Stage I would be regarded as satisfactorily completed if an adequate number of explants had survived without contamination, and was growing on. The objective is reproducibility, not 100% success.

Stage II: The production of suitable propagules

The object of Stage II is to bring about the production of new plant outgrowths or propagules, which, when separated from the culture are capable of giving rise to complete plants. According to the *in vitro* procedure that is being followed, multiplication can be brought about from newly-derived axillary or adventitious shoots, somatic embryos, or miniature storage or propagative organs. In some micropropagation methods, Stage II will include the prior induction of meristematic centres from which adventitious organs may develop. Some of the propagules produced at Stage II (especially shoots) can also be used as the basis for further cycles of multiplication in that they can usually be cultured again (subcultured) to increase their number.

Stage III: Preparation for growth in the natural environment

Shoots or plantlets derived from Stage II are small, and not yet capable of self-supporting growth in soil or compost. At Stage III, steps are taken to grow individual or clusters of plantlets, capable of carrying out photosynthesis, and survival without an artificial supply of carbohydrate. Some plantlets need to be specially treated at this stage so that they do not become stunted or dormant when taken out of the cultural environment. As originally proposed by Murashige, Stage III includes the *in vitro* rooting of shoots prior to their transfer to soil.

Rooting shoots is a very important part of any *in vitro* propagation scheme. A few species form adventitious roots on shoots during the course of Stage III culture, but usually it is necessary to adopt a separate rooting procedure using special media, or methods, to induce roots to form. Sometimes shoots may need to be specially elongated before rooting is attempted. To reduce the costs of micropropagation, many laboratories now remove unrooted shoots from the *in vitro* environment and root them outside the culture vessel. Therefore, in cultures where micropropagation relies on adventitious or axillary shoots, Stage III is often conveniently divided, as Debergh and Maene (1981) suggested, into:

1. Stage IIIa, the elongation of buds or shoots formed during Stage II, to provide shoots of a suitable size for Stage Mb;
2. Stage Mb, the rooting of Stage IIIa shoots *in vitro* or *extra vitrum*.

Stage IV: Transfer to the natural environment

Although not given a special numerical stage by Murashige, the methods whereby plantlets are transferred from the *in vitro* to the *ex vitro* external environment are extremely important. If not carried out carefully, transfer can result in significant loss of propagated material. There are two main reasons:

1. Shoots developed in culture have often been produced in high humidity and a low light '*intensity*'. This results in there being less leaf epicuticular wax or wax with an altered chemical composition, than on plants raised in growth chambers or greenhouses. In some plants, the stomata of leaves produced *in vitro* may also be atypical and incapable of complete closure under conditions of low relative humidity. Tissue cultured plants therefore lose water rapidly when moved to external conditions.
2. When supplied with sucrose (or some other carbohydrate) and kept in low light conditions, micropropagated plantlets are not fully

dependent on their own photosynthesis (they are *mixotrophic*). A stimulus which is not provided in the closed *in vitro* environment seems to be needed for them to change to being fully capable of producing their own requirements of carbon and reduced nitrogen (i.e. before they become capable of feeding themselves - *autotrophic*). The change only occurs after the plants have spent a period of several days *ex vitro*.

In practice, plantlets are removed from their Stage III containers, and if they have been grown on agar medium, the gel is carefully washed from the roots. The application of an anti-transpirant film to the leaves has been recommended at this stage, but in practice, seems to be seldom used. Plantlets are then transplanted into an adequate rooting medium (such as a peat:sand compost) and kept for several days in high humidity and reduced light intensity. A fog of water vapour is very effective for maintaining humidity. Alternatively, intermittent water misting may be applied automatically, or the plants placed inside a clear plastic enclosure and misted by hand. With some plants, an *in vitro* Stage III can be omitted; shoots from Stage II are rooted directly in high humidity, and, at the same time, gradually hardened to the exterior environment.

Micropropagation Methods

Propagation of Plants from Axillary Buds or Shoots

The production of plants from axillary buds or shoots has proved to be the most generally applicable and reliable method of true-to-type *in vitro* propagation. Two methods are commonly used:

1. Shoot culture
2. Single, or multiple, node culture.

Both depend on stimulating precocious axillary shoot growth by overcoming the dominance of shoot apical meristems.

Shoot (or shoot tip) culture

The term shoot culture is now preferred for cultures started from explants bearing an intact shoot meristem, whose purpose is shoot multiplication by the repeated formation of axillary branches. In this technique, newly formed shoots or shoot bases serve as explants for repeated proliferation; severed shoots (or shoot clumps) are finally rooted to form plantlets which can be grown *in vivo*. This is the most widely used method of micropropagation.

Explant size. Shoot cultures are conventionally started from the apices of lateral or main shoots, up to 20 mm in length, dissected from actively-growing shoots or dormant buds. Larger explants are also sometimes used with advantage: they may consist of a larger part of the shoot apex or be stem segments bearing one or more lateral buds; sometimes shoots from other *in vitro* cultures are employed. When apical or lateral buds were used almost exclusively as explants, the name '*shoot tip culture*' came to be widely used for cultures of this kind. As the use of larger explants has become more common, the term shoot culture has become more appropriate.

Large explants have advantages over smaller ones for initiating shoot cultures in that they:

1. Better survive the transfer to *in vitro* conditions,
2. More rapidly commence growth;
3. Contain more axillary buds.

However, the greater the size of the explant, the more difficult it may be to decontaminate from micro-organisms; in practice the size used will be the largest that can be gained in aseptic conditions. Shoot cultures are also frequently started directly from the shoots obtained from meristem tip cultures. Virus eradication then proceeds the shoot multiplication phase. Occasionally fragmented or macerated shoot tips are used. Meristem tip or meristem cultures are used for virus and bacteria elimination. Meristem cultures are initiated from much smaller explants and a single plantlet is usually produced from each. This terminology is very often abused.

Regulating shoot proliferation

The growth and proliferation of axillary shoots in shoot cultures is usually promoted by incorporating growth regulators (usually cytokinins) into the growth medium. Most often such a treatment effectively removes the dominance of apical meristems so that axillary shoots are produced, often in large numbers. These shoots are used as miniature cuttings for plant multiplication.

Removing the apex. In some plants, pinching out the main shoot axis is used as an alternative, or an adjunct, to the use of growth regulators for decreasing apical dominance. Pinching was found to be effective for some kinds of rose and for some apple cultivars. Pinching or '*tipping*' is usually done when plant material is removed for subculturing, for example removing the apical bud at the first subculture increased the branching of *Pistacia* shoot cultures. An effective kind of shoot tipping occurs when shoots are cropped as microcuttings. Standardi (1982) and Shen and Mullins (1984) obtained effective shoot proliferation of kiwi and pear varieties by transferring the basal shoot clump that is left at this stage, to fresh medium for further proliferation. (Note however that this practice can increase the likelihood of obtaining deviant plants). In just a few plants neither cytokinins nor pinching effectively remove apical dominance. Geneve *et al.* (1990) reported that seedling shoots of *Gymnocladus dioicus* produced 1-5 shoots, but only one grew to any appreciable length. If this shoot was removed, another took over.

Placing explants horizontally. In pear, pinching out the tips of shoots resulted in the growth of larger axillary shoots than in the controls, but the number of shoots was less. The most effective physical check to apical dominance was achieved by pinching the tips, and/or placing shoot explants horizontally on the medium. The treatment can be effective with many other woody plants: horizontal placement of shoot sections, consisting of 2-3 nodes, resulted in more axillary shoots being produced in cultures of *Acer rubrum*, *Amelanchier spicata*, *Betula nigra*, *Forsythia intermedia* and *Malus domestica*, than when explants were upright. Favourable results have also been reported with lilac and some apple cultivars.

Origin of shoots

Unfortunately not all the shoots arising in shoot cultures may originate from axillary buds. Frequently, adventitious shoots also arise, either directly from cultured shoot material, or indirectly from callus at the base of the subcultured shoot mass. For example, Nasir and Miles (1981) observed that in subcultures of an apple rootstock, some new shoots arose from callus at the base of the shoot clump; both adventitious and axillary shoots were produced in *Hosta* cultures; and shoot proliferation from some kinds of potato shoot tips was exclusively from organogenic callus.

The precise origin of shoots can sometimes only be determined from a careful anatomical examination. Hussey (1983) has termed cultures providing both adventitious and axillary shoots, '*mixed cultures*'. Adventitious shoots, particularly those arising indirectly from callus, are not desirable. For reasons, shoots of axillary origin will normally be genetically identical to the parent plant, whereas there is a probability that those regenerated from callus may differ in one or more characters. Genetically deviant plants may not occur with high frequency from newly initiated callus, but could begin to appear in significant number if shoot masses incorporating basal callus are simply chopped up to provide explants for subculture. The use of a strict protocol, using only axillary shoots, may present problems with some plants where the rate of shoot multiplication is comparatively slow. This has led to attempts by some workers to use a more relaxed regime and accept a proportion of adventitious shoots (e.g. with *Kalanchoe blossfeldiana*). The usual consequence is a degree of variation amongst ramets which may, or may not, be acceptable. The formation of callus and the subsequent development of adventitious shoots can often be controlled by modifying the growth regulators in the medium.

Fragmentation of a meristem tip, or its culture in a certain way, can lead to the formation of multiple adventitious shoots which can be used for plant propagation.

History

Although shoot culture has proved to be a widely applicable method of micropropagation, the appreciation of its potential value developed only slowly, and utilization largely depended on improvements in tissue culture technology. Robbins (1922) seems to have been the first person to have successfully cultured excised shoot tips on a medium containing sugar. Tip explants of between 1.75 and 3.75 mm were taken from pea, corn and cotton, and placed in a liquid medium. For some reason the cultures were maintained in the dark where they only produced shoots with small chlorotic leaves and numerous roots. Although it is tempting to suppose that the potential of shoot culture for plant propagation might have been appreciated at a much earlier date had the cultures been transferred to the light, the rapid rate of shoot multiplication achieved in modern use of this technique depends on later developments in plant science.

Only very slow progress in shoot culture was made during the next 20 years. As part of his pioneering work on plant tissue culture, White (1933) experimented with small meristem tips (0.1 mm or less) of chickweed (*Stellaria media*), but they were only maintained in hanging drops of nutrient solution. Leaf or flower primordia were observed to develop over a six-week period. Shoot culture of a kind was also carried out by La Rue (1936). His explants largely consisted of the basal and upper halves of seed embryos. Nevertheless, the apical plumular meristems of several plants were grown to produce entire plants. Whole plants were also obtained from axillary buds of the aquatic plant *Radicula aquatica*.

Significant shoot growth from vegetative shoot tip explants was first achieved by Loo, and reported in 1945 and 1946 a, b. *Asparagus* shoot tips 5-10 mm in length were supported on glass wool over a liquid medium and later grown on a solidified substrate. Loo (1945, 1946a, 1946b) made several significant observations showing that:

1. Growth depended on sucrose concentration, higher levels being necessary in the dark than in the light;
2. Explants, instead of being supported, could be grown satisfactorily on 0.5% agar;
3. *In vitro* shoot growth could apparently be continued indefmitely (35 transfers were made over 22 months);
4. shoot tip culture afforded a way to propagate plant material (clones were established from several excised shoot apices).

This work failed to progress further because no roots were formed on the *Asparagus* shoots in culture. Honours for establishing the principles of modern shoot culture must therefore be shared between Loo and Ball. Ball (1946) was the first person to produce rooted shoots from cultured shoot apices. His explants consisted of an apical meristem and 2-3 leaf primordia. There was no shoot multiplication but plantlets of nasturtium (*Tropaeolum majus*) and white lupine (*Lupinus alba*) were transferred to soil and grown successfully.

During several subsequent years, shoot apex (or meristem tip) culture was of interest only to plant pathologists who recognized its value for producing virus-tested plants. It was during studies of this kind that Morel made the significant discovery of protocorm formation from *Cymbidium* orchid shoot tips. Although they may be started from the same explants, cultures giving rise to protocorms are not typical shoot cultures.

The two major developments which made shoot culture feasible were the development of improved media for plant tissue culture and the discovery of the cytokinins as a class of plant growth regulators, with an ability to release lateral buds from dormancy. These developments were not immediately applied to shoot culture, and some years elapsed before it was appreciated that multiple shoots could be induced to form by appropriate growth regulator treatments.

Hackett and Anderson (1967) got either single shoots from carnation shoot apices, or else a proliferative tissue from which shoots were later regenerated. Walkey and Woolfitt (1968) reported a similar kind of direct or indirect shoot proliferation from *Nicotiana rustica* shoot tips. Vine and Jones (1969) were able to transfer large shoot tips of hop (*Humulus*) to culture, but shoots only rooted, and showed a high propensity for callus formation. Reports of plant multiplication using conventional shoot culture methods began to appear in the next decade.

Haramaki (1971) described the rapid multiplication of *Gloxinia* by shoot culture and by 1972 several reports of successful micropropagation by this method had appeared. Since then the number of papers on shoot culture published annually has increased dramatically and the method has been utilized increasingly for commercial plant propagation. Factors which have influenced the choice of shoot culture for practical micropropagation have been:

1. The way in which the method can be applied to a wide range of different plant species, using the same principles and basic methods;
2. The possibility of obtaining simultaneous virus control;
3. A general uniformity and '*trueness to type*' of the regenerated plants;
4. The relatively high rates of propagation which is possible in many species.

Methods

Primary explants. In most herbaceous plants, shoot tip explants may be derived from either apical or lateral buds of an intact plant, and consist of a meristematic stem apex with a subtended rudimentary stem bearing several leaf initials. In the axils of the more developed leaf primordia there will be axillary bud meristems. In some species (e.g. *Eucalyptus*) it is an advantage to commence shoot cultures with a piece of the stem of the mother plant bearing one or more buds (stem nodes). Shoot growth from the bud, and treatment of the culture, is thereafter the same as in conventional shoot tip culture. The use of nodal explants should not be confused with node culture in which a method of shoot multiplication is used that is different to that in shoot culture.

Shoot tips from trees, or other woody perennials, can be difficult to decontaminate. Because of this, Standardi and Catalano (1985) preferred to initiate shoot cultures of *Actinidia chinensis* from meristem tips which could be sterilized more easily. Shoot tips of woody plants are more liable than those of herbaceous species to release undesirable phenolic substances when first placed onto a growth medium. Buds taken from mature parts of the shrub or tree can also be reluctant to grow *in vitro* and seasonal factors may reinforce natural dormancy in buds from any source, so that cultures can only be readily initiated at certain times of the year. Shoot tip or lateral bud explants are usually most readily induced into growth if taken from juvenile shoots such as those of seedlings or young plants. The juvenile shoots which sometimes emerge from the base of mature plants or which arise form heavily pruned or coppiced bushes and trees, are alternative sources. However, developing techniques have made it possible to propagate some woody ornamentals, forest trees and fruit trees, using explants derived from mature shoots. De Fossard *et al.* (1977) could initiate cultures of *Eucalyptus ficifolia* with shoot tips from 36 year-old trees, but forest-gathered material was very difficult to decontaminate unless covered and protected for some period before excision (stage 0).

Secondary explants. Stage II subcultures are initiated from axillary shoots separated from primary shoot clusters. The place of the secondary explant within the primary shoot (cluster) can have a remarkable influence on the subsequent performance of the subcultures. A higher rate of shoot proliferation is often obtained from nodal explants or by subdivision of the basal shoot mass. Shoot tips were the best secondary explant for *Rosa* 'Fraser McClay', but with cherry ('F12/1') nodal explants gave more than twice as many shoots, and basal masses, three times as many as shoot tips. In Sitka spruce, cultures that had been apices in the previous subculture were able to proliferate buds at higher rates than those that had been axillary buds. The origin of an explant can also have a tremendous influence on the subsequent behaviour of the plant when established under field conditions. This was illustrated by Marks and Meyers (1994) for *Daphne odorata*.

To minimize the risk of genetic change in ramets, explants for subculture and shoots to be transferred to Stage III, should, as far as possible, be chosen from new shoots of axillary origin. It may be advisable to adjust the growth regulator content of the medium so that adventitious shoots are not formed, even though the rate of overali shoot multiplication is thereby reduced. In some circumstances callus arising at the base of an explant may be semi-organized and therefore capable of producing genetically-stable plants.

Stage II cultures are typically without roots, and shoots need to be detached and treated as miniature cuttings which, when rooted, will provide the new plants that are required. An alternative is to allow shoot clusters to elongate and to root singulated shoots under *ex vitro* conditions.

Media and growth regulators. A notable feature of shoot cultures of most plant species is the need for high cytokinin levels at Stage II to promote the growth of multiple axillary shoots.

Cytokinin growth regulators are usually extremely effective in removing the apical dominance of shoots. Their use can be combined with pinching the apex of shoots, or placing explants in an horizontal position. A cytokinin treatment can not only promote the formation of multiple shoots (axillary and/or adventitious), but also (if the compound used is unsuitable, or the concentration used is too high), cause the shoots formed to be too short for rooting and transfer.

Because or their nature, or the absence of an adequate method of culture, plants of some kinds fail to produce multiple shoots at Stage II and retain their apical dominance. In shoot cultures of *Gymnocladus dioicus*, for example, despite the formation of several axillary shoots in the presence of BA cytokinin, one shoot nearly always became dominant over the others. Most plants of this kind are best propagated by node culture.

Elongation

The length of the axillary shoots produced in shoot cultures varies considerably from one kind of plant to another. Species which have an elongated shoot system *in vivo* will produce axillary shoots which can be easily separated as microcuttings and then individually rooted. Apically dominant shoots which have not branched can be treated in the same way.

At the other extreme are plants with a natural rosette habit of growth, which tend to produce shoot clusters in culture. When these are micropropagated, it is difficult to separate individual shoots for use as secondary explants. It may then only be practical to divide the shoot mass into pieces and re-culture the fragments. Such shoot clusters can be induced to form roots when plants with a bushy habit are required (e.g. many species sold in pots for their attractive foliage). Otherwise it is necessary to specially elongate shoots before they are rooted. Shoot clusters are treated in such a way that axillary shoot formation is reduced, and shoot growth promoted. Individual shoots are then more readily handled and can be rooted as microcuttings.

Rooting and transfer. The cytokinin growth regulators added to shoot culture media at Stage II to promote axillary shoot growth, usually inhibit root formation. Single shoots or shoot clusters must therefore be moved to a different medium for rooting *in vitro* before being transferred as plantlets to the external environment. An alternative strategy for some plants is to root the plant material *ex vitro*. Treatments need to be varied according to the type of growth; the nature of the shoot proliferation produced during Stage II culture; and the plant habit required by the customer.

Current applications. Conventional shoot culture continues to be the most important method of micropropagation, although node culture is gaining in importance. It is very widely used by commercial tissue culture laboratories for the propagation of many herbaceous ornamentals and woody plants. The large numbers of manipulations required do, however, make the cost of each plantlet produced by this method comparatively expensive. Some success has been achieved in automating some stages of the process, in applying techniques for large-scale multiplication and in the use of robotics for plant separation and planting.

Shoot proliferation from meristem tips

Barlass and Skene (1978; 1980a,b; 1982a,b) have shown that new shoots can be formed adventitiously when shoot tips of grapevine or *Citrus* are cut into several pieces before culture. Tideman and Hawker (1982) also had success using fragmented apices with *Asclepias rotundifolia* but not with *Euphorbia peplus*. Usually leaf-like structures first develop from the individual fragments; these enlarge and shoots form from basal swellings. Axillary shoots often arise from the initial adventitious shoots.

Shoot cultures transferred to agitated liquid culture may form a proliferating mass of shoots. Although high rates of multiplication are possible, leafy shoots usually become *hyperhydric*. However, in some species at least, shoots can be reduced in size to little more than proliferating shoot initials which are then suitable for large-scale multiplication. A somewhat similar kind of culture consisting of superficial shoot meristems on a basal callus can sometimes be initiated from shoot tip explants or from the base of shoot cultures.

Single and multiple node culture (in vitro layering)

Single node culture is another *in vitro* technique which can be used for propagating some species of plants from axillary buds. As with shoot culture, the primary explant for single node culture is a shoot apex, a lateral bud or a piece of shoot bearing one or more buds (i.e. having one or more nodes). When shoot apices are used, it can be advantageous to initiate cultures with large explants (up to 20 mm), unless virus-tested cultures arc required, and small meristem-tips will be employed. Unbranched shoots are grown at Stage I until they are 5-10 cm in length and have several discrete and separated nodes. An environment that promotes etiolated shoot growth may be an advantage. Then at Stage II, instead of inducing axillary shoot growth with growth regulators (as in shoot culture), one of two manipulative methods is used to overcome apical dominance and promote lateral bud break:

1. Intact individual shoots may be placed on a fresh medium in an horizontal position. This method has been used by Wang (1977) to propagate potatoes, and has been termed *'in vitro* layering';
2. Each shoot may be cut into single-, or several-node pieces which are sub-cultured. Leaves are usually trimmed so that each second stage explant consists of a piece of stem bearing one or more lateral buds.
3. Each approach can be reiterated to propagate during stage II.

Unfortunately, *in vitro* layering seldom results in several axillary shoots of equal length, apical dominance usually causes the leading shoot, or shoots, to grow more rapidly than the rest. El Hasan and Debergh (1987) found that, even in potato, node culture was preferable. Node culture is therefore the simplest method of *in vitro* propagation, as it requires only that shoot growth should occur. Methods of rooting are the same as those employed for the microcuttings derived from shoot culture, except that prior elongation of shoots is unnecessary.

Note that "*node culture*" is distinct from shoot cultures started from the nodes of seedlings or mature plants.

Media and growth regulators

Media for single node culture are intrinsically the same as those suitable for shoot culture. As in shoot culture, optimum growth rate may depend on the selection of a medium particularly suited to the species being propagated, but adequate results can usually be produced from well-known formulations. It is often unnecessary to add growth regulators to the medium; for example, the shoots of some plants (e.g. *Chrysanthemum morifolium*) elongate satisfactorily without any being provided. If they are required, regulants at both Stages I and II will usually comprise an auxin and a cytokinin at rates sufficient to support active shoot growth, but not tissue proliferation or lateral bud growth. Sometimes gibberellic acid is advantageously added to the medium to make shoots longer and thus facilitate single node separation.

Current applications

Node culture is of value for propagating species that produce elongated shoots in culture (e.g. potato and *Alstroemeria*), especially if stimulation of lateral bud break is difficult to bring about with available cytokinins. Nowadays the technique becomes more and more popular in commercial micropropagation. The main reason is that it gives more guarantee for clonal stability. Indeed, although the rate of multiplication is generally less than that which can be brought about through shoot culture, there is less likelihood of associated callus development and the formation of adventitious shoots, so that Stage II subculture carries very little risk of induced genetic irregularity. For this reason, node culture has been increasingly recommended by research workers as the micropropagation method least likely to induces somaclonal variation.

Multiple shoots from seeds (MSS)

During the early 1980's it was discovered that it was possible to initiate multiple shoot cultures directly from seeds. Seeds are sterilized and then placed onto a basal medium containing a cytokinin. As germination occurs, clusters of axillary and/or adventitious shoots ('*multiple shoots*') grow out, and may be split up and serially sub-cultured on the same medium. High rates of shoot multiplication are possible. For instance, Hisajima (1982a) estimated that 10 million shoots of almond could be derived theoretically from one seed in a year. It is likely that multiple shoots can be initiated from the seeds of many species, particularly dicotyledons. The technique is effective in both herbaceous and woody species: soybean: sugar beet: almond: walnut: pumpkin and melon: cucumber and pumpkin: pea, peanut, mung bean, radish, *Zea mays* and rice. This technique does only make sense when elite seed is used or to gain preliminary information on the behaviour of a plant species under *in vitro* conditions.

Shoots from floral meristems

Meristems that would normally produce flowers or floral parts can sometimes be induced to give vegetative shoots *in vitro*. Success depends on the use of young inflorescences where the determination of individual flower meristems is not canalized. Meristems in older inflorescences are likely to give rise to floral structures. The exact origin of the shoots produced has not always been determined. In cauliflower and coconut they were thought to originate from actual flower meristems, but in sugar beet, from floral axillary buds. Some shoots formed from onion flower heads arose from various parts of the flower buds, but they were accompanied by other shoots which arose adventitiously over the entire receptacle surface. Shoots formed from young flower buds may therefore not always result from the reversion of floral meristems.

Propagation by Direct Organogenesis

Direct adventitious shoot initiation

In certain species, adventitious shoots which arise directly from the tissues of the explant (and not within previously-formed callus) can provide a reliable method for micropropagation. However, the induction of direct shoot regeneration depends on the nature of the plant organ from which the explant was derived, and is highly dependent on plant genotype. In responsive plants, adventitious shoots can be formed *in vitro* on pieces of tissue derived from various organs (e.g. leaves, stems, flower petals or roots); in others species, they occur on only a limited range of tissues such as bulb scales, seed embryos or seedling tissues. Direct morphogenesis is observed rarely, or is unknown, in many plant genera.

Direct shoot formation is sometimes accompanied by proliferation of unorganized cells, and a regenerative tissue that could be classed as callus, may ultimately appear. Its formation can usually be reduced by adjustment of the growth regulators in the medium. Because there is a risk of regenerating plants with a different genetic identity, use of the callus for further propagation is not recommended unless it has a highly organized nature. In some instances, the growth regulators used to initiate shoot buds directly on explants may not be conducive to continued bud growth. A closely packed mass of shoot primordia may then be mistaken for organized callus.

In those species where adult tissues have a high regenerative capacity, the main advantages of micropropagation by direct adventitious shoot regeneration are that:

1. Initiation of Stage I cultures and Stage II shoot multiplication, are more easily achieved than by shoot culture. It is, for example, simpler to transfer aseptically several pieces of *Saintpaulia leaf* petiole to culture medium, than to isolate an equivalent number of shoot meristems.
2. Rates of propagation can be high, particularly if numerous small shoots arise rapidly from each explant.

Stage I

Stage I consists of the establishment *in vitro* of suitable pieces of tissue, free from obvious contamination. As adventitious shoots are usually initiated on the tissue without transfer, Stages I and II are not generally discrete.

Stage II

Initially Stage II of this micropropagation method is recognized by the formation, growth and proliferation of adventitious shoots from the primary explant. Subsequently Stage II subcultures might, theoretically, be established from individual shoots by the techniques familiar in shoot culture. In practice, in plants such as *Saintpaulia*, both further adventitious and axillary shoots may develop in later stages of propagation. The result is a highly proliferative shoot mass and a very rapid rate of propagation. Subcultures are made by transferring shoot clumps (avoiding basal callus) to fresh media. In most commercial laboratories the micro-propagation of *Anthurium* species is initiated by adventitious shoot formation on leaf explants, followed by only axillary shoot development during the succeeding subcultures. Adventitious shoots sometimes arise directly from the leaves of plants during shoot culture. This often happens when leaves bend down to touch the semi-solid medium. Adventitious shoot formation of certain plants will take place in large vessels of aerated liquid medium, allowing the scale of propagation to be much increased.

Stage III

This is similar to the Stage III of most propagation systems. Individual shoots or shoot clumps are transferred to a nutrient medium with added growth regulators and ingredients that do not encourage further shoot proliferation and which promote rooting; alternatively shoots may be removed from culture and rooted *ex vitro*.

Some current applications

Several ornamental plants are at present propagated *in vitro* by direct shoot regeneration. Chief among these are plants of the family Gesneriaceae, (including *Achimenes*, *Saintpaulia*, *Sinningia* and *Streptocarpus*), where shoot buds can be freely regenerated directly on leaf explants without the formation of any intervening callus phase. Many other ornamentals and crop plants either are (or could be) propagated efficiently by this means, for example, begonias, *Epiphyllum*, cacti, *Gerbera*, *Hosta* and *Lilium*. Remember that this technique is more prone to yield off-types than shoot and node cultures, and that the technology is not applicable for the propagation of chimeras.

Regeneration from root pieces

In vitro shoot regeneration from root pieces is mainly reported from plants that possess thick fleshy roots such as those of the genera *Cichorium*, *Armoracia*, *Convolvulus*, and *Taraxacum*. It is, however, a method of propagation that is potentially applicable to a wide range of species. Shoots have, for instance, been induced to form directly on segments and apices of the roots of *Citrus* and *Poncirus* seedlings. Shoot regeneration from root pieces does not offer a continuous method of micropropagation unless there is a ready supply of aseptic root material (e.g. from isolated root cultures). Roots grown in soil *in vivo* are usually heavily contaminated and can be difficult to sterilize to provide an adequate number of uncontaminated cultures. They can however be used as an initial source of shoots which can be multiplied afterwards by shoot culture (e.g. *Robinia*).

Tissue maceration or fragmentation

The capacity of young fern tissue to regenerate adventitious shoots can be very high. Fern prothallus tissue (the gametophyte generation produced from germinating spores) has a high capacity for regeneration; a new prothallus can usually be grown from small isolated pieces of tissue, or even from single cells produced by maceration. Plants can also be regenerated from homogenized sporophyte tissue of some fern genera, and homogenization has been incorporated into tissue culture, or partial tissue culture techniques for the propagation of plants of this class.

Because a high proportion of the direct cost of micropropagation is attributable to the manual separation and transfer of explants and cultured material between media, the ability to regenerate plants from macerated or fragmented tissue would be extremely advantageous. Unfortunately there seem to be only a limited number of publications describing the formation of shoots directly from machine-macerated tissue of higher plants. One of them is the patent of Lindemann (1984), the claims of which may have been somewhat optimistic. Also Levin *et al.* (1997) reported on the regeneration of different plant species using a homogenization technology. However, shoot regeneration from fragmented shoot tips, or micropropagation of some plants by culturing shoot material or tissue fragments in fermentors, are somewhat comparable.

Organized calluses

In most callus cultures, shoots are produced from meristems which arise irregularly and may therefore be genetically altered. By contrast, so-called '*organized*' or '*semi-organized*' calluses are occasionally isolated in which there is a superficial layer of proliferating shoot meristems, overlaying an inner core of vacuolated cells acting as a mechanical and nutritional support. Calluses of this kind were termed organoid colonies by Hunault (1979): the names *meristemoids* and *nodules* have also been proposed. A meristemoid is defined as a cluster of isodiametric cells within a meristem or cultured tissue, with the potential for developmental (totipotential) growth. Meristemoids may give rise to plant organs (shoots, roots) or entire plants in culture.

Nodules also comprise meristematic cells, but they are distinct from meristemoids because they are independent spherical, dense cell clusters which form cohesive units, with analogy to both mineral nodules in geology and root nodules of legumes. Nodule culture has been extensively used for the propagation of *Cichorium intybus*.

The presence, in meristemoids, of an outer layer of shoot meristems seems to inhibit the unbridled proliferation of the unorganized central tissue. Geier (1988) has suggested that the control mechanisms which ensure the genetic stability of shoot meristems are still fully, or partly, active. Maintenance of a semi-organized tissue system depends on a suitable method of subculture and upon the use of growth regulator levels which do not promote excessive unorganized cell growth. Repeated selective transfer of unorganized portions of an organized *Anthurium scherzerianum* callus eventually resulted in the loss of caulogenesis. Conversely, by consistently removing the unorganized tissue when subculturing took place, shoot formation from the callus was increased.

Cultures consisting of superficial shoot meristems above a basal callus, seem to occur with high frequency amongst those initiated from meristem tip, or shoot tip, explants. Hackett and Anderson (1967) induced the formation of tissue of this type from carnation shoot tips by mutilating them with a razor blade before culture. Similar cultures were also obtained from seedling plumular tip explants of two (out of five tested) varieties of *Pisum sativum* placed on an agar medium. The calli were highly regenerative for 2-3 years by regular subculture to agar or shaken liquid medium. Maintenance was best achieved with an inoculum prepared by removing larger shoots and chopping the remainder of the callus and small shoots into a slurry. A callus, formed at the base of *Solanum curtilobum* meristem tips on filter paper bridges, gave rise to multiple adventitious shoots from its surface when transferred to shake culture in a liquid medium.

Callus with superficial proliferative meristems has also been induced by culture of shoot or meristem tips on a rotated liquid medium, in:

(a) *Nicotiana rustica*;

(b) *Chrysanthemum morifolium*;

(c) *Stevia rebaudiana*.

In *Stevia rebaudiana* (above), a slow rotation speed (2 rpm) was essential for initiation of an organized callus. A small callus formed upon the explant and in 2-3 weeks came to possess primary superficial shoot primordia which were globular and light green. Dark green aggregates of shoot primordia (termed '*secondary shoot primordia*") were developed within 6 weeks. If divided, the aggregations of shoot initials in both *Nicotiana* and *Stevia* could be increased by subculture or, if treated to a different cultural regime, could be made to develop into shoots with roots. Shoots were produced from *Chrysanthemum* callus upon subculture to an agar medium.

A spherical green dome-like structure was produced from meristem tips dissected from germinated *Eleusine coracana* (Gramineae) caryopses. When cut into four and subcultured, a green nodular structure was formed which grew to 5-10 mm in diameter. It was similar in appearance to a shoot dome, but much larger (a natural shoot dome is only 70-80 gm wide). The nodular structures were termed '*supradomes*' by Wazizuka and Yamaguchi (1987) because, unlike normal callus, superficial cells were arranged in an anticlinal plane and those beneath had a periclinal arrangement. Numerous multiple buds could be induced to form when the organized tissue was subcultured to a less complex medium.

Although proliferative meristematic tissue formed from shoot tips always appears to be accompanied by a basal callus, the superficial meristematic cells may well be derived directly from the cells of the apical shoot meristem of the explant, for they preserve the same commitment to immediate shoot formation. The presence of the apical meristem in the explant seems to be essential and culture of tissue immediately beneath it does not produce a callus with the same characteristics. Similar semi-organized callus can appear at the base of conventional shoot cultures. In the green granular callus mass which formed at the base of *Rhododendron* shoot tips, each granule represented a potential shoot. Organized caulogenic callus is thus closely comparable to embryogenic callus formed from pre-embryogenically determined cells.

Organized callus can be produced from explants other than shoot tips; in *Anthurium*, it has been derived from young leaf tissue and from spadix pieces. Organized callus has two characteristics which distinguish it from normal unorganized callus: the plants produced from it show very little genetic variation, and it can be subcultured for a very long period without losing its regenerative capacity. The callus of *Nicotiana* was able to produce plantlets over a ten-year period, while that of *Chrysanthemum* gave rise to plants continuously during four years.

The use of cultures with superficial proliferative meristems has not yet been widely used for micropropagation. There are three possible reasons:

1. The genetic variation which is almost invariably induced by shoot regeneration from normal callus, has cautioned against the use of any sort of callus culture for this purpose;
2. Organized callus may not always be readily distinguished from its unorganized counterpart;
3. Methods of initiating organized callus in a predictable fashion have not yet been fully elucidated.

There are examples of the initiation of organized callus from a sufficiently wide range of plant species (particularly from meristem tip explants) to suggest that it could be a method of general applicability. Multiplication may well be amenable to large-scale culture in fermentors.

Direct embryogenesis

Somatic embryos are often initiated directly upon explanted tissues. Of the occurrences, one of the most common is during the *in vitro* culture of explants associated with, or immediately derived from, the female gametophyte. The tendency for these tissues to give rise to adventitious somatic embryos is especially high in plants where sporophytic polyembryony occurs naturally, for example, some varieties of *Citrus* and other closely related genera.

Ovules, nucellar embryos, nucellus tissues and other somatic embryos are particularly liable to display direct embryogenesis. In *Carica* somatic embryos originated from the inner integument of ovules and in carrot tissue of the mericarp seed coat can give rise to somatic embryos directly.

The nucellus tissue of many plants has the capacity for direct embryogenesis *in vitro*. As explained, explants may also give rise to a proliferative tissue capable of embryogenesis. The high embryogenic competence of the nucellus is usually retained during subsequent cell generations *in vitro*, should the tissue be induced to form '*callus*' (or *cell suspensions*). It is not clear whether all cells of the nucellus are embryogenically committed. In *Citrus*, somatic embryos are formed from the nucellus even in cultivars that are normally monoembryonic (i.e. the seeds contain just one embryo derived from the zygote), whether the ovules have been fertilized or not. It has been suggested that only those cells destined to become zygotic proembryos can become somatic proembryos or give rise to embryogenic callus; somatic embryos have been shown to arise particularly from the micropylar end of *Citrus* nucellus.

Adventitious (adventive) embryos are commonly formed *in vitro* directly upon the zygotic embryos of monocotyledons, dicotyledons and gymnosperms, upon parts of young seedlings (especially hypocotyls and cotyledons) and upon somatic embryos at various stages of development (especially if their growth has been arrested). The stage of growth at which zygotic embryos may undergo adventive embryogenesis is species-dependent: in many plants it is only immature zygotic embryos which have this capacity. Unfortunately, as the phenotypic potential of seedlings is rarely known, using them as a source of clonal material is of limited value.

Embryogenic determination can be retained through a phase of protoplast culture. Protoplasts isolated from embryogenic suspensions, may give rise to somatic embryos directly, without any intervening callus phase. Treating protoplasts derived from leaf tissue of *Medicago sativa* with an electric field, induced them to produce somatic embryos directly upon culture.

Adventitious embryos arising on seedlings are sometimes produced from single epidermal cells. Zee and Wu (1979) described the formation of proembryoids within petiole tissue of Chinese celery seedlings, and Zee *et al.* (1979) showed that they arose from cortical cells adjacent to the vascular bundles which first became meristematic. Hypocotyl explants from seedlings of the leguminous tree *Albizia lebbek* showed signs of cracking after two weeks of culture and frequently young embryoids emerged. Stamp and Henshaw (1982) found that primary and secondary embryogenesis occurred in morphogenically active ridges produced on the surface of cotyledon pieces taken from mature cassava seeds.

Somatic embryos have been observed on the roots and shoots of *Hosta* cultures and on the needles and cultured shoots of various gymnosperm trees.

Protocorm formation in orchids

The seeds of orchids (like those of some other saprophytic or semi-parasitic plants) contain a small embryo of only about 0.1 mm diameter, without any associated endosperm storage tissue. Upon germination, the embryo enlarges to form a small, corm-like structure, called a *protocorm*, which possesses a quiescent shoot and root meristem at opposite poles. In nature, a protocorm becomes green and accumulates carbohydrate reserves through photosynthesis. Only when it has grown and has sufficient stored organic matter does it give rise to a shoot and a root. Normal seedling growth then continues utilizing the stored protocorm food reserves.

Bodies which, in their structure and growth into plantlets, appear to be identical with seedling protocorms (except that on synthetic media they may not be green), are formed during *in vitro* culture of different types of orchid organs and tissues. These somatic protocorms can appear to be dissimilar to seedling protocorms, and many workers on orchid propagation, have used terms such as '*protocorm-like bodies*' (PLBs) to describe them.

When a shoot tip of an orchid is transferred to culture on a suitable medium, it ceases to grow and to develop as a mature shoot apex; instead it behaves as though it were the apex of an embryo, i.e. it gives rise to a protocorm. Protocorm-like bodies also arise directly on some other orchid explants and proliferate from other PLBs in a fashion which is exactly comparable to the direct formation of somatic embryos.

Champagnat and Morel (1972) and Norstog (1979) considered the appearance of protocorms to be a manifestation of embryogenesis because they represent a specialized stage in embryo development and are normally derived directly from zygotic embryos.

Other protocorm-like structures. *In vitro* culture of small immature proembryos from developing barley seeds or from the fern *Todea barbara* has been noted to result in the formation of protocorm-like tissue masses from which root and shoots are regenerated after a period of irregular growth. Mapes (1973) recorded the appearance of such protocorm-like structures on shoot tips of pineapple, and Abo El-Nil and Zettler (1976) describe their direct formation on shoot tip explants of the yam *Colocasia esculenta*, or indirectly in subsequent callus cultures.

Embryogenesis from microspores or anther culture

Somatic embryos can be initiated directly from microspores. Usually it is necessary to culture the microspores within anthers, but occasionally it has been possible to induce embryogenesis from isolated microspores. Anther and microspore culture are described already, but because the plants produced by anther culture are likely to be dissimilar to their parents, we shall not consider the method in any detail, and reports of anther culture have been largely omitted.

Anther culture can result in callus formation; the callus may then give rise to plants through indirect embryogenesis or adventitious shoot formation.

Embryo proliferation

Accessory embryos on zygotic embryos. Occasionally new somatic embryos are formed directly on zygotic embryos that have been transferred to *in vitro* culture. Such adventitious embryos have been reported, for example, in: *Cuscuta reflexa*; barley; *Rex aquifolium*; *Thuja orientalis*; *Trifolium repens*; *Zamia integrifolia*; *Theobroma cacao*; *Linum usitatissimum*; *Vitis vinifera*.

When direct embryogenesis occurs on pre-formed embryonic tissue, the newly formed embryos are sometimes termed direct secondary embryos or accessory embryos.

Accessory embryos on somatic embryos. The *in vitro* induction of somatic embryogenesis starts a highly repetitive process, lacking some of the controls which must exist in nature during the formation of zygotic embryos. This results in the frequent development of small additional embryos on somatic embryos which have arisen directly on explants, or indirectly in callus and suspension cultures. Accessory embryos can occur along the whole axis of the original embryo, or grow preferentially from certain sites (e.g. the hypocotyl region or the scutellum of monocot embryoids). In walnut, accessory embryos appear to arise from single cells of the epidermis of somatic embryos.

Sometimes the term polyembryony is used to describe the formation of accessory, or secondary, embryos. The process has also been called *repetitive embryogenesis* or *recurrent somatic embryogenesis*. Such additional embryos are liable to be developed during all kinds of *in vitro* embryogenesis. They have been noted for example, on the somatic embryos formed in:

1. Anther cultures: *Atropa belladonna*; *Brassica napus*; *Carica papaya*; *Citrus aurantifolia*; *Datura innoxia*; *Vitis hybrids*;
2. Suspension cultures: *Daucus carota*; *Ranunculus scleratus*;
3. Callus cultures: *Aesculus hippocastanum*; alfalfa (when individual embryoids were transferred to a fresh medium); carrot; Citrus; parsley; *Pennisetum purpureum*; *Ranunculus sceleratus*, *Theobroma cacao*.

Protocorms arising directly on explanted shoot tips or leaf pieces of orchids, frequently produce other adventive 'daughter' protocorms in culture, in a fashion that is similar to the adventive formation of somatic embryos. Somatic embryos formed in callus of oil palm have been reported to give rise to protocorm-like bodies, which regenerated shoots repeatedly as subculture was continued.

Practical uses in propagation

From a quantitative point of view, indirect embryogenesis does provide an efficient method of micropropagation; the same is not true of direct embryogenesis when it is unaccompanied by the proliferation of embryogenic tissue. Although plants can be regenerated from embryos directly initiated *in vitro*, and may be present in sufficient numbers for limited plant production in breeding programmes, the numbers of primary embryos per explant will usually be inadequate for large scale cloning. To increase the number of somatic embryos formed directly on immature zygotic embryos of sunflower, Freyssinet and Freyssinet (1988) cut larger zygotic embryos into four equal pieces.

Additional embryos are generally unwanted: they are frequently joined one to another as twins or larger groups so that abnormal seedlings with multiple shoots develop from them. The presence of accessory embryos can also impede the growth of the primary somatic embryo. Growth then becomes asynchronous and normal seedlings may not be obtained unless the adventive embryos are removed.

However it has been suggested that accessory embryos might be used for micropropagating some species. Perhaps this is a method of micropropagation which will be developed more in the future? Examples of where it has been successful are:

1. *Helianthus annuus*;
2. *Juglans regia*;
3. *Medicago sativa*.

As mentioned earlier embryogenesis has a great potential for mass propagation, however, all adventitious techniques do still have the associated problem of the lack of clonal stability. Therefore the commercial application of this technology remains limited except, perhaps, where embryos arise directly from parental tissue.

Propagation of orchids

Morel (1960) noticed that when protocorms of *Cymbidium* were divided new protocorms were formed from the pieces, whereas if they were not divided, original and regenerated protocorms developed into new plantlets. Morel (1960, 1964) suggested that meristem or shoot tip explants could be used to establish cultures for the clonal propagation of orchids, providing thereby the basis of the method which is now used for many orchid genera. Rates of propagation are improved through the use of slightly more complex media than used by Morel, and by including growth regulators. However, many commercial micropropagation laboratories do not favour the use of protocorms for micropropagation because of the lack of clonal fidelity.

Some orchids not only form protocorms on apical meristems, but also directly on explants such as leaves, or flower stalks, or they may be formed from callus or callus via suspension cultures.

Propagation by Indirect Organogenesis

Propagation by all methods of indirect organogenesis carries a risk that the regenerated plants will differ genetically from each other and from the stock plant. Propagation by indirect organogenesis is described here for the sake of completeness; because of its potential as a propagation method, if the occurrence of genetic variation can be controlled; and because it is necessary for the regeneration and propagation of plants, which have been genetically transformed.

Indirect adventitious shoots from callus

Because they are not formed on tissues of the original mother plant, shoots (or other organs) are said to be regenerated indirectly when they are formed on previously unorganized callus, or in cell cultures. Separate root and shoot initials are characteristically formed in callus cultures and are only observed occasionally in suspensions where they are typically produced in large cell aggregates. Somatic embryogenesis occurs in both callus and suspension cultures. Adjustment of the growth regulators in the culture medium can bring about shoot or root formation in callus from a very large number of species. Inception of roots and shoots is most frequent in tissues that have been recently isolated, and morphogenic capacity generally declines with time as the tissues are subcultured. Nevertheless, some callus cultures maintain their regenerative ability over long periods.

Callus cultures vary in their morphogenic potential or competence. Because of this, the callus which originates from some plants, or from some kinds of explant, may not be responsive to techniques and media which frequently result in morphogenesis. The tissue may be non-morphogenic, or may only produce roots, from which plants cannot be regenerated. In some cases callus lines with different appearances (texture, colour, etc) and/or morphogenic capacities can be isolated from the same explant. These differences may reflect the epigenetic potential of the cells, or be caused by the appearance of genetic variability amongst the cells of the culture and support Street's (1979) suggestion that primary explants may be composed of cells or tissues capable of morphogenesis (competent cells) and others that are incapable (non-competent cells). Another possibility is that the operator is not able to create the appropriate conditions to express the full potential of the plant material he is working with.

Morphogenic and non-morphogenic callus lines, selected from primary callus, can retain their characteristics over many years. Special treatments, such as, a change of medium, an altered cultural environment, or an adjustment of the growth regulators added to the medium, may induce shoots or roots to form in some apparently non-morphogenic calluses; but generally, treatments to reverse a non-regenerative condition are unsuccessful.

In practice, the speed and efficiency with which plantlets can be regenerated from callus depends upon:

1. The interval between culture initiation and the onset of morphogenesis;
2. The choice of the appropriate type of callus;
3. The frequency and rate of shoot bud initiation;
4. Whether shoot regeneration can be readily re-induced when the callus is subcultured;
5. The number of subcultures that are possible without loss of morphogenesis;
6. Whether newly-initiated buds can be grown into shoots capable of being isolated and subsequently rooted.

Normal callus cultures produce shoots relatively slowly, but from some plants, and certain explants, under conditions that are not yet fully understood, callus can be initiated which has an especially high ability to regenerate shoots or somatic embryos.

Stage I

In most herbaceous broad leafed plants, it is possible to initiate morphogenically competent callus cultures from explants derived from many different tissues. Leaf, stem or root segments, pieces of storage tissue (e.g. tubers), seed embryos, shoot tips and seedling tissues have been used at various times. In monocotyledons there is a narrower range of suitable organs; embryos, very young leaf tissue, stem nodes and immature inflorescences being the most common sources. Initiation of callus cultures of many tree species, including gymnosperms, is frequently limited to explants derived from tissues near the vascular bundles or the cambium of stem or root sections. Explants containing actively dividing cells may be necessary if callus possessing a high level of morphogenic competence is to be isolated.

Callus growth is usually initiated by placing the chosen explant on a semi-solid medium into which auxin has been incorporated at a relatively high level, with or without a cytokinin. One or more transfers on the same medium may be necessary before the callus is separated from the parental tissue for subculture. Because more than one kind of callus may arise from a single explant, successful propagation can depend on being able to recognize and subculture only the type (or types) which will eventually be able to give rise to shoots or somatic embryos. In the absence of previous experience, samples of each type of callus may have to be carried forward for testing on inductive media. Translucent, watery callus is seldom morphogenic, whereas nodular callus frequently is.

Organized adventitious shoots are usually induced to form in callus or suspension cultures by reducing the auxin level in the medium and/ or increasing the concentration of cytokinin. To grow callus-derived shoots into plantlets capable of survival in the soil, they must be rooted as micro-cuttings. Root production by callus is of little consequence for micropropagation purposes; even if roots are formed concurrently with adventitious shoots, the vascular connections between roots and shoots, through the callus tissue, are almost invariably insufficient for the development of a functional plantlet.

Stage II

Once a morphogenic callus has been isolated, propagation is carried out either by callus subdivision, or by the preparation of cell suspensions. The success of each technique depends on the subcultured tissues or cells continuing to regenerate shoots.

Callus subdivision. Callus is cut into smaller pieces which increase in size when subcultured in a liquid or an agar-solidified medium. The callus can either be subdivided further, or shoot regeneration allowed to occur. This may take place on the same medium, or the callus may need to be transferred to another shoot-inducing medium. The organogenic capacity of callus is easily lost on repeated subculture. Use of high growth regulator levels can encourage the proliferation of non-regenerative callus which will displace tissues having the competence to form new shoots (e.g. in *Pelargonium*).

The preparation of cell suspensions. Compared to the relatively rapid rates of propagation that are possible with shoot culture of some kinds of plants, propagation from morphogenically competent callus can be slow initially. Krikorian and Kann (1979) quoted a minimum of 135 days from the excision of daylily explants to the potting of plantlets. The rate at which propagation can proceed after that depends on the rate at which callus can be grown and subdivided. Providing a shoot-forming capacity is retained, a much faster rate of multiplication can be achieved by initiating a suspension culture from competent callus. After being increased by culture, the cells or cell aggregates can then be plated to produce new regenerative callus colonies. This is not an easy operation, as growth regulators favouring the formation of a dispersed cell suspension can cause the cells to lose their morphogenic capacity. There is also the problem that, by prolonging the period before shoot regeneration, genetic variability within the cell line will be increased.

Genetic stability

In some crop plants, the genetic differences between plants derived from callus and suspension cultures are considerable, and are sufficient to have attracted the interest of plant breeders as a new source of selectable variability. However, plants obtained from callus lines with a high degree of morphogenic competence, appear to be much more uniform genetically. Care must be taken though to see that primary explants are not taken from plant tissue likely to be endopolyploid. Subsequent exposure to high levels of growth substances such as 2,4-D should also be avoided as far as possible. Genetic stability of plants from highly competent callus cultures may be assisted by the continual presence of superficial meristems. As mentioned previously, these probably repress shoot formation from cells within the callus mass.

Morphogenic cereal cultures

The shoot forming capacity of some callus cultures has been attributed to the proliferation of meristematic centres derived from the tissues of the explant. King et al. (1978) have suggested that the small number of shoots produced by certain cereal tissue cultures arises in this way (e.g. in wheat, rice, oat and maize). Cure and Mott (1978) noticed that aberrant root-like structures existed within cereal cultures from which shoots arose. Such primordia, whether of root or shoot origin, are thought to proliferate adventitiously *in vitro,* surrounded by less organized tissues. Regenerative capacity is usually lost rapidly when the shoot primordia are diluted during subculture. Cereal callus of this kind does not have the same kind of inherent morphogenic capacity found in other types of callus cultures. Despite these observations, experience shows that morphogenesis can occur from previously unorganized cereal callus.

Current applications

In the past, several ornamental plants (e.g. *Freesia* and *Pelargonium*) have been micropropagated from adventitious shoots produced indirectly from callus. It had been hoped to extend the technique to other species possessing a strong natural tendency towards diploidy (e.g. some forest trees) where plantlets produced *in vitro* might have a normal karyotype, but it is now realized that the genetic changes which are almost universally induced in the genotype of cells during callus and cell culture make cloning by this technique inadvisable except where new genotypes are required for selection or further plant breeding. Another possibility is that mutated somatic cells, already present in the mother plant, are given the opportunity to develop into a plantlet.

Indirectly-initiated somatic embryos

Indirect formation of somatic embryos (or adventitious somatic embryogenesis) from callus or suspension cultures is observed more frequently than direct embryogenesis. Frequently callus which is wholly or partly embryogenic can be induced during the initial culture of explants derived from young meristematic tissues (see below), but induction is less common in cultures which have been kept and transferred for some period without organogenesis.

There are important requirements for the successful induction of embryogenic callus and suspension cultures:

1. The plant genotype must be capable of embryogenesis on the chosen system of induction (medium plus added growth regulators). In some genera most genotypes are competent, but in others there may be a wide variation in competence even between different varieties or cultivars within a species.
2. In most practical situations, cultures should be grown in the presence of an auxin for the induction (and initiation) of embryogenesis (Stage I).
3. The level of sugar (e.g. sucrose or glucose) in the medium may need to be within critical concentrations, and no embryos may be formed at all if the sugar concentration is too high.
4. After the beginning of embryogenesis, it is usually (but not invariably) necessary for Stage I tissues or cells to be subcultured to a medium containing a reduced auxin concentration, or containing no auxin at all (Stage II).

5. There may be an optimum length of time during which the Stage I routine should be maintained. An extended period before subculture can result in the failure to obtain embryogenesis at Stage II. Maintenance of the cultures on high auxin usually causes embryo development to be arrested or a loss of embryogenic capability.
6. A supply of reduced nitrogen is required. This may be supplied in the form of NH_4^+ ion and/or as an amino acid such as glutamine or alanine.

Embryogenesis in primary callus cultures

Callus capable of producing somatic embryos (embryogenic callus) is most reliably obtained from an explant during the initial phase of culture, and is frequently produced in conjunction with non-morphogenic tissue. Embryogenic callus can usually be distinguished by its nodular appearance, and is frequently produced preferentially from one part of an explant (e.g. the scutellum of a monocotyledon embryo), probably because only the cells of that part of the explant were embryogenically pre-determined. These may be the same tissues, which in another cultural environment are capable of producing embryos directly. According to this hypothesis, although competent and non-competent cells may produce callus, only that which grows from competent cells will give rise to somatic embryos.

The expression of competence depends on the use of a suitable medium for the culture, containing requisite growth regulators at the correct concentration. The formation of somatic embryos in *Lolium multiflorum*, for example, was medium dependent. On the most suitable medium, immature inflorescence explants produced three types of callus, only one of which spontaneously formed embryo-like structures. Unless such different callus types are separated, cells of different regenerative capabilities may become mixed. Morphogenically competent cells could then be lost by competition in the combined callus tissue that results.

Stage I. Selection of an appropriate explant is most important. Embryogenic callus has been commonly gained from seed embryos, nucelli or other highly meristematic tissues such as parts of seedlings, the youngest parts of newly initiated leaves and inflorescence primordia. Within an inflorescence, staminoids (*Theobroma cacao*) and filaments (*Aesculus hippocastanum*) have been reported to be adequate sources of explants. The initiation of embryogenic callus from root tissue is rare but has been reported in some monocotyledons e.g. rice; oil palm, Italian ryegrass and *Allium carinatum*. Callus is usually commenced on a semi-solid medium incorporating a relatively high level of an auxin. Only a few tissues with a high natural embryogenic capacity do not require the addition of endogenous auxin for the development of embryogenic callus. Occasionally, primary callus arising from an explant may show no morphogenic capacity, but can be induced to give rise to new embryogenic tissue during later (secondary) subcultures by transfer to an inductive medium. Ahee *et al.* (1981) have used this method to propagate oil palms. On the medium used, calluses arising on the veins of young leaf fragments had no morphogenic capability. However, when primary calluses were subcultured onto appropriate media (unspecified), some of them gave rise to tissue that was different in structure and form, and grew at a much faster rate. These '*fast-growing calluses*' could be induced to produce structures resembling embryoids, and afterwards plantlets, upon further subculture to other media.

One highly embryogenic tissue that has been extensively studied is that of the nucellus of the polyembryonic '*Shamouti*' orange. Here it seems that cells at just one end of the embryo sac (the micropylar end) are embryogenically predetermined and retain this capacity in subsequent cell generations. On subculture, proliferation of the nucellus cells proceeds without the addition of growth regulators to the medium, and results in the formation of an habituated callus. A tissue is said to have become habituated when it will grow without a growth regulator, or some other organic substance which is normally necessary, being added to the medium. Addition of auxins to the growth medium is inhibitory to the growth of auxin-habituated '*Shamouti*' orange tissue, which has been thought to be composed (at least initially) of numerous pro-embryos and not of undifferentiated cells. Embryogenic callus has also been obtained from the nucellus tissue of other plants, mainly tropical fruit species.

Stage II. As a general rule, somatic embryos formed on a medium containing a relatively high concentration of an auxin, will only develop further if the callus culture is transferred to a second medium from which auxin has been omitted, another '*less active*" auxin has been substituted, or the level of the original auxin much reduced. This treatment is occasionally ineffective and sometimes adding a cytokinin helps to ensure embryo growth. A further essential requirement is the need for a supply of reduced nitrogen in the form of an ammonium salt or amino acid. No change of nitrogen source is required if MS medium was used for Stage I, but if, for example, White's medium were used for Stage I, it would need to be supplemented with reduced nitrogen, or the culture transferred to MS.

Callus subculture. Once obtained, embryogenic callus can continue to give rise to somatic embryos during many subcultures over long periods. The continued production of somatic embryos in these circumstances depends either on the continued proliferation of pro-embryogenic nodules, and/or the *de novo* formation of embryogenic tissue from young somatic embryos during each subculture. Inocula for subcultures must be carefully selected. In wheat, callus with continued embryogenesis was only re-initiated from inocula taken close to somatic embryos; tissue from the same culture which did not contain embryos was not embryogenic in the next passage.

Sometimes the number of embryos produced per unit weight of callus rises during a few passages and then slowly falls, the capacity to form somatic embryos eventually being irrevocably lost. Callus derived from the nucellus of '*Shamouti*' oranges increased in its capacity to form somatic embryos when subcultured at 10-15 week intervals, while transfer at 4-5 week intervals, reduced embryogenesis.

Somatic embryos can be formed relatively freely in callus tissue, but where they are to be used for large scale propagation, their numbers can often be increased more rapidly and conveniently by initiating an embryogenic suspension culture from the primary callus.

Embryogenesis in suspension cultures

Cultures from embryogenic callus (*Stage I*). Suspension cultures can sometimes be initiated from embryogenic callus tissue, and the cells still retain the capacity to regenerate somatic embryos freely. Obtaining such cultures is not always a simple matter, for the auxin levels that are often used to promote cell dispersion may result in the loss of morphogenic capability. Embryogenic cell suspensions are most commonly initiated from embryogenic callus that is placed in liquid medium on a shaker. Vasil and Vasil (1981a,b) and Lu and Vasil (1981a,b) have reported producing cultures of this type from pearl millet and guinea grass respectively. Suspensions were initiated and subcultured in MS medium containing 1-2.5 mg/l 2.4-D and 2.5-5% coconut milk, and came to be composed of a mixture of embryogenic cells (small, highly cytoplasmic and often containing starch) and non-embryogenic cells (large and vacuolated). Embryoids were induced to develop into somatic seedlings when plated onto an agar medium without growth regulators, or with lower levels of auxin than used at the previous stage.

Cultures from non-embryogenic sources. Embryogenesis can be induced in cell suspensions of some plants when the cultures are produced from non-morphogenic callus and have been maintained without morphogenesis for one or more transfers. Induction occurs most readily in recently isolated suspensions and usually becomes much less probable with increasing culture age. Loss of regenerative ability is often associated with the appearance of some cells with abnormal chromosome numbers, but it can also be due to culture on an inappropriate medium.

Embryogenesis in suspension cultures seems to require media at Stage I and Stage II with similar compositions to those necessary for somatic embryo formation in callus cultures. Somatic embryos can be formed in suspension cultures in very large numbers. Reinert *et al.* demonstrated that the continued capacity of carrot cell suspensions to form embryos depended on an adequate supply of nitrogen. Embryogenesis ceased on a medium containing little nitrogen, but it was re-induced for several transfers after the culture was returned to a high-nitrogen medium.

In a few kinds of plants it is possible to induce embryogenesis in previously unorganized suspension cultures. Success is so far recorded only in members of the families Apiaceae (Umbelliferae), Cruciferae and Scrophulariaceae. This is not therefore a method of propagation which can be readily utilized. There is a greater chance of obtaining an embryogenic suspension culture from embryogenic callus.

Abnormal embryos and plantlets

Unfortunately embryogenesis in both callus and suspension cultures is seldom synchronous so that embryoids at different stages of development are usually present in a Stage II culture from the onset. This presents a major drawback for plant propagation which could otherwise be very rapid, especially from suspensions. A proportion of the seedlings developing from somatic embryos can also be atypical: abnormalities include the possession of multiple or malformed cotyledons, more than one shoot or root axis, and the presence of secondary adventive embryos. Embryos with three cotyledons have been observed to give rise to well-formed plantlets. Abnormal somatic embryos do however produce secondary embryos, which are usually of normal morphology.

Pretova and Williams (1986) suggested that embryo proliferation, or '*cleavage*', and the formation of accessory cotyledons and root poles, to be homologous with the production of discrete complete embryoids. They suggested that production of accessory cotyledons and somatic embryos on the hypocotyl, and additional root poles near the base of existing embryos, may represent either a gradient along these organs in the early stages of determination, or, be caused by a factor affecting cell to cell co-ordination.

Differential filtering and sedimentation to separate embryos at different stages of development can improve the uniformity of embryo populations in suspension cultures. More recently image analysis has been used to select embryos in specific developmental stages. In addition cultures can be maintained on media containing high levels of sucrose, and/or low levels of abscisic acid. Both approaches, as well as the addition of imazalil to the culture medium, limit the number of abnormalities and give a higher degree of synchronization. High levels of sucrose and abscisic acid induce reversible dormancy in somatic embryos and thus might be used to temporarily suspend growth should this be advantageous in a planned micropropagation programme.

Dormancy is however not always reversible. Indeed somatic embryos can remain dormant, and conversion to plantlets can be problematic. Different types of approaches can be used to overcome this problem, i.e. desiccation, supplementing the medium with osmotic agents [e.g. *polyethylene glycol* (PEG), mannitol].

Genetic stability

Plants regenerated through somatic embryogenesis are usually morphologically and cytologically normal, but sometimes a proportion of aberrant plants is obtained. Genetically abnormal plants are more likely to occur where embryogenesis is initiated in callus or suspension cultures after a period of unorganized growth or when embryogenic cultures are maintained for several months. A proportion of albino plants lacking chlorophyll is characteristically produced in anther culture of cereals and grasses and during embryogenesis from other monocot explants. Dale *et al.* (1980) found that plants produced from embryogenic callus cultures of Italian ryegrass were more likely to be devoid of chlorophyll the longer the cultures were maintained. After one year, some cultures produced only albinos. Embryo-like structures (although still present on the surface of the callus) tended to be distorted.

Current applications

Few plant species are at present propagated on a large scale via embryogenesis *in vitro*. This method of morphogenesis does however offer advantages which suggest that it will be used increasingly for plant cloning in the future:

1. In some monocotyledons (e.g. cereals, date palm and oil palm) it provides a method of micropropagation where shoot culture has

not been successful (but note however that in some attempts to clone oil palms through embryogenesis, the resulting plants have been very variable);

2. Providing embryogenic cell suspensions can be established, plantlets can theoretically be produced in large numbers and at much lower cost because plantlets do not have to be handled and subcultured individually;
3. Somatic embryos probably provide the only way for tissue culture methods of plant propagation to be economically deployed on extensively planted field crops and forest trees.

Storage Organ Formation

Many ornamental and crop species are normally propagated, stored and planted in the form of vegetative storage organs. It is therefore not surprising that, where such organs are produced *in vitro,* they often provide a convenient means of micropropagation and/or genotype storage. Characteristic, though small, storage structures can be induced to form in cultures of several plant species, for example:

Bulbils	*Amaryllis*, hyacinth, lily, onion *Narcissus*
Cormlets	*Gladiolus*
Miniature tubers	Potato, yams

Protocorm formation as a method of propagation has been considered under Direct Embryogenesis.

Methods of obtaining storage organs vary according to the kind of tissue being cultured. Some storage organs formed *in vitro* can be planted ex *vitro* directly into the soil.

Production of Bulbils and Cormlets

Species that naturally produce bulbs can be induced to form small bulbs (bulbils or bulblets) in culture. Bulbils can be produced from axillary buds, but frequently they are formed from adventitious buds developed on pieces of leaf, on inflorescence stalks, or on ovaries, and particularly on detached pieces of bulb scale.

Both axillary and adventitious shoots and bulbils are formed on bulb scale pieces *in vitro.* Strong dominance of the main shoot apex often prevents the formation of axillary buds at the bases of bulb scales *in vivo,* but buds capable of giving rise to bulblets (or to shoots upon which bulbs will be formed later) are freely produced when bulb scales or bulb sections are cultured. In some species it is important to include part of the basal plate in the explant. Depending on the kind of bulb being cultured, explants for continued Stage II propagation may consist of scales taken from bulblets, swollen shoot bases, or bulblets which have been trimmed and split. Propagules for transfer from Stage III to the external environment can be plantlets, plantlets with a bulblet at the base, or dormant bulblets.

Instead of producing storage organs composed of swollen leaf bases (bulbs), some monocotyledons store food reserves in swollen stem bases (corms). Small corms (cormlets) of *Gladiolus* may be formed directly on explanted tissue or on callus in culture, or they are produced on rooted plantlets grown in culture jars until the leaves senesce. Cormlets formed *in vitro* can be planted in soil or used to start new *in vitro* cultures.

Miniature tubers

Under appropriate environmental conditions, plants that naturally produce tubers can be induced to produce miniature versions of these storage organs, in a medium containing high cytokinin levels. Tubers normally formed on underground stolons are produced *in vitro* in axillary positions along *in vitro* shoots. Two crops where miniature tubers have been utilized for propagation are potato and yams. Methods for inducing the *in vitro* tuberization of potatoes were first described by Lin *et al.* (1978) and Hussey and Stacey (1981a,b), and since by several other workers. Potato tubers form best in darkness, but those of *Dioscorea* mainly appear at the base of stem node cuttings in the light.

Miniature tubers have the great advantage that they can be readily removed from culture flasks in a dormant condition and stored *ex vitro* without precautions against sepsis. When planted in soil they behave as normal tubers and produce plants from axillary shoots. If they are produced *in vitro* from virus-tested shoots, miniature tubers provide an ideal method of propagating and distributing virus-tested stock to growers.

MICROGRAFTING

The transfer of small shoot apices onto rootstocks (termed micrografting), can be carried out *in vivo* or *in vitro*. Navarro (1988) lists four uses which have been found for the technique:

1. Obtaining plants free from specific virus and virus-like diseases;
2. A method for separating virus and virus-like organisms in mixed infections;
3. For studying graft incompatibility between scions and rootstocks, and the histological and physiological aspects of grafting;

4. A minimum risk method for importing plant material through quarantine.

Micrografting is thus indirectly useful in micropropagation: the necessary techniques are usually too time consuming and the proportion of successful takes is generally too low, for it to be of direct application.

The rootstocks used for micrografting are commonly newly-germinated seedlings, but it is also possible to use rooted cuttings or micropropagated shoots. Before micrografting can be carried out, it is necessary to prepare suitable rootstock material. When seedling rootstocks are used, and all stages of grafting are conducted *in vitro,* seeds are surface sterilized and germinated aseptically in vessels containing nutrient salts (e.g. those of MS medium). The seedlings may be supported on agar medium or on a porous substrate, such as sterile vermiculite, which allows the growth of a branched root system. Micrografting is then effected by cutting off the tops of the seedling rootstocks and placing small shoot apices onto the exposed surface. When grafts are successful, rootstock and scion grow together to produce a plant. It is usually necessary to examine freshly grafted seedlings on a regular basis and remove any adventitious shoot arising on or below the graft union.

Shoot tips to be grafted onto seedling rootstocks (i.e. the scions) are carefully excised from preferred plant material. In *Citrus* there has been most success when scions have been placed directly in one of two positions:

(a) into inverted T-shaped incisions immediately below the cut surface of a decapitated rootstock, or,

(b) onto the cambium layer or vascular ring of the cut surface.

Shoot or meristem tips intended for grafting can be taken from apical or axillary buds of actively growing shoots in the greenhouse or field, or may be removed from shoots growing *in vitro.* Once transferred, the survival of micrografted apices is partly dependent on their size. Very small apices must be used for virus elimination, making the technique difficult and unreliable. Tips 0.1-0.2 mm in length have been grafted for virus elimination from vines; with peach, slightly larger apices (0.5-1 mm) have been employed. The excision and transfer of very small shoot apices requires precise micro-manipulation under a binocular microscope. If large shoot apices are to be grafted, their bases are often cut into a wedge which is then inserted into a vertical cut on the rootstock. Several techniques have been found to increase the proportion of successful graft unions:

1. Tissue blackening, which commonly results in the death of very small scions, can be reduced by soaking explants in an anti-oxidant solution, and/or placing a drop of solution onto the severed rootstock immediately before inserting the scion. A solution of 2 g/1 sodium diethyldithiocarbamate (DIECA) has been used for this purpose. Navarro (1988) advocates rapid manipulations to prevent phenolic oxidation and says that it is more effective than anti-oxidants.
2. Apices to be grafted may be placed either directly onto a decapitated rootstock, or cultured for a short period before being transferred. There is often a better 'take' and more rapid growth if they are pre-cultured for a short while supported on paper above an MS mineral salt medium containing growth regulators. Jonard *et al.* (1983) found that adding a cytokinin to the medium (e.g. 0.1 mg/l zeatin if the apex is cultured for 48 h; 0.01 mg/l zeatin if the culture is continued for 48-240 h) was particularly effective in encouraging the rapid formation of leafy shoots once the graft has been made. An alternative is to place scion shoot tips into a growth regulator solution for a short period before grafting: a 5-10 minute immersion in either 10 mg/l 2,4-D or 1 mg/l BA, doubled the number of successful micrografts of *Citrus*. Starrantino and Caruso (1988) got a greater percentage of viable grafts when they dipped both shoot tips and the cut apex of young rootstocks in 0.5 mg/l BA for 20 min before the two were united. Yet another method is to place cytokinin (e.g. 2 mg/l BA or, for peach, 10 mg/l zeatin) in a drop of water or agar gel between the scion and the rootstock.
3. A greater proportion of graft unions may result from growing isolated meristem tips to a larger size before they are implanted. Isolated scion tips of peach have been cultured *in vitro* by the initial stages of meristem tip culture for a period of about two weeks until they have grown from 0.5-1 mm to ca. 10 mm.
4. Desiccation is a major cause of the failure of graft unions. To prevent drying, Pliego-Alfaro and Murashige (1987) applied a layer of moist nutrient agar gel to connect the graft area with the medium. The gel had to be progressively removed from the top downwards during weeks 1-3 after grafting, or poor unions resulted.

When, as has been most common, micrografting has been carried out *in vitro,* considerable care needs to be taken over transferring grafted plants to the external environment. However, several authors have found that sterility is not necessary and that shoot apices can be united onto rootstocks grown *in vivo*. The proportion of completed grafts may be less than under aseptic conditions, but problems with eventual transfer are eliminated. A variety of scion material has been utilized for *in vivo* grafting, including directly-excised 0.1-0.2 mm tips of *Citrus limon*, and meristem-cultured scion tips of peach. Once again, graft unions may be improved if cut surfaces are anointed with a DIECA (1 g/l) plus cytokinin (10 g/l) solution. Plants are probably best kept in a growth room for a period after grafting, and desiccation of the graft union prevented by enclosing each plant in a plastic bag or by placing an elastic strip around the graft.

Grafting mature shoots of woody perennials onto juvenile rootstocks is known to induce juvenile characteristics in the resulting shoots, particularly if it is repeated successively. Half of the plants resulting from micrografting adult lateral buds of avocado onto seedling rootstock were found to have some juvenile symptoms. But in *Citrus*, micrografting does not seem to induce juvenile characteristics, providing shoot tips are taken from an adult source: plants are thornless and come into flower rapidly.

From having been developed as a method of producing virus-tested *Citrus*, micrografting is now widely used for the improvement of plants of this and related genera. It has also been practised on a wide range of other plants, primarily for virus-elimination.

5

Cultural Media

Inorganic Medium Components

Plant tissues and organs are grown *in vitro* on artificial media, which supply the nutrients necessary for growth. The success of plant tissue culture as a means of plant propagation is greatly influenced by the nature of the culture medium used. For healthy and vigorous growth, intact plants need to take up from the soil: (i) relatively large amounts of some inorganic elements (the so-called *major plant nutrients*): ions of nitrogen (N), potassium (K), calcium (Ca), phosphorus (P), magnesium (Mg) and sulphur (S); and, (ii) small quantities of other elements (minor plant nutrients or trace elements): iron (Fe), nickel (Ni), chlorine (Cl), manganese (Mn), zinc (Zn), boron (B), copper (Cu), and molybdenum (Mo).

According to Epstein (1971), an element can be considered to be essential for plant growth if

1. A plant fails to complete its life cycle without it;
2. Its action is specific and cannot be replaced completely by any other element;
3. Its effect on the organism is direct, not indirect on the environment;
4. It is a constituent of a molecule that is known to be essential.

The elements listed above are - together with carbon (C), oxygen (O) and hydrogen (H) - the 17 essential elements. Certain others, such as cobalt (Co), aluminium (Al), sodium (Na) and iodine (I), are essential or beneficial for some species but their widespread essentiality has still to be established. The most commonly used medium is the formulation of Murashige and Skoog (1962). This medium was developed for optimal growth of tobacco callus and the development involved a large number of dose-response curves for the various essential minerals. The relatively low levels of Ca, P and Mg in MS are evident. The most striking differences are the high levels of Cl and Mo and the low level of Cu. Each plant species has its own characteristic elementary composition which can be used to adapt the medium formulation. These media result often in a much improved growth. A major problem in changing the mineral composition of a medium is precipitation, which may often occur only after autoclaving because of the endothermic nature of the process.

Plant tissue culture media provide not only these inorganic nutrients, but usually a carbohydrate (sucrose is most common) to replace the carbon which the plant normally fixes from the atmosphere by photosynthesis. To improve growth, many media also include trace amounts of certain organic compounds, notably vitamins, and plant growth regulators.

In early media, '*undefined*' components such as fruit juices, yeast extracts and protein hydrolysates, were frequently used in place of defined vitamins or amino acids, or even as further supplements. As it is important that a medium should be the same each time it is prepared, materials, which can vary in their composition are best avoided if at all possible, although improved results are sometimes obtained by their addition. Coconut milk, for instance, is still frequently used, and banana homogenate has been a popular addition to media for orchid culture.

Plant tissue culture media are therefore made up from solutions of the following components:

1. Macronutrients (always employed);
2. Micronutrients (nearly always employed but occasionally just one element, iron, has been used);
3. Sugar (nearly always added, but omitted for some specialized purposes);
4. Plant growth substances (nearly always added)
5. Vitamins (generally incorporated, although the actual number of compounds added, varies greatly);
6. A solidifying agent (used when a semi-solid medium is required. Agar or a gellan gum are the most common choices).
7. Amino acids and other nitrogen supplements;
8. Undefined supplements such as coconut milk etc. (which, when used, contribute some of the five components above and also plant growth substances or regulants);

9. Buffers (have seldom been used, but the addition of organic acids or buffers could be beneficial in some circumstances).

Finally, it should be noted that minerals may also have a signalling role altering developmental patterns. This is most obvious in root architecture which is logical as roots have a principal function in ion uptake and the root system should be such that uptake is optimal. So growth and branching of roots should be affected by mineral concentrations in the soil. Ramage and Williams (2002) also argue that minerals appear to have an important role in the regulation of plant morphogenesis as opposed to just growth.

Uptake of Inorganic Nutrients

Plants absorb the inorganic nutrients they require from soils almost entirely as ions. An ion is an atom, or a group of atoms, which has gained either a positive charge (a cation) or a negative charge (an anion). Inorganic nutrients are added to plant culture media as salts. In weak aqueous solutions, such as plant media, salts dissociate into cations and anions. Thus calcium, magnesium and potassium are absorbed by plant cells (normally those of the root) as the respective cations Ca^{2+}, Mg^{2+} and K^+; nitrogen is mainly absorbed in the form nitrate (the anion, NO_3^-) although uptake of ammonium (the cation, NH_4^+) may also occur, phosphorus as the phosphate ions HPO_4^{2-} and $H_2PO_4^-$; and sulphur as the sulphate ion SO_4^{2-}. In tissue culture, uptake is generally proportional to the medium concentration up to a concentration of twice MS. For specific elements this may be different. For example, Leiffert *et al,* (1995) found only a small increase in Zn uptake with increasing medium concentration indicating that the concentration of Zn in the cultured tissues was adequate, not requiring further uptake. Selective uptake also suggests active uptake.

In the whole plant, nutrients are either taken up passively, or through active mechanisms involving the expenditure of energy. Active uptake is generally less dependent on ionic concentration than passive uptake. Both systems are however influenced by the concentration of other elements, pH, temperature, and the biochemical or physiological status of the plant tissues. These factors can in turn be controlled by the solution presented to the roots, or they may dictate the ionic balance of an ideal solution. For example, Mg^{2+} competes with other cations for uptake. Under conditions of high K^+ or Ca^{2+} concentrations, Mg deficiency can result, and vice versa. Active uptake of phosphate falls off if the pH of the solution should become slightly alkaline when the $(H_2PO_4)^-$ ion becomes changed to $(HPO_4)^{2-}$. There is some evidence that ammonium is utilized more readily than nitrate at low temperatures and that uptake may be enhanced by high carbohydrate levels within plant cells. Calcium is not absorbed efficiently and concentrations within plant tissues tend to be proportional to those in the soil. Plants are comparatively insensitive to sulphate ions, but high concentrations of dissolved phosphate can depress growth, probably through competitively reducing the uptake of the minor elements Zn, Fe and Cu. Although the biochemistry and physiology of nutrient uptake in tissue cultures may be similar, it is unlikely to be identical.

In vivo, plants take up mineral ions with their roots. No studies have been made on how uptake of nutrients occurs in shoot cultures. For IAA, it has been shown that most uptake is *via* the cut surface and that only a small fraction is taken up via the epidermis. The same likely holds for minerals. It should be noted, though, that in tissue culture the stomata are always open in the portion of the explant exposed to the gaseous phase and the same may apply for tissues that are exposed to semi-solid or liquid medium. Uptake via the stomata is well possible.

Once taken up, transport within the plant occurs in the mass flow via the xylem. In *in vivo* plants there are two mechanisms for driving the water flow, root pressure and water potential gradient between at one end the soil and at the other end the atmosphere. Under normal conditions, the latter is the far more important, but water flow resulting from root pressure is in itself sufficient for long-distance mineral supply. Plants without roots are often cultured *in vitro* where the atmosphere is very humid, and the flow driven by a difference in water potential consequently reduced. In spite of this, in tissue culture there still seems to be sufficient water flow which may be favoured by the stomata being continuously open. There are no indications that the structure of the xylem is altered in such a way as to reduce transport of ions.

When explants are first placed onto a nutrient medium, there may be an initial leakage of ions from damaged cells, especially metallic cations (Na^+, Ca^{2+}, K^+, Mg^{2+}) for the first 1-2 days, so that the concentration in the plant tissues actually decreases. Cells then commence active absorption and the internal concentration slowly rises. Phosphate and nitrogen (particularly ammonium) are absorbed more rapidly than other ions. In liquid medium, almost all phosphorus and ammonium are taken up in the first two weeks of culture (e.g. by 5 microshoots of *Dahlia* in 50 ml stationary liquid medium). After uptake, phophorus is massively redistributed to tissues that are formed after the initial two weeks. Nitrate in a medium very similar to that of Wood and Braun (1961) B medium, was reduced by *Catharanthus roseus* suspensions from 24 mM to 5 mM in 15 days, while Na^+, K^+, and SO_4^{2-}, fell to only just over half the original concentrations in the same time. Carnation shoot cultures were found to use 31-75% Mg^{2+}, and 29-41% Ca^{2+} in MS medium during a 4 week period.

Unintended Alterations

Nutrients, and especially micronutrients, may also be added via impurities, and especially via agar. Such impurities may well be beneficial. This is particularly true of Ni, which has recently been shown to be an essential element but was not known to be when most medium formulations were established. This element is usually not included in the inorganic constituents but can be provided by impurities. Agar provides a large addition of sodium but levels of sulphur and copper are also significantly increased. Increases in the other elements in MS, are less than 20 %. Gelrite contains fewer organic impurities but inorganic ones occur at high concentrations. It should be noted that the data are from determinations done more than 15 years ago and that the production process of gelrite has been improved ever since. Gelrite is being used in medicines as an ophthalmic vehicle. Furthermore, minerals are absorbed to a significant percentage by agar and by activated charcoal but whether this has a significant effect has not been examined.

MACRONUTRIENTS

Nitrogen

Forms of nitrogen

Nitrogen is essential to plant life. It is a constituent of both proteins and nucleic acids and also occurs in chlorophyll. Most animals cannot assimilate inorganic nitrogen or synthesize many of the amino acids unless assisted by bacteria (e.g. in the rumen of cattle). Nitrogen is available in the atmosphere as N_2 but only legumes have the capacity to utilize this nitrogen using *Rhizobium* bacteria in the root nodules. In most plants, nitrate (NO_3^-) is the sole source of nitrogen. After uptake, NO_3^- is reduced to NH_4^+ prior to incorporation into organic molecules. The relevance of nitrogen is illustrated by the vast amounts of nitrogen reserves in seeds (as storage proteins).

Both growth and morphogenesis in tissue cultures are markedly influenced by the availability of nitrogen and the form in which it is presented. Compared to the nitrate ion, NO_3^- (which is a highly oxidized form of nitrogen), the ammonium ion, NH_4^+, is the most highly '*reduced*' form. Plants utilize reduced nitrogen for their metabolism and internally, nitrogen exists almost entirely in the reduced form. As a source of reduced nitrogen, plant cultures are especially able to use primary amines:

(a) $R-NH_2$ and amides: R-CO-NH-R- (where R and R- are functional groups)

(b) The primary amines which are most commonly employed in culture media are ammonia (NH_4^+) and, occasionally, amino acids.

(c) Amides are less commonly added to culture media: thosc which can be used by plants are particularly

(d) $NH_2-CO-NH_2$ (urea)

and ureides, which include allantoin and allantoic acid. Allantoin or allantoic acid are sometimes more efficient nitrogen sources than urea.

Most media contain more nitrate than ammonium ions, but as plant tissue culture media are usually not deliberately buffered, the adopted concentrations of ammonium and nitrate ions have probably been more due to practical pH control, than to the requirement of the plant tissues for one form of nitrogen or another. Uptake of nitrate only takes place effectively in an acid pH, but is accompanied by extrusion of anions from the plant, leading to the medium gradually becoming less acid. By contrast, uptake of ammonium results in the cells excreting protons (H^+) into the medium, making it more acid. The exchange of ions preserves the charge balance of the tissues and may also assist in the disposal of an excess of protons or hydroxyl

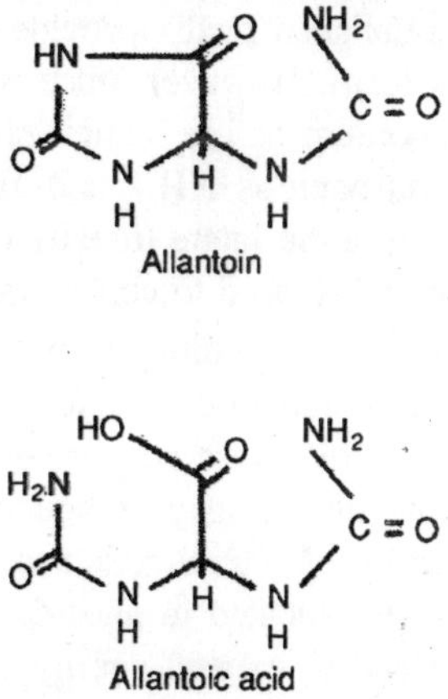

Allantoin

Allantoic acid

Fig. 5.1. The structures of allantoin and allantoic acid.

(OH^-) ions generated during metabolism. Uptake of nitrogen by cell suspension cultures of *Nicotiana tabacum* is an active (energy-dependent) process and is dependent on a supply of oxygen.

Plant culture media are usually started at pH 5.4-5.8. However, in one containing both nitrate and ammonium ions, a rapid uptake of ammonium into plant tissue causes the pH to fall to ca. 4.2-4.6. As this happens, further ammonium uptake is inhibited, but uptake of nitrate ion is stimulated, causing the pH to rise again. In unbuffered media, efficient nitrogen uptake can therefore depend on the presence of both ions. Unless otherwise stated, comments in this section on the roles of nitrate and ammonium refer to observations on unbuffered media.

There is generally a close correlation in tissue cultures between uptake of nitrogen, cell growth and the conversion of nitrogen to organic materials. A readily available supply of nitrogen seems to be important to maintain cultured cells in an undifferentiated state. The depletion of nitrogen in batch cultures, triggers an increase in the metabolism of some nitrogen-free compounds based on phenylpropanes (such as lignin), which are associated with the differentiation of secondarily-thickened cells. However, the growth in culture of differentiated cotton fibres composed largely of cellulose, is nitrate-dependent; the presence of some reduced nitrogen in the culture medium decreased the proportion of cultured embryos which produced fibres, and particularly in the absence of boron, promoted the cells of the embryos to revert to callus formation.

Nitrate ions (NO_3)

Nitrate ions are an important source of nitrogen for most plant cultures, and nearly all published media provide the majority of their available nitrogen in this form. However, once within the cell, nitrate has to be reduced to ammonium before being utilized biosynthetically. Why not simply supply nitrogen as NH_4^+ and avoid the use of NO_3^- altogether? The reason lies in the latent toxicity of the ammonium ion in high concentration, and in the need to control the pH of the medium.

Conversion of nitrate to ammonium is brought about firstly by one, or possibly two, nitrate reductase enzymes, which reduce NO_3^- to nitrite (NO_2^-). One nitrate reductase enzyme is thought to be located in the cytoplasm, while the second may be bound to membranes. The NO_2^- produced by the action of nitrate reductase is reduced to NH_4^+ by a nitrite reductase enzyme located in plastids. Reduction of nitrate to ammonia requires the cell to expend energy. The ammonium ions produced are incorporated into amino acids and other nitrogen-containing compounds. Nitrate and nitrite reductase enzymes are substrate induced, and their activity is regulated directly by the level of nitrate-nitrite ions within cultured cells, but also apparently by the products of the assimilation of reduced nitrogen.

Unlike the ammonium ion, nitrate is not toxic and, in many plants, much is transported to the shoots for assimilation. On the other hand, the nitrite ion can become toxic should it accumulate within plant tissues or in the medium, for example when growth conditions are not favourable to high nitrite reductase activity and when nitrate is the only nitrogen source. In *Pinus pinaster*, nitrate reductase is induced by the presence of KNO_3, and plants regenerated *in vitro* exhibit an ability to reduce nitrate similar to that of seedlings.

For most types of culture, the nitrate ion needs to be presented together with a reduced form of nitrogen (usually the NH_4^+ ion), and tissues may fail to grown a medium with NO_3^- as the only nitrogen source.

Reduced nitrogen

In the natural environment and under most cropping conditions, plant roots usually encounter little reduced nitrogen, because bacteria rapidly oxidize available sources. An exception is forest soils in mountainous regions of the northern hemisphere, where nitrates are not usually available. If NH_4^+, and other reduced nitrogen compounds are available, (and this is particularly the case in the aseptic *in vitro* environment), they can be taken up and effectively utilized by plants. In fact the uptake of reduced nitrogen gives a plant an ergonomic advantage because the conversion of nitrate to ammonium ions (an energy-requiring process) is not necessary.

The free ammonium ion can cause toxicity, which, at least in whole plants, can lead to an increase in ethylene evolution. Shoots grown on an unbuffered medium containing a high proportion of ammonium ions may become stunted or *hyperhydric*. These effects can sometimes be reversed by transfer to a medium containing a high proportion of NO_3^- or to one where NO_3^- is the only N source. Hyperhydricity is the *in vitro* formation of abnormal organs, which are brittle and have a water-soaked appearance.

Growth of plant cultures may also be impaired in media containing high concentrations of NH_4^+ even when high concentrations of NO_3^- are present at the same time. Growth inhibition may not only be due to depressed pH, but may reflect a toxicity induced by the accumulation of excess ammonium ions. In normal circumstances the toxic effect of ammonium is avoided by conversion of the ion into amino acids. There are two routes by which this takes place, the most important of which, under normal circumstances, is that by which L-glutamic acid is produced from glutamine through the action of glutamine synthetase (GS) and glutamate synthetase (GOGAT) enzymes. Compounds, which block the action of GS can be used as herbicides. The reaction of α-ketoglutaric acid with NH_4^+ is usually less important, but seems to have increased significance when there is an excess of ammonium ions. Detoxification and ammonium assimilation may then be limited by the availability of α-ketoglutaric acid, but this may be increased *in vitro* by adding to the medium one or more acids which are Krebs' (tri-carboxylic acid) cycle intermediates. Their addition can stimulate growth of some cultures on media containing high levels of NH_4^+.

In comparison with media having only nitrate as the nitrogen source, the presence of the ammonium ion in media usually leads to rapid amino acid and protein synthesis, and this takes place at the expense of the synthesis of carbohydrate compounds. This diversion of cellular metabolism can be disadvantageous in some shoot cultures, and can contribute towards the formation of hyperhydric shoots. Hyperhydricity no longer occurs when NH_4^+ is eliminated from the medium or greatly reduced. It is possible that adding an organic acid to the medium might also alleviate the symptoms on some plants.

A supply of reduced nitrogen in addition to nitrate, appears to be beneficial for at least two processes involved with cell division:

1. The formation of a cell wall. Without a complete cell wall, protoplasts require a factor capable of inducing wall formation. Freshly-isolated protoplasts may contain sufficient of this substance to promote wall formation for just a few divisions. The wall-forming factor is only effective when NH_4^+ is present in the medium: glutamine does not substitute for NH_4^+:
2. The activity of growth regulators.

There are several reports in the literature that, with constant amounts of NO_3^-, ammonium sulphate has not provided such a good source of NH_4^+ as ammonium nitrate or ammonium chloride. Possibly the reason is that a medium containing ammonium sulphate has a greater tendency to become acid, than one containing less sulphate ions. This would result if the presence of sulphate ions accelerated the uptake of NH_4^+, or slowed the uptake of NO_3^-. Ammonium sulphate has been used as the only source of the ammonium ion in some media used for the culture of legumes, including B5.

Ammonium as the sole nitrogen source

pH adjustment

If plant tissues are presented with a medium containing only NH_4^+ nitrogen, the pH falls steadily as the ion is taken up (for example, a decrease of 0.9 pH units in 15 days in *Asparagus* callus - Hunault (1985) or 0.7 pH units with potato shoots - Avila *et al.*, (1998). Growth and morphogenesis is possible in suspension cultures containing only NH_4^+ ions, providing the pH of the medium is frequently adjusted by the addition of a base, or the medium is buffered. In wild carrot, the induction of embryogenesis required the medium to be adjusted to pH 5.4 at 8 hourly intervals. Without adjustment, the pH of media containing only NH_4^+ falls rapidly to a point where cells cannot grow.

Buffering

Ammonium can also serve as the only nitrogen source when the medium is buffered. Tobacco cells could be grown on a medium containing NH_4^+ nitrogen if the organic acid ion, succinate, was added to the medium. Gamborg and Shyluk (1970) found that cultured cells could be grown without frequent pH adjustment on a medium containing only NH_4^+ nitrogen, when a carboxylic acid was present. The organic acids appeared to minimize the acidification of the medium through NH_4^+ uptake. Similarly *Asparagus* internode callus grew just as well on NH_4^+ as the only nitrogen source as on a medium containing both NH_4^+ and NO_3^-, but only when organic acids (such as citrate, or malate) or MES buffer were added to the medium. When media were buffered with MES, the best callus growth occurred when the pH was 5.5.

Additional effect of organic acids

Although Krebs' cycle organic acids can act as buffers, they may also act as substrates for amino acid synthesis from NH_4^+. To be assimilated into amino acids via the GDH enzyme, the ammonium ion must react with α-ketoglutaric acid, which is produced by the Krebs' cycle. Its availability may govern the rate at which ammonium can be metabolized by this route. The rate of assimilation might be expected to be improved by supplying the plant with α-ketoglutarate directly, or by supplying acids which are intermediates in the Krebs' cycle (citrate, iso-citrate, succinate, fumarate or malate), for then the natural production of α-ketoglutarate should increase.

This hypothesis was confirmed by Behrend and Mateles (1976) who concluded that succinate, or other Krebs' cycle acids, acted mainly as a nutrient, replacing α-ketoglutarate as it was withdrawn from the cycle during NH_4^+ metabolism and amino acid synthesis. Depletion of α-ketoglutarate causes the cycle to cease unless it, or another intermediate, is replaced. The optimum molar ratio of NH_4^+ to succinate, was 1.5 (e.g. 10 mM NH_4^+: 15 mM succinate).

Chaleff (1983) thought that the growth of rice callus on Chaleff R3 (NH_4) medium, containing 34 mM of only ammonium nitrogen, (Chaleff R3 NH_4 medium) was enabled by the presence of 20 mM succinate or α-ketoglutarate, partly by the buffering capacity of the acids, and partly by their metabolism within the plant, where they may serve as substrate for amino acid synthesis. Similar conclusions have been reached by other workers; Dougall and Weyrauch (1980); Hunault (1985); Molnar (1988b), who have found that compounds such as ammonium malate and ammonium citrate are effective nitrogen sources.

Orange juice promotes the growth of *Citrus* callus. Einset (1978) thought that this was not due to the effect of citric acid, but Erner and Reuveni (1981) showed that citric acid, particularly at concentrations above the 5.2 mM found in the juice used by Einset, does indeed promote the growth of *Citrus* callus; it had a more pronounced effect than other Krebs' cycle acids, perhaps due to the distinctive biochemistry of the genus.

Organic acids not only enhance ammonium assimilation when NH_4^+ provides the only source of nitrogen, but may sometimes also do so when nitrate ions are in attendance. The weight of rice anther callus was increased on Chaleff (1983) R3 medium, if 20 mM succinate or α-ketoglutarate was added. Similarly the rate of growth of *Brassica nigra* suspensions on MS medium, was improved either by adding amino acids, or 15 mM succinate. An equivalent improvement (apparently due entirely to buffering) only occurred through adding 300 mM MES buffer. However, the presence of organic acids may be detrimental to morphogenesis. In Chaleff 's experiment, the presence of succinate in R3 medium markedly decreased the frequency of anther callus formation.

Photosynthesis

Although plants grown on nutrient solutions containing only NH_4^+ nitrogen have been found to possess abnormally high levels of PEP enzyme (the enzyme facilitating CO_2 fixation in photosynthesis), media containing high levels of NH_4^+ tend to inhibit chlorophyll synthesis and photosynthesis.

Urea

Plants are able to absorb urea, but like the ammonium ion, it is not a substance that is normally available in soils in the natural environment. It is however produced as a by-product of nitrogen metabolism; small quantities are found in many higher plants, which are able to utilize urea as a source of nitrogen, providing it is first converted to ammonium ions by the enzyme urease. In legumes and potato, urease requires the microelement nickel for activity. In conifers, the epidermal cells of cotyledons and cotyledons are capable of urease induction and ammonium ion formation.

Urea can be used as the sole nitrogen source for cultures, but growth is less rapid than when ammonium and nitrate ions are supplied; urease enzyme increases after cultures have been maintained for several passages on a urea-based medium. Although the metabolism of urea, like that of other reduced nitrogen compounds, causes the production of excess hydrogen ions, less are predicted to be secreted into the medium than during the utilization of NH_4^+, so that urea is less suitable than ammonium to balance the pH of media containing NO_3^-. Nitrate ions are utilized in preference to urea when both nitrogen sources are available. Urea is able to serve as a reduced nitrogen source during embryogenesis, but has been used in relatively few culture media, and of these, none has been widely adopted.

Media with nitrate and ammonium ions

Most intact plants, tissues and organs take up nitrogen more effectively, and grow more rapidly, on nutrient solutions containing both nitrate and ammonium ions, than they do on solutions containing just one of these sources. Although in most media, reduced nitrogen is present in lower concentration than nitrate, some morphogenic events depend on its presence, and it can be used in plant cultures in a regulatory role. Adventitious organs may also develop abnormally if NH_4^+ is missing.

Possible explanations, which have been put forward for the regulatory effect of NH_4^+ are:

1. The reduction and assimilation of NO_3^- is assisted by the presence of NH_4^+ or the products of its assimilation. When grown on a medium containing a small amount of NH_4^+ nitrogen in addition to nitrate, suspension cultured cells of '*Paul's Scarlet*' rose accumulated twice as much protein as when grown on a medium containing only nitrate, even though ammonium finally accounted for only 10% of the total protein nitrogen. Dougall (1977) considered

this to be an oversimplified interpretation, moreover nitrate reductase activity is effectively increased by the presence of NO_3^- and in some plants, a high concentration of NH_4^+ inhibits nitrate reductase activity.

2. Ammonium ions effectively buffer plant nutrient media in the presence of nitrate and so enhance nitrate uptake.

Cultures of some plants are capable of growing with only NO_3^- nitrogen (e.g. cell cultures of *Reseda luteoli*, soybean, wheat, flax and horse radish; callus of *Medicago sativa*, although yields are generally better when the medium is supplemented with NH_4^+. Craven *et al.*, (1972) with carrot, and Mohanty and Fletcher (1978) with *Rosa 'Paul's Scarlet'*, found that the presence of NH_4^+ was particularly important during the first few days of a suspension culture. After that cells increase in cell number and dry weight more rapidly on NO_3^- nitrogen alone.

The response of plant cultures to nitrate and ammonium ions depends to a large extent on the enzymes, and the manner in which their activities are increased or inhibited in different tissues by the presence of the ions. These factors vary according to the degree of differentiation of the tissue, its physiological age, and its genotype. For example, the high level of NH_4^+ in MS medium inhibited the activity of glutamate synthetase enzyme in soybean suspension cultures, while in *N. tabacum*, a peak of *glutamate dehydrogenase* (GDH) appeared to exist at 10 mM NH_4^+. The activity of GDH and NADH-dependent GOGAT developed rapidly in cultured tobacco cells, while nitrate reductase and ferridoxin-dependent GOGAT activity increased more slowly during growth. By contrast, in sunflower cultures, the specific activity of GDH and ferridoxin-dependent GOGAT only reached a maximum at the end of growth, and the presence of 15 mM NH_4^+ inhibited the activity of nitrate reductase, indicating that the cells were entirely dependent on the reduced nitrogen in the medium.

In consequence of such variation, the relative concentrations of ammonium and nitrate in media may need to be altered for different cultures.

Correct balance of ions

When trying to find media formulations suitable for different plant species and different kinds of cultures, two important factors to be considered are:

1. The total concentration of nitrogen in the medium;
2. The ratio of nitrate to ammonium ions.

There is a high proportion of NH_4^+ nitrogen in MS medium [ratio of NO_3^- to NH_4^+, 66:34] and the quantity of total nitrogen is much higher than that in the majority of other media. For some cultures, the total amount of nitrogen is too high and the balance between the two forms in the medium is not optimal.

It is noticeable that in several media used for legume culture, there is a greater proportion of NO_3^- to NH_4^+ ions than in MS medium. Evans *et al.*, (1976) found that soybean leaf callus grew more rapidly and formed more adventitious roots when the ammonium nitrate in MS medium was replaced with 500 mg/l ammonium sulphate. A reduction in the ammonium level of their medium for the callus culture of red clover, was also found to be necessary by Phillips and Collins (1979, 1980) to obtain optimum growth rates of suspension cultured cells. A similar adjustment can be beneficial in other genera. Eriksson (1965) was able to enhance the growth rate of cell cultures of *Haplopappus gracilis* when he reduced the ammonium nitrate concentration to 75% of that in MS medium, and doubled the potassium dihydrogen phosphate level.

It will be seen that the balance between NO_3^- and NH_4^+ in these different experiments has varied widely. This implies that the ratio between the two ions either needs to be specifically adjusted for each plant species, or that the total nitrogen content of the medium is the most important determinant of growth or morphogenesis. Only occasionally is an even higher concentration of nitrate than that in MS medium beneficial. Trolinder and Goodin (1988) found that the best growth of globular somatic embryos of *Gossypium* was on MS medium with an extra 1.9 g/l KNO_3.

There are reports that adjustments to the nitrogen content and ratio of NO_3^- to NH_4^+, can be advantageous in media containing low concentrations of salts. Biedermann (1987) found that even quite small adjustments could be made advantageously to the NO_3^- content of a low salt medium for the shoot culture of different *Magnolia* species and hybrids, but too great a proportion of NO_3^- was toxic.

Nitrate-ammonium ratio for various purposes

Root growth

Root growth is often depressed by NH_4^+ and promoted by NO_3^-. Media for isolated root culture contain no NH_4^+, or very little. Although roots are able to take up nitrate ions from solutions, which become progressively more alkaline as assimilation proceeds, the same may not be true of cells, tissues and organs *in vitro*.

Shoot cultures

Media containing only nitrate nitrogen are used for the shoot culture of some plants, for example strawberry, which can be cultured with 10.9 mM NO_3^- alone; supplementing the medium with 6 mM NH_4^+ causes phytotoxicity. However, shoot cultures of strawberry grown without NH_4^+ can become chlorotic: adding a small amount of NH_4NO_3 (or another source of reduced nitrogen) to the medium at the last proliferation stage, or to the rooting medium, may then give more fully developed plants with green leaves. On some occasions it is necessary to eliminate or reduce NH_4^+ from the medium for shoot cultures to prevent hyperhydricity.

Nitrogen supply and morphogenesis

Morphogenesis is influenced by the total amount of nitrogen provided in the medium and, for most purposes, a supply of both reduced nitrogen and nitrate seems to be necessary. The requirement for both forms of nitrogen in a particular plant species can only be determined by a carefully controlled experiment: simply leaving out one component of a normal medium gives an incomplete picture. For example, cotyledons of lettuce failed to initiate buds when NH_4NO_3 was omitted from Miller (1961) salts and instead formed masses of callus: was this result due to the elimination of NH_4^+, or to reducing the total nitrogen content of the medium to one third of its original value?

Importance of the nitrate/ammonium balance

The importance of the relative proportions of NO_3^- and NH_4^+ has been demonstrated during indirect morphogenesis and the growth of regenerated plants. Grimes and Hodges (1990) found that although the initial cellular events which led to plant regeneration from embryo callus of indica rice, were supported in media in which total nitrogen ranged from 25 to 45 mM and the NO_3^- to NH_4^+ ratio varied from 50:50 to 85:15, differentiation and growth were affected by very small alterations to the NO_3^- to NH_4^+ ratio. Changing it from 80:20 (N6) medium, to 75:20, brought about a 3-fold increase in plant height and root growth, whereas lowering it below 75:25, resulted in short shoots with thick roots. Atypical growth, resulting from an unsuitable balance of nitrate and ammonium, has also been noted in other plants. It gave rise to abnormal leaves in *Adiantum capillus-veneris*; the absence of ammonium in the medium caused newly initiated roots of *Carica papaya* to be abnormally thickened, and to have few lateral branches. Shoot cultures may survive low temperature storage more effectively when maintained on a medium containing less NH_4NO_3 than in MS medium.

By comparing different strengths of Heller (1953; 1955) and MS media, and varying the NH_4NO_3 and $NaNO_3$ levels in both, David (1972) was led to the conclusion that the principal ingredient in MS favouring differentiation in Maritime pine explants is NH_4NO_3. However, in embryonic explants of *Pinus strobus*, adventitious shoot formation was better on Schenk and Hildebrandt (1972) medium than on MS (which induced more callus formation). The difference in the ammonium level of the two media was mainly responsible.

Morphogenesis influenced by total available nitrogen

Others have found that the total nitrogen content of culture media influences morphogenesis more than the relative ammonium concentration. Results of Margara and Leydecker (1978) indicated that adventitious shoot formation from rapeseed callus was optimal in media containing 30-45 mM total nitrogen. The percentage of explants forming shoots was reduced on media containing smaller or greater amounts (e.g. on MS medium). Increasing the ratio of NH_4^+ to total N in media, from 0.20 to 0.33 was also detrimental. Similarly, Gertsson (1988) found that a small number of adventitious shoots was obtained on petiole segments of *Senecio hybridus* when the total nitrogen in MS medium was increased to 75 mM, but that an increased number of shoots was produced when the total nitrogen was reduced to 30 mM (while keeping the same ratio of NO_3^- to NH_4^+). Shoot production was more than doubled if, at the same time as the total N was reduced, the potassium ion concentration was fixed at 15 mM, instead of 20 mM.

The total amount of nitrogen in a medium was shown by Roest and Bokelmann (1975) to affect the number of adventitious shoots formed directly on *Chrysanthemum* pedicels. The combined amount of KNO_3 + NH_4NO_3 in MS medium (60 mM), was adjusted, while the ratio of NO_3^- to NH_4^+ (66:34) was unchanged. From 30-120 mM total nitrogen was optimal. However there was clearly a strong effect of genotype, because the cultivar '*Bravo*' was much more sensitive to increased nitrogen than '*Super Yellow*'.

Nitrogen x sugar interaction

The enhancement of morphogenesis caused by high nitrogen levels may not be apparent unless there is an adequate sucrose concentration in the medium. In *Dendrobium*, the uptake of NO_3^- is slower than that of NH_4^+. Uptake is dependent on the nature and concentration of the sugar in the medium, being slower in the presence of fructose than when sucrose or glucose are supplied. The rate of growth of *Rosa* 'Paul's Scarlet' suspensions was influenced by the ratio of NO_3^- to sucrose in the medium. A high ratio favoured the accumulation of reduced nitrogen, but not the most rapid rate of cell growth.

Embryogenesis and embryo growth

It is accepted that the presence of some reduced nitrogen is necessary for somatic embryogenesis in cell and callus cultures; but although reduced nitrogen compounds are beneficial to somatic embryo induction, apparently they are not essential until the stage of embryo development. A relatively high level of both nitrate and ammonium ions then seems to be required. Some workers have also noted enhanced embryogenesis and/or improved embryo growth when media have been supplemented with amino acids in addition to NO_3^- and NH_4^+. Street (1979) thought that an optimum level of NH_4^+ for embryogenesis was about 10 mM (from NH_4Cl) in the presence of 12-40 mM NO_3^- (from KNO_3): that is: [NO_3 to NH_4^+ ratio, from 55:45 to 80:20; Total N 22-50 mM].

Walker and Sato (1981) obtained no embryogenesis in alfalfa callus in the absence of either ammonium or nitrate ions. Miller's medium (12.5 mM NH_4^+) supported a high rate of embryogenesis: [NO_3^- to NH_4^+ ratio, 68:32; Total N 39.1 mM], but only a small number of embryos were produced on Schenk and Hildebrandt (1972) medium (SH): [NO_3^- to NH_4^+ ratio, 90:10, Total N 2732 mM], unless it was supplemented with NH_4^+ from either ammonium carbamate, ammonium chloride or ammonium sulphate. An optimal level of ammonium in SH medium was 12.5 mM: [NO_3^- to NH_4^+ ratio, then 66:34, total N, 37.22], although embryogenesis was still at a high level with 100 mM ammonium ion: [NO_3^- to NH_4^+ ratio, then, 20:80; total N, 124.72 mM].

By contrast, *Coffea arabica* leaf callus, which was formed on a medium containing MS salts, was induced to become embryogenic by first being cultured on a medium with MS salts (and high auxin): [NO_3^- to NH_4^+ ratio, 66:34; total N, 30.0 mM], and then moved to another medium containing MS salts with an extra 2850 mg/l KNO_3: [NO_3^- to NH_4^+ ratio, 82:18; total N, 58.2 mM] (and low auxin)

Zygotic embryos

The presence of some reduced nitrogen in the growth medium is also required for the continued growth of zygotic embryos in culture. Nitrate alone is insufficient. An optimum concentration of NH_4^+ for the development of barley embryos in culture was 6.4 mM. A similar provision seems to be necessary in most plants for the *in vitro* growth of somatic embryos.

Flower bud formation and growth

Nitrate was essential for the formation of adventitious buds on leaf segments of *Begonia franconis*. The greatest proportion of flower buds was obtained with 5 mM NO_3^- and 1.5 mM NH_4^+. Above this level, NH_4^+ promoted vegetative sprouts. The best *in vitro* growth of *Begonia franconis* flower buds detached from young inflorescences, occurred on a medium with 10-15 mM total nitrogen (NO_3^- to NH_4^+ ratio, 50:50 to 67:33). Detached flower buds of *Cleome iberidella* were found to grow best *in vitro* with 25 mM total nitrogen (NO_3^- to NH_4^+ ratio, 80:20), but the complete omission of NH_4NO_3 from MS medium, where the salts had been diluted to their original concentration, promoted the development (but not the initiation) of adventitious floral buds of *Torenia fournieri*.

Effect on the action of growth regulants

The ratio of NO_3^- to NH_4^+ present in the culture medium has been found to affect the activity of plant growth substances and plant growth regulators. The mechanisms by which this occurs are not fully elucidated. It has been noted, for example, that cells will divide with less added cytokinin when the proportion of reduced nitrogen is reasonably high. To induce tobacco protoplasts to divide, it was necessary to add 0.5-2 mg/l BAP to a medium containing only NO_3^- nitrogen. The presence of glutamine or NH_4^+ in the medium together with NO_3^-, reduced the cytokinin requirement, and division proceeded without any added cytokinin when urea, NH_4^+, or glutamine were the sole N-sources of the medium. Sargent and King (1974) found that soybean cells were dependent on cytokinin when cultured in a medium containing NO_3^- nitrogen, but independent of cytokinin when NH_4^+ was present as well.

The relative proportion of nitrate and ammonium ions also affects the response of cells to auxin growth regulators in terms of both cell division and morphogenesis. It is possible that this is through the control of intracellular pH. Carrot cultures that produce somatic embryos when transferred from a high- to a low-auxin medium, can also be induced into embryogenesis in a high-auxin medium, if it contains adequate reduced nitrogen. Only root initials are formed in high-auxin media which do not contain reduced nitrogen. The number of plants regenerated from rice callus grown on Chu *et al.*, (1975) medium containing 0.5

mg/l 2,4-D, depended on the ratio of NO_3^- to NH_4^+. It was high in the unaltered medium (ratio 4:1), but considerably less if, with the same total N, the ratio of the two ions was changed to 1:1.

Cells of *Antirrhinum majus* regenerated from isolated protoplasts were stimulated to divide with a reduced quantity of auxin in a medium containing 39.77 mM total nitrogen [NO_3^- to NH_4^+ ratio, 39:77 (2.98)] by adding further ammonium ion to give a total nitrogen content of 54.72 mM [NO_3^- to NH_4^+ ratio, 54:46 (1.19)] or, alternatively, 400 mg/l of casein hydrolysate.

In experiments of Koetje *et al.*, (1989) and Grimes and Hodges (1990), when the NO_3^- to NH_4^+ ratio in N6 medium was 80:20, there was a strong dose response curve to the auxin 2,4-D with 0.5 mg/l being the best concentration to induce embryogenesis in *Oryza sativa* callus; if the medium was modified, so that the NO_3^- to NH_4^+ ratio was 66:34 or 50:50, 2,4-D was less effective, and there was little difference in the number of plants regenerated between 0.5 and 3 mg/l 2,4-D. The ratio of NO_3^- to NH_4^+ therefore seemed either to alter the sensitivity of cells to the auxin, or to affect its uptake or rate of metabolism.

Walker and Sato (1981) also found that the proportion of ammonium ion in the medium can influence the way in which growth regulants control morphogenesis. Having been placed for 3 days on a medium which would normally induce root formation (medium containing 5 mM 2,4-D and 50 mM kinetin), suspension cultured cells were subsequently plated on a modification of the same medium (which contains 24.8 mM NO_3^-) without regulants, in which the concentration of NH_4^+ had been adjusted to various levels. Media containing high levels of ammonium ion would have tended to become acid, especially as the extra ammonium was added as ammonium sulphate. Possibly this affected the uptake or action of the regulants?

Addition of amino acids

Amino acids can be added to plant media to satisfy the requirement of cultures for reduced nitrogen, but as they are expensive to purchase, they will only be used in media for mass propagation where this results in improved results. For most tissue culture purposes, the addition of amino acids may be unnecessary, providing media contain adequate amounts, and correct proportions, of nitrate and ammonium ions. For example, Murashige and Skoog (1962) found that when cultures were grown on media such as Heller (1953; 1955), Nitsch and Nitsch (1956) N1, and Hildebrandt *et al.*, (1946) Tobacco, containing sub-optimal amounts of inorganic chemicals, a casein hydrolysate (consisting mainly of a mixture of amino acids) substantially increased the yield of tobacco callus, whereas it gave only marginal increases in yield when added to their revised MS medium. Arginine (0.287 mM) increased the growth of sugar cane callus and suspension cultures grown on Nickell and Maretzski (1969) medium but was without effect on cultures of this plant grown on a medium based on Scowcroft and Adamson (1976) CS5 macronutrients.

It is noticeable from the literature that organic supplements (particularly amino acids) have been especially beneficial for growth or morphogenesis when cells or tissues were cultured on media such as White (1943a), which do not contain ammonium ions. White (1937) and Bonner and Addicott (1937), for example, used known amino acids to replace the variable mixture provided by yeast extract. For the culture of *Picea glauca* callus, Reinert and White (1956) supplemented Risser and White (1964) medium with 17 supplementary amino acids, and similar, or greater numbers, were used by Torrey and Reinert (1961) and Filner (1965) in White (1943) medium for the culture of carrot, *Convolvulus arvensis*, *Haplopappus gracilis* and tobacco tissues.

Dependence on the nitrate to ammonium ratio

Grimes and Hodges (1990) found that when both NO_3^- and NH_4^+ are present in the medium, the response to organic nitrogen depends on the ratio of these two ions. Twice as many plants were regenerated from embryogenic rice callus when 1 g/l CH was added to Chu *et al.*, (1975) N6 medium, providing the proportion of NO_3^- to NH_4^+ was also changed to 1 (50:50). There was little response to CH with the same amount of total N in the medium, if the NO_3^-/NH_4^+ ratio was 4 (i.e. 80%:20%, as in the original medium), or more.

Amino acids as the sole N source

As most of the inorganic nitrogen supplied in culture media is converted by plant tissues to amino acids, which are then assimilated into proteins, it should be possible to culture plants on media in which amino acids are the only nitrogen source. This has been demonstrated: for example, *Nicotiana tabacum* callus can be cultured on MS salts lacking NO_3^- and NH_4^+ (but with an extra 20.6 mM K^+), if 0.1 mM glycine, 1 mM arginine, 2 mM aspartic acid and 6 mM glutamine are added; wild carrot suspensions can be grown on a medium containing glutamine or casein hydrolysate as the sole nitrogen source. Amino acids provide plant cells with an immediately available source of nitrogen, and uptake can be much more rapid than that of inorganic nitrogen in the same medium. Only the L- form of amino acids is biologically active.

Amino acids can also provide reduced nitrogen in culture media in place of NH_4^+ and as a supplement to NO_3^-. However they are usually employed as minor additions to media containing both NH_4^+ and NO_3^-. Uptake of amino acids into cultured tissues causes a decrease in the pH of the medium, similar to that which occurs when NH_4^+ ions are absorbed.

Sugar-based amines such as glucosamine and galactosamine can also serve as a source of reduced nitrogen in morphogenesis.

Biologically-active amino acids

Amino acids are classified according to their stereoisomers and according to the relative positions of the amino group and the acidic radical. Only the L- isomers of the a-amino acids are important for plant tissue culture media. They have the general structure:

$$\begin{array}{c} NH_2 \\ | \\ \text{R-CH- COOH} \end{array}$$

β-Amino acids are present in plants but tend to result from secondary metabolism. They have the general structure:

$$\begin{array}{c} NH_2 \\ | \\ \text{R-CHCH}_2\text{- COOH} \end{array}$$

(where R = functional groups)

Unfortunately the particular amino acid, or mixture of amino acids, which promotes growth or morphogenesis in one species, may not do so in another. For instance, L-α-alanine, glutamine, asparagine, aspartic acid, glutamic acid, arginine and proline could serve as a source of reduced nitrogen in a medium containing 20 mM NO_3^-, and were effective in promoting embryogenesis in *Daucus carota* callus and suspensions, but lysine, valine, histidine, leucine and methionine were ineffective.

Competitive inhibition

Some amino acids are growth inhibitory at fairly low concentrations and this is particularly observed when mixtures of two or more amino acids are added to media. Inhibition is thought to be due to the competitive interaction of one compound with another. In oat embryo cultures, phenylalanine and L-tyrosine antagonize each other, as do L-leucine and DL-valine; DL-isoleucine and DL-valine, and L-arginine and L-lysine. Lysine and threonine often exert a co-operative inhibition when present together, but do not inhibit growth when added to a medium singly.

Glycine

Glycine is an ingredient of many media. It has usually been added in small amounts, and has been included by some workers amongst the vitamin ingredients. Despite frequent use, it is difficult to find hard evidence that glycine is really essential for so many tissue cultures, but possibly it helps to protect cell membranes from osmotic and temperature stress.

White (1939) showed that isolated tomato roots grew better when his medium was supplemented with glycine rather than yeast extract and that glycine could replace the mixture of nine amino acids that had been used earlier. It was employed as an organic component by Skoog (1944) and continued to be used in his laboratory until the experiments of Murashige and Skoog (1962). They adopted the kinds and amounts of organic growth factors specified by White (1943a) and so retained 2 mg/l glycine in their medium without further testing.

Linsmaier and Skoog (1965), furthering the study of medium components to organic ingredients, omitted glycine from MS medium and discovered that low concentrations of it had no visible effect on the growth of tobacco callus, while at 20 mg/l it depressed growth. No doubt the success of MS medium has caused the 2 mg/l glycine of Skoog (1944) to be copied in many subsequent experiments. Many workers overlook the later Linsmaier and Skoog paper.

Casein and other protein hydrolysates

Proteins, which have been hydrolysed by acid, or enzymes, and so broken down into smaller molecules, are less costly than identified amino acids. The degree of degradation varies: some protein hydrolysates consist of mixtures of amino acids together with other nitrogenous compounds such as peptide fragments, vitamins, and elements which might (if they can form inorganic ions, or are associated with organic compounds that can be taken up by plant tissues) be able to serve as macro- or micro-elements. Peptones are prepared from one or several proteins in a similar fashion but generally consist of low molecular weight proteins. Although protein hydrolysates are a convenient source of substances which may promote plant growth, they are by nature relatively undefined supplements. The proportion of individual amino acids in different hydrolysates depends on the nature of the source protein and the method by which the product has been prepared.

The hydrolysate most often used in culture media is that of the milk protein, casein, although lactalbumin hydrolysate has been employed. Peptones and tryptone have been used less frequently, but there are reports of their having been added to media with advantage. Casein hydrolysates can be a source of calcium, phosphate, several microelements, vitamins and, most importantly, a mixture of up to 18 amino acids. Several *casein hydrolysates* (CH) are available commercially but their value for plant tissue culture can vary considerably. Acid hydrolysis can denature some amino acids and so products prepared by enzymatic hydolysis are to be preferred. The best can be excellent sources of reduced nitrogen, as they can contain a relatively large amount of glutamine.

Casein hydrolysate produces an improvement in the growth of *Cardamine pratensis* and *Silene alba* suspensions, only if the medium is deficient in phosphorus. Glutamine has the same effect; it is the most common amino acid in CH, and its synthesis requires ATP. For these reasons, Bister-Miel *et al.*, (1985) concluded that CH overcomes the shortage of glutamine when there is insufficient phosphorus for adequate biosynthesis.

However several investigators have concluded that casein hydrolysate itself is more effective for plant culture than the addition of the major amino acids which it provides. This has led to speculation that CH might contain some unknown growth-promoting factor. In prepared mixtures of amino acids resembling those in CH, competitive inhibition between some of the constituents is often observed. For instance, the induction of embryogenesis in carrot cell suspensions on a medium containing glutamine as the only nitrogen source, was partly inhibited by the further addition of L-amino acids similar in composition to those in CH. This suppression was mainly caused by the L-tyrosine in the mixture.

There may be a limit to the amount of CH, which can be safely added to a medium. Anstis and Northcote (1973) reported that the brand of CH known as 'N-Z-amine', can produce toxic substances if concentrated solutions are heated, or if solutions are frozen and thawed several times. Possibly these are reasons why mixtures of amino acids occasionally provide more valuable supplements than CH. *Nicotiana tabacum* callus grew better on a nitrogen-free MS medium when a mixture of the amino acids L-glutamine (6 mM), L-aspartic acid (2 mM), L-arginine (1 mM) and glycine (0.1 mM) was added, rather than 2 g/l casein hydrolysate (which would have provided about 2 mM glutamine, 0.6 mM aspartic acid, 0.2 mM arginine and 0.3 mM glycine).

Beneficial effects of amino acid additions

Improved growth

The growth rate of cell suspensions is frequently increased by the addition of casein hydrolysate or one or more amino acids (particularly glutamine) to media containing both nitrate and ammonium ions. Some workers have included a mixture of several amino acids in their medium without commenting on how they improved growth. In other cases the benefit resulting from a specific compound has been clearly shown. The lag phase of growth in suspensions of *Pseudotsuga menziesii* cultured on Cheng (1977; 1978) medium was eliminated by the addition of 50 mM glutamine, and the final dry weight of cells was 4 times that produced on the unammended medium. Similarly the rate of growth of *Actinidia chinensis* suspensions was improved by the addition of 5 mM glutamine and those of *Prunus amygdalus* cv. 'Ferragnes' could not be maintained unless 0.2% casamino acids was added to the medium. Molnar (1988b) found that the growth of *Brassica nigra* cell suspensions was improved by adding 1-4 g/l CH or a mixture of 4 mM alanine, 4 mM glutamine and 1 mM glutamic acid. In this case the medium contained MS salts (but less iron) and B5 vitamins.

Amino acid supplements have also been used to boost the rate of growth of callus cultures. For instance, Short and Torrey (1972) added 5 amino acids and urea to a medium containing MS salts for the culture of pea root callus, and Sandstedt and Skoog (1960) found that aspartic and glutamic acids promoted the growth of tobacco callus as much as a mixture of several amino acids (such as found in yeast extract). Glutamic acid seemed to be primarily responsible for the growth promotion of sweet clover callus caused by casein hydrolysate on a medium containing 26.6 mM NO_3^-, 12.5 mM NH_4^+ and 2.0 mM PO_4^{3-}. Amino acids are often added to media for protoplast culture. It was essential to add 2 mM glutamine and 2 mM asparagine to a medium containing MS salts, to obtain cell division, colony growth and plantlet differentiation from *Trigonella* protoplasts.

Shoot cultures

Many shoot cultures are grown on MS medium containing glycine, although in most cases the amino acid is probably not an essential ingredient. Usually it is unnecessary to add amino acids to media supporting shoot cultures, but methionine may represent a special case. Druart (1988) found that adding 50-100 mg/l L-methionine to the medium seemed to stimulate cytokinin activity and caused cultures of *Prunus glandulosa* var. *sinensis* to have high propagation rates through several subcultures. This promotive effect of L-methionine was thought to be due to it acting as a precursor of ethylene. Glutamine inhibited the growth of apical domes excised from *Coleus blumei* shoots and 50-100 mg/l glutamic acid inhibited shoot growth, the formation of axillary buds and shoot proliferation in cultures of woody plants.

L-Citrulline is an important intermediate in nitrogen metabolism in the genus *Alnus*. The addition of 1.66 mM (4.99 mM NH_2) to WPM medium improved the growth of *A. cordata* and *A. subcordata* shoot cultures.

Contaminants grow more rapidly on media containing amino acids. Casein hydrolysate is therefore sometimes added to the media for Stage I shoot cultures so that infected explants can be rejected quickly. The health of shoots grown from seedling shoot tips of *Feijoa* (*Acca*) *sellowiana* was improved when 500 mg/l CH was added to Boxus (1974) medium (which does not contain ammonium ions).

Organogenesis

The presence of amino acids can enhance morphogenesis, either when they provide the only source of reduced nitrogen, or when they are used as a supplement to a medium containing both NO_3^- and NH_4^+. In a medium containing 25 mM nitrate, but no NH_4^+, direct adventitious shoot formation on cauliflower peduncle explants was induced by the addition of a mixture of the amino acids asparagine, proline, tyrosine and phenylalanine, each at a concentration of only 0.1 mM. A high rate of adventitious shoot regeneration and embryogenesis, from *Beta vulgaris* petioles or petiole callus, was achieved on a medium comprised of several amino acids and a complex vitamin mixture with MS salts. The addition of CH to MS medium was found to be essential for shoot formation from callus.

Adding only 1-10 mg/l of either L-leucine or L-isoleucine to Gamborg *et al.*, (1968) B5 medium, decreased callus growth of *Brassica oleracea* var *capitata*, but increased adventitious shoot formation. Basu *et al.*, (1989) thought that this might be due to these amino acids being negative effectors of *threonine deaminase* (TD) enzyme, the activity of which was diminished in their presence. Threonine, methionine and pyruvic acid, which increased callus growth in this species, enhanced TD activity.

There are several examples of the amino acid L-asparagine being able to stimulate morphogenesis. This may be because it too can be a precursor of ethylene, the biosynthesis of which may be increased by greater substrate availability. Kamada and Harada (1977) found that the addition of 5 mM L-asparagine stimulated both callus and bud formation in stem segments of *Torenia fournieri*, while alanine (and, to a lesser extent, glutamic acid) increased flower bud formation from *Torenia* internode segments when both an auxin and a cytokinin were present. An increase in the number of adventitious buds formed on the cotyledons and hypocotyl of *Chamaecyparis obtusa* seedlings occurred when 1.37 mM glutamine and 1.51 mM asparagine were added together to Campbell and Durzan (1975) medium, but not when they were supplied on their own. L-asparagine was also added to MS medium by Green and Phillips (1975), to obtain plant regeneration from tissue cultures of maize; adding it to Finer and Nagasawa (1988) 10A4ON medium caused there to be more embryogenic clumps in *Glycine max* suspension cultures.

Amino acid additions do not invariably enhance morphogenesis. Supplementing Linsmaier and Skoog (1965) medium with 0.5-5 mM glutamine, caused callus of *Zamia latifolia* to show greatly decreased organogenesis and ammending Linsmaier and Skoog (1965) medium with 100 mg/l CH, prevented adventitious shoot formation from stem internode callus of apple and cherry rootstocks.

Embryogenesis

The presence the ammonium ion is usually sufficient for the induction of embryogenesis in callus or suspension cultures containing NO_3^-, but on media where NH_4^+ is lacking, casein hydrolysate, or an amino acid such as alanine, or glutamine, is often promotory. For embryogenesis in carrot cultures, Wetherell and Dougall (1976) have shown that in a medium containing potassium nitrate, reduced nitrogen in the form of ammonium chloride matched the effectiveness of an equivalent concentration of nitrogen from casein hydrolysate. Casein hydrolysate could be replaced by glutamine, glutamic acid, urea or alanine. Suspensions of wild carrot cells grew and produced somatic embryos on a medium containing either glutamine or CH as the sole nitrogen source.

There have also been many reports of embryogenesis being promoted by the addition of casein hydrolysate, or one or more specific amino acids, when both NO_3^- and NH_4^+ were available in the medium. In many cases, embryogenic callus and/or embryo formation did not occur without the presence of the amino acid source, suggesting that without amino acid, the medium was deficient in NH_4^+ or total nitrogen. Armstrong and Green (1985) found that the frequency of friable callus and somatic embryo formation from immature embryos of *Zea mays* increased almost linearly with the addition of up to 25 mM proline to Chu *et al.*, (1975) N6 medium, but there was no benefit from adding proline to MS medium (containing 150 mg/l asparagine hydrate).

The growth of somatic embryos can also be affected by the availability of reduced nitrogen. That of *Coronilla varia* embryos was poor on Gamborg *et al.*, (1968) B5 medium (Total N 26.74 mM; NH_4^+ 2.02 mM), unless 10 mM asparagine or 20 mM NH_4Cl was added to the medium, or unless the embryos were moved to Saunders and Bingham (1972) BOi2Y medium, which has 37.81 mM total inorganic N, 12.49 mM NH_4^+ and 2000 mg/l casein hydrolysate (approx. 9.9 mM NH_4^+ equivalence). However, the germination of *Triticum aestivum* somatic embryos was completely prevented by adding 800 mg/l CH to MS medium.

Culture of immature cotyledons

Young storage cotyledons isolated from immature zygotic embryos accumulate protein efficiently when cultured with amino acids in a medium without nitrate and ammonium ions. A medium such as that of Millerd *et al.* (1975), or of Thompson *et al.*, (1977) is normally used, but where the effect of different amino acids on protein assimilation is being studied, the amino acid content of the medium is varied. Glutamine is often found to be the most efficient nitrogen source for this purpose, but protein increase from culture with asparagine and glutamate (glutamic acid) is usually also significant.

Causes of the stimulatory effect of amino acids

We may conclude therefore, that for many cultural purposes, amino acids are not essential media components; but their addition as identified pure compounds, or more cheaply through casein hydrolysates, can be an easy way of ensuring against medium deficiency, or of providing a source of nitrogen that is immediately available to cultured cells or tissues. An observation by Murashige and Skoog (1962) that the presence of casein hydrolysate allowed vigorous organ development over a broader range of IAA and kinetin levels, may be of significance.

In gram moles per litre, amino acids can be a much more efficient source of reduced nitrogen than ammonium compounds. For instance, the mixture of amino acids provided by 400 mg/l of casein hydrolysate (containing at most as much reduced nitrogen as 3.3 mM NH_4^+) was as effective as 14.95 mM NH_4Cl in stimulating the division of protoplast-derived cells of *Antirrhinum*.

Why should this be, and why can additions of amino acids (sometimes in comparatively small amounts) stimulate growth or morphogenesis when added to media, which already contain large amounts of NH_4^+? Some hypotheses, which have been advanced are:

Conservation of ATP - alleviating phosphate deficiency

Durzan (1982) pointed out that when plant tissues take up the ammonium ion, they consume adenosine tri-phosphate (ATP) in converting it to amino acids. If suitable amino acids are available from the medium, some of ATP may be conserved. Bister-Miel *et al.,* (1985) noted that CH promoted growth in cultures where phosphate became growth-limiting. They suggested that amino acids compensated for phosphate deficiency. With the plant well supplied with amino acids, some of the phosphate, which is normally used for ATP production can be diverted to other uses. Several authors have pointed out that CH itself is also a source of phosphate. For example, Bridson (1978) and some chemical catalogues, show that some casein hydrolysates normally contain about 1.3 g P_2O_5 per 100 g. The addition of 2 g/l of CH will therefore increase the phosphate content of MS medium by 11% and that of White (1954) medium by 44% (assuming complete phosphate availability).

Enhanced nitrogen assimilation

Glutamine and glutamic acid are directly involved in the assimilation of NH_4^+. A direct supply of these amino acids should therefore enhance the utilization of both nitrate and ammonium nitrogen and its conversion into amino acids.

A replacement for toxic ammonium ions

Certain plant tissues are particularly sensitive to NH_4^+. Ochatt and Caso (1986) and Ochatt and Power (1988a, b) found that protoplasts of *Pyrus* spp. would not tolerate the ion, and that to obtain sustained cell division it was necessary to eliminate it from MS medium, and use 50 mg/l casein hydrolysate as a source of reduced nitrogen. CH can however be extremely toxic to freshly isolated protoplasts of some species and varieties of plants. Conifer tissues too are unable to cope with high concentrations of NH_4^+, but cultures can be supplied with equivalent levels of reduced nitrogen in the form of amino acids without the occurrence of toxicity. In soybean suspension cultures, the high level of ammonium in MS medium has been shown to inhibit isocitrate dehydrogenase (a Krebs' cycle enzyme) and glutamine synthetase, which contribute to the conversion of NH_4^+ to glutamine.

Adjustment of intracellular pH

As intracellular pH is important for the activation of sea urchin eggs, and amino acids can promote embryogenesis, Nuti Ronchi *et al.*, (1984) speculated that the uptake and assimilation of amino acids might help to regulate cellular pH in plants.

As mentioned before, there is commonly a minimum inoculation density below which growth cannot be initiated *in vitro*. This minimum varies according to both the source of the cells and the nature of the medium. It can usually be lowered by employing a 'conditioned' medium (i.e. a fresh medium into which the products of another medium in which cells are actively growing, have been added). Alternatively, initial growth at low densities can be supported by the close presence of other actively growing plant cells ('*nurse cultures*'). Compounds responsible for this effect must be freely diffusible from living cells and could include growth substances, reducing sugars, vitamins and amino acids. Addition of such supplements has been found to overcome the inhibited growth of some cells at low densities.

Phosphate

Phosphorus is a vital element in plant biochemistry. It occurs in numerous macromolecules such as nucleic acids, phospholipids and co-enzymes. It functions in energy transfer via the pyrophophate bond in ATP. Phosphate groups attached to different sugars provide energy in respiration and photosynthesis and phosphate bound to proteins regulates their activity. Phosphorus is absorbed into plants in the form of the primary or secondary orthophosphate anions $H_2PO_4^-$ and HPO_4^{2-} by an active process, which requires the expenditure of respiratory energy. Phosphate, in contrast to nitrate and sulphate, is not reduced in plants, but remains in the highly oxidized form. It is used in plants as the fully oxidized orthophosphate (PO_4^{3-}) form.

In culture media the element is provided as soluble potassium mono- and di-hydrogen phosphates. The di- and mono-valent phosphate anions respectively provided by these chemicals are interconvertible in solution depending on pH. Monovalent $H_2PO_4^-$ predominates at pH values below 7, characteristic of most tissue culture media, and it is this ion, which is most readily absorbed into plants. Conversion of $H_2PO_4^-$ into divalent HPO_4^{2-} begins to occur as solutions become more alkaline. The divalent ion is said to be only sparingly available to plants but Hagen and Hopkins (1955) and Jacobsen *et al.*, (1958) thought that its absorption could be significant, because even though the ion is normally at a relatively low concentration in nutrient solutions, its affinity with the site of absorption is greater than that of the monovalent form. Trivalent PO_4^{3-}, which appears in alkaline solutions, is not generally absorbed by plants.

In some early tissue culture media, all, or part of the phosphorus was supplied as sparingly-soluble phosphates. A slow rate of phosphorus availability seems to be possible from such compounds. The optimum rate of uptake of phosphate (HPO_4^{2-}) into cultured *Petunia* cells occurred at pH 4 but Zink and Veliky (1979) did not observe any decline in the absorption of phosphate by *Ipomoea* suspension cultures at pH 6.5, when HPO_4^{2-} and $H_2PO_4^-$ were present in approximately equal concentrations. Plant tissue cultures secrete phosphatase enzymes into the medium, which could release phosphate ions from organic phosphates.

In the cytoplasm, phosphate is maintained at a constant concentration of 5-10 mM, more or less independent of the external concentration. Phosphate in the vacuole fluctuates according to the external concentration but does not increase above 25 mM. When there is a high supply of phosphate and it is taken up at rates that exceed the demand, a number of processes act to prevent toxic phosphate concentrations, among others storage into inorganic compounds such as phytic acid. High concentrations of dissolved phosphate can depress growth, possibly because calcium and some microelements are precipitated from solution and/or their uptake reduced. In *Arabidopsis thaliana*, four different phosphate transporter genes have been isolated (APT1-4). *In vivo,* the genes are predominantly expressed in the roots and their expression is constitutive or induced by phosphate starvation. Overexpression of APT1 gene in tobacco cell cultures increased the rate of phosphate uptake.

Although the concentration of phosphate introduced into plant culture media has been as high as 19.8 mM, the average level is 1.7 mM and most media contain about 1.3 mM. However many reports indicate that such typical levels may be too low for some purposes. When phosphate is depleted from MS medium, there is an increase in free amino acids in *Catharanthus roseus* cells, because protein synthesis has ceased and degradation of proteins is occurring. Phosphate (starting concentration 2.64 mM) and sucrose were the only nutrients completely depleted in *Catharanthus roseus* batch suspension cultures, and the period of growth could be prolonged by increasing the levels of both. MS medium contains only 1.25 mM phosphate which may be insufficient for suspension cultures of some plants. The phosphate in MS medium is insufficient for *Cardamine pratensis* suspension cultures, all having been absorbed in 5 days: it is however adequate for *Silene alba* suspensions.

The phosphate in MS medium is also inadequate for static cultures of some plants, or where a large amount of tissue or organs are supported on a small amount of medium (for example where many separate shoots are explanted together in a static shoot culture). The concentration of the ion is then likely to be reduced almost to zero over several weeks. Insufficient levels of phosphate were present from MS during culture of *Hemerocallis*, *Iris* and *Delphinium*. Although growth can continue for a short while after the medium is depleted of phosphate, for some purposes it has been found to be beneficial to increase the phosphate concentration of MS to 1.86 mM, 2.48 mM, 3.1 mM or 3.71 mM, for example, to induce adventitious shoot formation from callus, or to increase the rate of shoot multiplication in shoot cultures. It should be noted that there is *in vivo* a significant retranslocation of phosphate from older leaves to the growing shoot. Retranslocation also occurs in tissue culture. In *Dahlia* culture in liquid medium, phosphate is almost completely taken up after 2 weeks. In spite of this, the concentration in tissues formed after the exhaustion is 'normal'. The depletion of phosphate early during culture has also a major effect on the pH of tissue culture media in which added phosphate is the major buffering component. When phosphate levels are increased to obtain a more rapid rate of growth of a culture, it can be advisable to investigate the simultaneous enhancement of the level of myo-inositol in the medium.

Potassium

Potassium is the major cation (positive ion) within plants reaching in the cytoplasm and chloroplasts concentrations of 100 - 200 mM. The biphasic uptake kinetics suggest two uptake systems: a high-affinity and a low-affinity one. K^+ is not metabolized. It contributes significantly to the osmotic potential of cells. K^+ counterbalances the negative charge of inorganic and organic anions. It functions in cell extension through the regulation of turgor, it has a major role in stomatal movements and functions in long-distance nutrient flow. Potassium ions are transported quickly across cell membranes and two of their major roles are regulating the pH and osmotic environment within cells. Potassium, calcium, sodium and chloride ions conserve their electrical charges within the plant, unlike the cation NH_4^+ and the anions NO_3^-, SO_4^{2-}, and $H_2PO_4^-$, which are rapidly incorporated into organic molecules. In intact plants, potassium ions are thought to cycle. They move, associated with cations (particularly NO_3^-), upwards from the roots in the xylem. As nitrate is reduced to ammonia and assimilated, carboxylic acid ions (RCO_3^-, malate) are produced.

These become associated with the released K^+ ions and are transported in the phloem to the roots, where they are decarboxylated, releasing K^+ for further anion transport. Carboxylate transported to the roots gives rise to OH ion, which is excreted into the soil (or medium) to counterbalance NO_3^- uptake. Potassium ions will clearly have a similar role in cultured tissues, but obvious transport mechanisms will usually be absent.

Many proteins show a high specificity for potassium which, acting as a cofactor, alters their configuration so that they become active enzymes. Potassium ions also neutralize organic anions produced in the cytoplasm, and so stabilize the pH and osmotic potential of the cell. In whole plants, deficiency of potassium results in loss of cell turgor, flaccid tissues and an increased susceptibility to drought, salinity, frost damage and fungal attack.

A high potassium to calcium ratio is said to be characteristic of the juvenile stage in woody plants. Potassium deficiency in plant culture media is said to lead to hyperhydricity, and a decrease in the rate of absorption of phosphate. However quite wide variations in the potassium content of MS medium had little effect on the growth or proliferation of cultured peach shoots.

Lavee and Hoffman (1971) reported that the optimum rate of callus growth of two apple clones was achieved in a medium containing 3.5 mM K^+: when the concentration was much higher than this, or when it was less than 1.4 mM, the callus grew less vigorously. However, the growth rate of wild carrot suspensions was said by Brown *et al.*, (1976) to be at, or near, the maximum when K^+ concentration was 1 mM: for embryogenesis 10-50 mM K^+ was required. Uptake of potassium into plants is reduced in the absence of calcium.

Within a large sample of different macronutrient compositions, it is found that authors have tended to relate the concentration of potassium to the level of nitrate. This is correlated with a coefficient of 0.78, $P<0.001$. The average concentration of potassium in these media was 13.6 mM and the most common value (median), 10.5 mM. Murashige and Skoog (1962) medium contains 20.04 mM K^+.

Sodium

Sodium ions (Na^+) are taken up into plants, but in most cases they are not required for growth and development and many plants actively secrete them from their roots to maintain a low internal concentration. The element can function as an osmotic stabilizer in halophytic plants; these have become adapted so that, in saline soils with low water potential, they can accumulate abnormally high concentrations of Na^+ ion in vacuoles, and thereby maintain sufficient turgor for growth.

Sodium does appear to have a beneficial nutritional effect on some plants and is therefore considered as a functional element. Small amounts of sodium chloride (e.g. 230 mg/l) can stimulate the growth of plants in the families Chenopodiaceae and Compositae even when there is no limitation on the availability of K^+. In other plants such as wheat, oats, cotton and cauliflower, sodium can partially replace potassium, but is not essential.

Sodium only appears to be essential to those salt-tolerant plants, which have a C4 (crassulacean acid) metabolism. Examples are *Bryophyllum tubiflorum* (Crassulaceae) and *Mesembryanthemum crystallimum* (Aizoaceae). In these plants the element is necessary for CO_2 fixation in photosynthesis.

Most macronutrient formulations do not contain any sodium at all, and the average concentration in 615 different preparations was 1.9 mM. Even if the element is not deliberately added as a macronutrient, small amounts are incorporated in most media from the salts added to provide micronutrients. Plant macronutrient preparations containing high concentrations of both sodium and chloride ions are not well formulated.

Magnesium

Magnesium is an essential component of the chlorophyll molecule and is also required non-specifically for the activity of many enzymes, especially those involved in the transfer of phosphate. ATP synthesis has an absolute requirement for magnesium and it is a bridging element in the aggregation of ribosome subunits. Magnesium is the central atom in the porphyrin structure of the chlorophyll molecule. Within plants, the magnesium ion is mobile and diffuses freely and thus, like potassium, serves as a cation balancing and neutralizing anions and organic acids. Macklon and Sim (1976) estimated there to be 2.1 mM Mg^{2+} in the cytoplasm of *Allium cepa* roots while McClendon (1976) put the general cytoplasmic requirement of plants as high as 16 mM. Plant culture media invariably contain relatively low concentrations of magnesium (average 6.8 mM, median 5.3 mM). Very often $MgSO_4$ is used as the unique source of both magnesium and sulphate ions.

Walker and Sato (1981) found there to be a large reduction in the number of somatic embryos formed from *Medicago sativa* callus when Mg^{2+} was omitted from the medium. In sympathy with this finding, Kintzios *et al.*, (2004) observed in tissue culture of melon that the highest level of magnesium occurred in direct somatic embryogenic cultures and the lowest level in callus cultures.

Sulphur

The sulphur utilized by plants is mainly absorbed as SO_4^{2-}, which is the usual source of the element in plant culture media. Uptake is coupled to nitrogen assimilation, and is said to be independent of pH. It results in the excretion of OH^-ions by the plant, making the medium more alkaline. However, according to Mengel and Kirkby (1982), plants are relatively insensitive to high sulphate levels and only when the concentration is in the region of 50 mM, is growth adversely affected. Although sulphur is mainly absorbed by plants in the oxidized form, that which is incorporated into chemical compounds is mainly as reduced -SH, -S- or -S-S- groups. The sulphur-containing amino acids cysteine and methionine become incorporated into proteins. Sulphur is used by plants in lipid synthesis and in regulating the structure of proteins through the formation of S-S bridges. The element also acts as a ligand joining ions of iron, zinc and copper to metalloproteins and enzymes. The reactive sites of some enzymes are -SH groups. Sulphur is therefore an essential element and deficiency results in a lack of protein synthesis. Sulphur-deficient plants are rigid, brittle and thin-stemmed. Important sulphur compounds are glutathione, which acts in detoxification of oxygen radicals, and the proteins thioredoxin and ferrodoxin that are involved in redox chemistry.

Growth and protein synthesis in tobacco cell suspensions were reduced on a medium containing only 0.6 mM SO_4^{2-} instead of 1.73 mM and when the supply of S in the medium was used up, large amounts of soluble nitrogen accumulated in the cells. Most media contain from 2-5 meq/l SO_4^{2-} (1 - 2.5 mM).

Calcium

As a major cation, calcium helps to balance anions within the plant, but unlike potassium and magnesium, it is not readily mobile.

Because of its capacity to link biological molecules together with coordinate bonds, the element is involved in the structure and physiological properties of cell membranes and the middle lamella of cell walls. The enzyme β-(1→3)-glucan synthase depends on calcium ions, and cellulose synthesis by cultured cells does not occur unless there are at least micro-molar quantities of Ca^{2+} in the medium. Many other plant enzymes are also calcium-dependent and calcium is a cofactor in the enzymes responsible for the hydrolysis of ATP.

Although calcium can be present in millimolar concentrations within the plant as a whole, calcium ions are pumped out of the cytoplasm of cells. Ca^{2+} is sequestered in the vacuole, complexes with calcium-binding proteins and may precipitate into calcium oxalate crystals to maintain the concentration at around only 0.1 mM. The active removal of Ca^{2+} is necessary to prevent the precipitation of phosphate (and the consequent disruption of phosphate-dependent metabolism) and interference with the function of Mg^{2+}. The uniquely low intra-cellular concentration of Ca^{2+} allows plants to use calcium as a chemical 'second messenger'. Regulatory mechanisms are initiated when Ca^{2+} binds with the protein calmodulin, which is thus enabled to modify enzyme activities. A temporary increase in Ca^{2+} concentration to 1 or 10 mM does not significantly alter the ionic environment within the cell, but is yet sufficient to trigger fundamental cell processes such as polarized growth, response to gravity and plant growth substances, cytoplasmic streaming, and mitosis. Physiological and developmental processes, which are initiated through the action of phytochrome are also dependent on the presence of Ca^{2+}. A short-term increase in cytosolic free Ca^{2+} has been observed for osmoadaptation, phytoalexin synthesis, thermotolerance and induction of free-radical scavengers.

Large quantities of calcium can be deposited outside the protoplast, in cell vacuoles and in cell walls. Calcification strengthens plant cell walls and is thought to increase the resistance of a plant to infection. By forming insoluble salts with organic acids, calcium immobilizes some potentially damaging by-products. The element gives protection against the effects of heavy metals and conveys some resistance to excessively saline conditions and low pH.

The Ca^{2+} ion is involved in *in vitro* morphogenesis and is required for many of the responses induced by plant growth substances, particularly auxins and cytokinins. In the moss *Funaria*, cytokinin causes an increase in membrane-associated Ca^{2+} specifically in those areas which are undergoing differentiation to become a bud. Protocorm formation from callus of *Dendrobium fibriatum* was poor on Mitra *et al.,* (1976) A medium when calcium was omitted and in *Torenia* stem segments, adventitious bud formation induced by cytokinin seems to be mediated, at least in part, by an increase in the level of Ca^{2+} within cells. Exogenous Ca^{2+} enhanced the formation of meristemoids and the first phases of outgrowth into organs in tobacco pith explants. In carrot, somatic embryogenesis coincides with a rise of free cytosolic Ca^{2+} and applied Ca^{2+} increases the number of somatic embryos.

Shoot tip necrosis

Calcium deficiency in plants results in poor root growth and in the blackening and curling of the margins of apical leaves, often followed by a cessation of growth and death of the shoot tip. The latter symptoms are similar to aluminium toxicity. Tip necrosis has been especially observed in shoot cultures, sometimes associated with hyperhydricity. It often occurs after several subcultures have been accomplished (e.g. in *Cercis canadensis*). After death of the tip, shoots often produce lateral branches, and in extreme cases the tips of these will also die and branch again. The cause of tip necrosis has not always been determined (e.g. in *Pistacia* shoot cultures, where shoots showing symptoms may die after planting out). The occurrence of necrosis was reduced in *Pistacia* and *Prunus tenella* by more frequent subculturing, but this is a costly and time-consuming practice. In *Pictacia*, calcium reduced necrosis.

Tip necrosis was found in *Psidium guajava* shoot cultures after prolonged subculturing, if shoots were allowed to grow longer than 3 cm, and was common in rapidly growing cultures; it occurred on *Sequoiadendron giganteum* shoots only when they were grown on relatively dilute media. Necrosis of *Rosa hybrida* 'White Dream', was cured by adding 0.1 mg/l GA_3 to the medium.

Analysis of necrotic apices has shown them to be deficient in calcium, and a shortage of this element has been associated with tip necrosis in *Amelanchier*, *Betula*, *Populus*, *Sequoia*, *Ulmus*, *Cydonia* and other woody plants, although the extent of damage is variable even between genotypes within a species. As calcium is not remobilized within plant tissues, actively growing shoots need a constant fresh supply of ions in the transpiration stream. An inadequate supply of calcium can result from limited uptake of the ion, and inadequate transport, the latter being caused by the absence of transpiration due to the high humidity in the culture vessel. A remedy can sometimes be obtained by reducing the culture temperature so that the rate of shoot growth matches calcium supply, using vessels which promote better gas exchange (thereby increasing the transpiration and xylem transport), or by increasing the concentration of calcium in the medium. The last two remedies can have drawbacks: the medium will dry out if there is too free gas exchange; adding extra calcium ions to the medium is not always effective (e.g. in cultures of *Castanea sativa*); and can introduce undesirable anions. Chloride toxicity can result if too much calcium chloride is added to the medium. To solve this difficulty, McCown *et al.*, added 6 mM calcium gluconate to Lloyd and McCown (1981) WPM medium to correct Ca^{2+} deficiency, without altering the concentrations of the customary anions. There is a limit to the concentration of calcium, which can be employed in tissue culture media because several of its salts have limited solubility.

Chloride

The chloride ion (Cl^-) has been found to be essential for plant growth, but seems to be involved in few biological reactions and only very small quantities are really necessary. Rains (1976) listed chlorine as a micronutrient. Chloride is required for the water-splitting protein complex of Photosystem II, and it can function in osmoregulation in particular in stomatal guard cells. The chloride ion is freely transported and many plants can tolerate the presence of high concentrations without showing toxicity. The chief role of chloride seems to be in the maintenance of turgor and in balancing rapid changes in the level of free cations such as K^+, Mg^{2+} and Na^+. Plants deprived of Cl^- are liable to wilting.

In isolated chloroplasts, chloride (together with Mn^{2+}) ions are required for oxygen evolution in photosystem II of photosynthesis, although there has been some doubt whether this requirement exists *in vivo*. Chloride ions are best taken into plants at slightly acid pH.

The most common concentration of chloride in culture media is 3 mM, the average 6 mM. MS medium contains 6 mM Cl^-; Quoirin and Lepoivre (1977) medium, only 0.123 μM. Some species are sensitive to chloride ions. McCown and Sellmer (1987) reported that too high a concentration, seemed to cause woody species to have yellow leaves and weak stems: sometimes tissues collapsed and died. An excess of Cl^- has been thought to be one cause of the induction of hyperhydricity, and omission of the ion does seem to prevent the development of these symptoms in *Prunus*. Pevalek-Kozlina and Jelaska (1987) deliberately omitted chloride ions from WPM medium for the shoot culture of *Prunus avium* and obtained infrequent hyperhydricity in only one genotype. The presence of 7 mM Cl^- can be toxic to pine suspension cultures.

As chlorine has only a relatively small nutritional significance, steps are sometimes taken to reduce the concentration of chloride ion in culture media, but in order to adjust the concentration of other ions, it is then often necessary to make a marked increase in SO_4^{2-}. For example, using ammonium sulphate instead of ammonium chloride to supply NH_4^+ in Eeuwens (1976) Y3 medium, would increase the sulphate level from 2 to 12 meq/l (from 1 to 6 mM).

Micronutrients

Plant requirements for microelements have only been elucidated over the past 50-60 years. Before the end of the last century, it had been realized that too little iron caused chlorophyll deficiency in plants, but the importance of other elements took many years to prove conclusively. Maze, for example, used hydroponic techniques during the years 1914-1919 to show that zinc, manganese and boron improved the growth of maize plants. Sommer and Lipman (1926) also showed the essentiality of boron, and Sommer (1931) of copper, but uncertainty over which elements were really indispensible to growth still existed in 1933 when Hoagland and Snyder proposed two supplementary nutrient solutions for water culture which in total contained 26 elements. It took several further years to prove that molybdenum and cobalt in very small amounts, were most important for healthy plant growth. Early plant tissue culture work was to both profit from, and contribute to the findings of previous hydroponic studies. Our understanding has been enhanced by investigations into the biochemical role of minor elements.

Early Use in Plant Tissue Culture Media

At the time of the early plant tissue culture experiments, uncertainty still existed over the nature of the essential microelements. Many tissues were undoubtedly grown successfully because they were cultured on media prepared from impure chemicals or solidified with agar, which acted as a micronutrient source.

In the first instance, the advantage of adding various micronutrients to culture media was mainly evaluated by the capability of individual elements to improve the growth of undifferentiated callus or isolated root cultures. Knudson (1922) incorporated Fe and Mn in his very successful media for the non-symbiotic germination of orchid seeds, and, following a recommendation by Berthelot (1934), Gautheret (1939) and Nobecourt (1937) included in their media (in addition to iron) copper, cobalt, nickel, titanium and beryllium. Zinc was found to be necessary for the normal development of tomato root systems, and without Cu, roots ceased to grow. Hannay and Street (1954) showed that Mo and Mn were also essential for root growth.

An advantage adding five micronutrients to tissue culture media was perhaps first well demonstrated by Heller in 1953 who found that carrot callus could be maintained for an increased number of passages when Fe, B, Mn, Zn and Cu were present.

Micronutrients from Trace Impurities

Micronutrients tend to be added to modern media by the addition of fairly standard chemicals. Street (1977) rightly emphasized that even analytical grade chemicals contain traces of impurities that will provide a hidden supply of micronutrients to a medium. An illustration of this, comes from the work of Dalton *et al.*, (1983) who found traces of silicon (Si) in a precipitate from MS medium which had been made up with analytical grade laboratory chemicals. Gelling agents contain inorganic elements but whether cultures can utilize them is unclear. Amounts of contaminating substances in chemicals would have been greater in times past, so that an early medium such as Knudson (1922; 1943) B, prepared today with highly purified chemicals, will not have quite the same composition as when it was first used by Knudson in 1922; the addition of some micronutrients might improve the results obtained from a present-day formulation of such early media.

Optimum Micro-element Concentrations

Most modern culture media use the microelements of Gamborg *et al.*, (1968) B5 medium, or the more concentrated mixtures in MS or Bourgin and Nitsch (1967) H media. Several research workers have continued to use Heller (1953) micro-nutrient formulation, even though higher levels are now normally recommended. Quoirin and Lepoivre (1977) showed clearly that in conjunction with MS or their Quoirin and Lepoivre (1977) B macro-elements, the concentration of Mn in Heller's salts should be increased by 100-fold to obtain the most effective growth of *Prunus* meristems.

Cell growth and morphogenesis of some species may even be promoted by increasing the level of micronutrients above that recommended by Murashige and Skoog (1962). The induction and maintenance of callus and growth of cell suspensions of juvenile and mature organs of both Douglas fir and loblolly pine, was said to be improved on Litvay *et al.*, (1981) LM medium in which Mg, B, Zn,

Mo, Co and I were at 5 times the concentration of MS micro-elements, and Mn and Cu at 1.25 and 20 times respectively. Other authors to have employed high micronutrient levels are Barwale *et al.*, (1986) who found that the induction of adventitious shoots from callus of 54 genotypes of *Glycine max* was assisted by adding four times the normal concentration of minor salts to MS medium.

A further example of where more concentrated micro-elements seemed to promote morphogenesis is provided by the work of Wang, *et al.*, (1980, 1981). Embryogenesis could be induced most effectively in callus derived from *Hevea brasiliensis* anthers, by doubling the concentration of microelements in MS medium, while at the same time reducing the level of macronutrients to 60-80% of the original.

Despite these reports, few research workers seem to have accepted the need for such high micronutrient levels. To diminish the occurrence of hyperhydricity in shoot culture of carnation, Dencso (1987) reduced the level of micronutrients to those in MS medium, but this mixture was inadequate for *Gerbera* shoot cultures and the rate of propagation was less than that with the normal MS formulation.

The need for macronutrient concentrations to be optimized as the first step in media development seems to be emphasized by results of Eeuwens (1976). In an experiment with factorial combinations of the macro- and micro-nutrient components of four media, his Eeuwens (1976) Y3 micronutrients gave a considerable improvement in the growth of coconut callus, compared with other micro-element mixtures, when they were used with Y3 and MS macronutrients, but not when used with those of White (1942) or Heller (1953; 1955).

Cellular Differentiation and Morphogenesis

Welander (1977) obtained evidence, which suggested that plant cells are more demanding for minor elements when undergoing morphogenesis. Petiole explants of *Begonia hiemalis* produced callus on media without micronutrients, but would only produce adventitious shoot buds directly when micronutrients were added to the macronutrient formulation. The presence of iron is particularly important for adventitious shoot and root formation.

That mineral nutrition can influence cellular differentiation in combination with growth hormones, was shown by Beasley *et al.*, (1974). Cotton ovules cultured on a basic medium containing 5.0 mM IAA and 0.5 mM GA_3, required 2 mM calcium (normally present in the medium) for the ovules to develop fibres. Magnesium and boron were essential for fibre elongation.

Roles of Micronutrients

Manganese

The essential micronutrient metals Fe, Mn, Zn, B, Cu, Co and Mo are components of plant cell proteins of metabolic and physiological importance. At least five of these elements are, for instance, necessary for chlorophyll synthesis and chloroplast function. Micronutrients have roles in the functioning of the genetic apparatus and several are involved with the activity of growth substances.

Manganese (Mn) is one of the most important microelements and has been included in the majority of plant tissue culture media. It is generally added in similar concentrations to those of iron and boron, i.e. between 25-150 mM. Manganese has similar chemical properties to Mg^{2+} and is apparently able to replace magnesium in some enzyme systems. However there is normally 50- to 100-fold more Mg^{2+} than Mn^{2+} within plant tissues, and so it is unlikely that there is frequent substitution between the two elements.

The most probable role for Mn is in definition of the structure of metalloproteins involved in respiration and photosynthesis. It is known to be required for the activity of several enzymes, which include decarboxylases, dehydrogenases, kinases and oxidases and superoxide dismutase enzymes. Manganese is necessary for the maintenance of chloroplast ultra-structure. Because Mn(II) can be oxidized to Mn(IV), manganese plays an important role in redox reactions. The evolution of oxygen during photosystem II of the photosynthetic process, is dependent on a Mn-containing enzyme and is proportional to Mn content. Mn is toxic at high concentration.

In tissue cultures, omission of Mn ions from Doerschug and Miller (1967) medium reduced the number of buds initiated on lettuce cotyledons. A high level of manganese could compensate for the lack of molybdenum in the growth of excised tomato roots (and vice versa). Natural auxin levels are thought to be reduced in the presence of Mn^{2+} because the activity of IAA-oxidase is increased. This is possibly due to Mn^{2+} or Mn-containing enzymes inactivating oxidase inhibitors, or because manganous ions are one of the cofactors for IAA oxidases in plant cells. Manganese complexed with EDTA increased the oxidation of naturally-occurring IAA, but not the synthetic auxins NAA or 2,4-D. However, Chee (1986) has suggested that, at least in blue light, Mn^{2+} tends to cause the maintenance of, or increase in, IAA levels within tissues by inactivating a co-factor of IAA oxidase. When the Mn^{2+} level in MS medium was reduced from 100 mM to 5 mM, the production in blue light, of axillary shoots by *Vitis* shoot cultures was increased.

Zinc

Zinc is a component of stable metallo-enzymes with many diverse functions, making it difficult to predict the unifying chemical property of the element, which is responsible for its essentiality. Zinc is required in more than 300 enzymes including alcohol dehydrogenase, carbonic anhydrase, superoxide dismutase and RNA-polymerase. Zinc forms tetrahedral complexes with N-, O-, and S-ligands. In bacteria, Zn is present in RNA and DNA polymerase enzymes, deficiency resulting in a sharp decrease in RNA levels. DNA polymerase is concerned with the repair of incorrectly formed pieces of DNA in DNA replication, and RNA polymerase locates the point on the DNA genome at which initiation of RNA synthesis is to take place. Divalent Mg^{2+}, Mn^{2+} or Co^{2+} are also required for activation of these enzymes.

Zinc deficient plants suffer from reduced enzyme activities and a consequent diminution in protein, nucleic acid and chlorphyll synthesis. Molybdenum-and zinc-deficient plants have a decreased chlorophyll content and poorly developed chloroplasts. Plants deprived of zinc often have short internodes and small leaves.

The concentration of Zn^{2+} in MS medium is 30 1.1M but amounts added to culture media have often varied widely between 0.1-70 μM and experimental results to demonstrate the most appropriate level are limited. When Eriksson (1965) added 15 mg/l $Na_2ZnEDTA.2H_2O$ (40 μM Zn^{2+}) to *Haplopappus gracilis* cell cultures, he obtained a 15% increase in cell dry weight which was thought to be due to the presence of zinc rather than the chelating agent. Zinc was also shown to increase growth of a rice suspension. The highest concentration tested, 520 μM, resulted in the fastest rate of growth and it was suggested that zinc had increased auxin activity. Zinc is required for adventitious root formation in *Eucalyptus*. In cassava, additional zinc promotes somatic embryogenesis and rooting. However, very high concentrations of zinc are found to be inhibitory, and the microelement has been noted to prevent root growth at a concentration higher than 50 μM.

There is a close relationship between the zinc nutrition of plants and their auxin content. It has been suggested that zinc is a component of an enzyme concerned with the synthesis of the IAA precursor, tryptophan. The importance of Zn for tryptophan synthesis is especially noticeable in crown gall callus which normally produces sufficient endogenous auxin to maintain growth on a medium without synthetic auxins, but which becomes auxin-deficient and ceases to grow in the absence of Zn.

Boron

Boron is involved in plasma membrane integrity and functioning, probably by influencing membrane proteins, and cell wall intactness. Reviews have been provided by Lewis (1980) and by Blevins and Lukaszewski (1998). The element is required for the metabolism of phenolic acids, and for lignin biosynthesis: it is probably a component, or co-factor of the enzyme which converts *p*-coumaric acid to caffeate and 5-hydroxyferulate. Boron is necessary for the maintenance of meristematic activity, most likely because it is involved in the synthesis of N-bases (uracil in particular); these are required for RNA synthesis. It is also thought to be involved in the maintenance of membrane structure and function, possibly by stabilizing natural metal chelates which are important in wall and membrane structure and function. Boron is concerned with regulating the activities of phenolase enzymes; these bring about the biosynthesis of phenylpropane compounds, which are polymerized to form lignin. Lignin biosynthesis does not take place in the absence of boron. Boron also mediates the action of phytochrome and the response of plants to gravity.

Use in culture media

In the soil, boron occurs in the form of boric acid and it is this compound, which is generally employed as the source of the element in tissue cultures. Uptake of boric acid occurs most readily at acid pH, possibly in the undissociated form or as $H_2BO_3^-$. A wide range of boron concentrations has been used in media, the most usual being between from 50 and 100 μM: MS medium contains 100 μM. Bowen (1979) found boron to be toxic to sugarcane suspensions above 2 mg/l (185 μM), but there are a few reports of higher concentrations being employed. High concentrations of boron may have a regulatory function; for example, 1.6-6.5 mM have been used in simple media to stimulate pollen germination.

Boric acid reacts with some organic compounds having two adjacent *cis*-hydroxyl groups. This includes *o*-diphenols, hexahydric alcohols such as mannitol and sorbitol (commonly used in plant tissue culture as osmotic agents), and several other sugars, but excludes sucrose which forms only a weak association. Once the element is complexed it appears to be unavailable to plants. This led Lewis (1980) to suggest that because boric acid was required for lignin biosynthesis, vascular plants were led, during evolution, to use sucrose exclusively for the transport of carbohydrate reserves.

Although the addition of sugar alcohols and alternative sugars to sucrose can be beneficial during plant tissue culture, it is clearly necessary to return to a sucrose-based medium for long-term culture, or boron deficiency may result.

Deficiency symptoms

Boron is thought to promote the destruction of natural auxin and increase its translocation. Endogenous IAA levels increase in the absence of boron and translocation is reduced, the compound probably being retained at the site of synthesis. Plants suffering from boron deficiency have restricted root systems and a reduced capacity to absorb $H_2PO_4^-$ and some other ions. High levels of auxins can have the same effect on growth and ion uptake. Neales (1959, 1964) showed that isolated roots stopped growing unless a minimum concentration of boron was present (although the necessity for the element was not so apparent when cultures were grown in borosilicate glass vessels). Inhibition of root elongation in the absence of boron has been shown to be due to the cessation of mitosis and the inhibition of DNA synthesis. Boron deficiency also results in depressed cytokinin synthesis. Cell division is inhibited in the absence of boron, apparently because there is a decrease in nuclear RNA synthesis. However, deficiency often leads to increased cambial growth in intact dicotyledonous plants.

One of the changes seen in some plants grown under boron deficiency is the outgrowth of lateral buds resulting in plants with a bushy or rosette appearance. In pea, this was associated with a sharp decrease in IAA-export from the apex. It is generally accepted that the outgrowth of lateral buds is inhibited by polar auxin transport in the stem and that disruption of this transport by decapitation or auxin transport inhibitors results in the outgrowth of lateral buds.

Cotton ovules which otherwise develop fibres when cultured, commence extensive callus formation when placed on a medium deficient in boron. On the other hand, the growth rate of callus cultures of *Helianthus annuus* and *Daucus carota*, and cell cultures of sugar cane, was much reduced when boron was not present in the growth medium. Boron influences the development of the suspensor of somatic embryos in *Larix deciduas*. Boron had no influence on the induction of embryogenesis in *Daucus carota* but altered the development of embryos: root development was promoted at low concentrations and shoot development at high. This coincided with a high and low auxin-cytokinin ratio, respectively.

Adventitious root formation

Boron is thought to promote the destruction of auxin. Although auxin is required for the formation of adventitious root initials, boron is necessary in light grown-plants for the growth of these primordia; possibly boron enhances the destruction of auxin in these circumstances, which in high concentrations is inhibitory to root growth. An interaction between boron and auxins in the rooting of cuttings has been noticed in several species and a supply of exogenous borate has been shown to be essential. However, excessive boron concentrations lead to a reduction in the number of roots formed. Boron deficiency had no observed effect on the rooting of *Eucalyptus* microcuttings.

Copper and Molybdenum

Copper is an essential micronutrient, even though plants normally contain only a few parts per million of the element. Two kinds of copper ions exist; they are the monovalent cuprous [Cu(I)] ion, and the divalent cupric [Cu(II)] ion: the former is easily oxidized to the latter; the latter is easily reduced. The element becomes attached to enzymes, many of which bind to, and react with oxygen. They include the cytochrome oxidase enzyme system, responsible for oxidative respiration, and superoxide dismutase (an enzyme which contains both copper and zinc atoms). Detrimental superoxide radicals, which are formed from molecular oxygen during electron transfer reactions, are reacted by superoxide dismutase and thereby converted to water. Copper atoms occur in plastocyanin, a pigment participating in electron transfer.

Several copper-dependent enzymes are involved in the oxidation and hydroxylation of phenolic compounds, such as ABA and dopamine. The hydroxylation of monophenols by copper-containing enzymes leads to the construction of important polymeric constituents of plants, such as lignin. These same enzymes can lead to the blackening of freshly isolated explants. Copper is a constituent of ascorbic acid oxidase and the characteristic growth regulatory effects of ethylene are thought to depend on its metabolism by an enzyme, which contains copper atoms.

High concentrations of copper can be toxic. Most culture media include ca. 0.1-1.0 μM Cu^{2+}. Ions are usually added through copper sulphate, although occasionally cupric chloride or cupric nitrate have been employed. In hydroponic culture of *Trifolium pratense*, uptake of copper into the plant depended on the amount of nitrate in solution. Uptake was considerably reduced when NO_3^- was depleted. The concentration of Cu in tissue culture media is very small relative to the level in plants. It is therefore not surprising that various authors report strong increases of growth when Cu is added at 1-5 μM.

Plants utilize hexavalent molybdenum and absorb the element as the molybdate ion (MoO_4^{2-}). This is normally added to culture media as sodium molybdate at concentrations up to 1 mM. Considerably higher levels have occasionally been introduced apparently without adverse effect, although Teasdale (1987) found pine suspension cultures were injured by 50 mM. Molybdenum is a component of several plant enzymes, two being nitrate reductase and nitrogenase, in which it is a cofactor together with iron: it is therefore essential for nitrogen utilization. Tissues and organs presented with NO_3^- in a molybdenum-deficient medium can show symptoms of nitrate toxicity because the ion is not reduced to ammonia.

Cobalt

Cobalt is not regarded as an essential element. Nevertheless, it was found to have been included in approximately half of a large sample of published plant culture media. Murashige and Skoog (1962) included Co in their medium because it had been shown to be required by lower plants and that it might have a role in regulating morphogenesis in higher plants. However, no stimulatory effect on the growth of tobacco callus was observed by adding cobalt chloride to the medium at several concentrations from 0.1 μM and above, and at 80.0 and 160 μM the compound was toxic.

Similarly Schenk and Hildebrandt (1972) obtained no clear evidence for a Co requirement in tests on a wide variety of plants, but retained the element in their medium because they occasionally observed an apparent stimulation to the callus growth of some monocotyledons. *Pinus* suspension cultures do not require cobalt. The concentration most commonly added to a medium is ca. 0.1 μM, although ten times this amount has sometimes been used. Cobalt is the metal component of Vitamin B12 which is concerned with nucleic acid synthesis, but evidence that the element has any marked stimulatory effect on growth or morphogenesis in plant tissue cultures is hard to find. Cobalt may replace nickel in urease and thereby render it inactive, e.g., in potato.

Advantage from adding cobalt to plant culture media might be derived from the fact that the element can have a protective action against metal chelate toxicity and it is able to inhibit oxidative reactions catalyzed by ions of copper and iron. The Co^{2+} ion can inhibit ethylene synthesis.

Aluminium and Nickel

Several workers, following Heller (1953), have included aluminium and nickel in their micronutrient formulations. However, the general benefit of adding the former metal does not seem to have been adequately demonstrated.

It was believed that in most plants Ni^{2+} is not absolutely required for normal growth and development. However, more recently, it has been found by careful experimentation that nickel is essential. The ion is a component of urease enzymes, which convert urea to ammonia. It has been shown to be an essential micronutrient for some legumes and to activate urease in potato microshoot cultures. In tissue cultures the presence of 0.1 mM Ni^{2+} strongly stimulates the growth of soybean cells in a medium containing only urea as a nitrogen source. Slow growth occurs on urea without the deliberate addition of nickel, possibly supported by trace amounts of the element remaining in the cells. Cells and tissues are not normally grown with urea as a nitrogen source, and as urease is the only enzyme, which has been shown to have a nickel component, it could be argued that nickel is not essential. However, without it soybean plants grown hydroponically, accumulate toxic concentrations of urea (2.5%) in necrotic lesions on their leaf tips, whether supplied with inorganic nitrogen, or with nitrogen compounds obtained from bacterial symbiotic nitrogen fixation. These symptoms can be alleviated in plants growing in hydroponic culture by adding 1 mg/l Ni to the nutrient solution. Absence of nickel in a hydroponic solution also results in reduced early growth and delayed nodulation.

Despite these findings nickel has not been added deliberately to tissue culture media. However, it should be noted that agar contains relatively high levels of nickel and the possibility of urea toxicity may have been avoided because, in tissue cultures, urea diffuses into the medium. Quoirin and Lepoivre (1977) showed that at the concentrations recommended by Heller, Al^{3+} and Ni^{2+} were without effect on the growth of *Prunus* meristems and were inhibitory at higher levels. If it is thought that Ni should be added to a culture medium, 0.1 mM is probably sufficient.

Aluminium has been said to be necessary for the growth of some ferns, but is not generally added to tissue culture media for fern propagation.

Iodine

Iodine is not recognized as an essential element for the nutrition of plants, although it may be necessary for the growth of some algae, and small amounts do accumulate in higher plants (ca. 12 and 3 mol/kg dry weight in terrestrial and aquatic plants respectively). However, the iodide ion has been added to many tissue culture media (e.g. to 65% of micronutrient formulations).

The practice of including iodide in plant culture media began with the report by White (1938) that it improved the growth of tomato roots cultured *in vitro.* Hannay (1956) obtained similar results and found that root growth declined in the absence of iodine which could be supplied not only from potassium iodide, but also from iodoacetate or methylene iodide, compounds which would only provide iodide ions very slowly in solution by hydrolysis. Street (1966) thought that these results indicated that iodine could be an essential nutrient element, but an alternative hypothesis is that any beneficial effect may result from the ability of iodide ions to act as a reducing agent. Oxidants convert iodide ions to free iodine. Eeuwens (1976) introduced potassium iodide into his Y3 medium at 0.05 mM, as it prevented the browning of coconut palm tissue cultures. The presence of 0.06 μM potassium iodide slightly improved the survival and growth of cultured *Prunus* meristems.

Although Gautheret (1942) and White (1943) had recommended the addition of iodine to media for callus culture, Hildebrandt *et al.*, (1946) obtained no statistically significant benefit from adding potassium iodide to tumour callus cultures of tobacco and sunflower. However, as the average weight of tobacco callus was 11% less without it, the compound was included (at different levels) in both of the media they devised. Once again iodine also had no appreciable effect on tobacco callus yield in the experiments of Murashige and Skoog (1962), but was nevertheless included in their final medium. Other workers have omitted iodine from MS medium or from new media formulations without any apparent ill effects. However, Teasdale *et al.*, (1986); Teasdale, (1987) reported a definite requirement of *Pinus taeda* suspensions for 25 mM KI when they were grown on Litvay *et al.*, (1981) LM medium.

There seems, at least in some plants, to be an interaction between iodine and light. Eriksson (1965) left KI out of his modification of MS medium, finding that it was toxic to *Haplopappus gracilis* cells cultured in darkness: shoot production in *Vitis* shoot cultures kept in blue light was reduced when iodine was present in the medium, but the growth of roots on rooted shoots was increased. Chee thought that these results supported the hypothesis that iodine enhanced the destruction and/or the lateral transport of IAA auxin. This seems to be inconsistent with the suggestion that r acts mainly as a reducing agent.

Silicon

Silicon (Si) is the second most abundant element on the surface of the earth. Si has been demonstrated to be beneficial for the growth of plants and to alleviate biotic and abiotic stress. The silicate ion is not normally added to tissue culture media, although it is likely to be present in low concentrations. Deliberate addition to the medium might, however, improve the growth of some plants. Adatia and Besford (1986) found that cucumber plants depleted silicate from a hydroponic solution and in consequence their leaves were more rigid, had a higher fresh weight per unit area and a higher chlorophyll content than the controls. The resistance of the plants to powdery mildew was also much increased.

Iron

Chelating agents

Some organic compounds are capable of forming complexes with metal cations, in which the metal is held with fairly tight chemical bonds. The complexes formed may be linear or ring-shaped, in which case the complex is called a *chelate*. Metals can be bound (or sequestered) by a chelating agent and held in solution under conditions where free ions would react with anions to form insoluble compounds, and some complexes can be more chemically reactive than the metals themselves. For example, Cu^{2+} complexed with amino acids is more active biologically than the free ion.

Chelating agents vary in their sequestering capacity (or avidity) according to chemical structure and their degree of ionization, which changes with the pH of the solution. Copper is chelated by amino acids at relatively high pH, but in conditions of greater acidity, it is more liable to be complexed with organic acid ligands. The higher the stability of a complex, the higher the avidity of the complexing agent. One, and in many cases, two or three molecules of a complexing agent may associate with one metal ion, depending on its valency.

Despite tight bonding, there is always an equilibrium between different chelate complexes and between ions in solution. Complexing agents also associate with some metal ions more readily than with

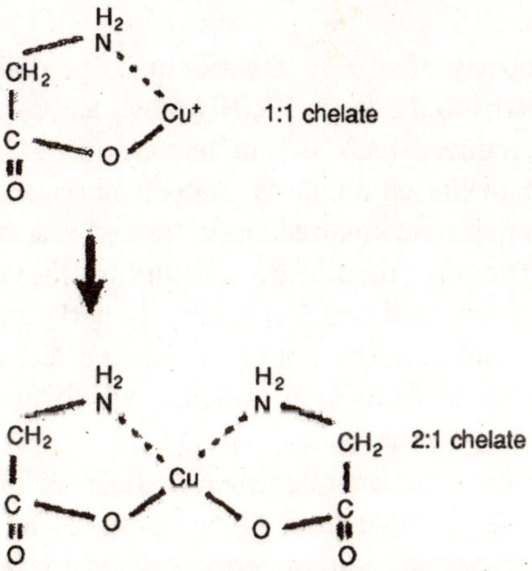

Fig. 5.2. Copper chelated with amino acid, glycone.

others. In general Fe^{3+} (for agents able to complex with trivalent ions) complexes have a higher stability than those of Cu^{2+}, then (in descending order), Ni^{2+}, Al^{3+} (where possible), Zn^{2+}, Co^{2+}, Fe^{2+}, Mn^{2+} and Ca^{2+}.

For a chelated metal ion to be utilized by a plant there must be some mechanism whereby the complex can be broken. This could occur if it is absorbed directly and the ion displaced by another more avid binding agent, or if the complex is biochemically denatured. Metals in very stable complexes can be unavailable to plants; copper in EDTA chelates may be an example. High concentrations of avid chelating agents are phytotoxic, probably because they competitively withdraw essential elements from enzymes.

Naturally-occurring compounds act as chelating agents. Within the plant very many constituents such as proteins, peptides, porphyrins, carboxylic acids and amino acids have this property: some of those with high avidity are metal-containing enzymes. Amino acids are able to complex with divalent metals. Grasses are thought to secrete a chelating agent from their roots to assist the uptake of iron. There are also synthetic chelating agents with high avidities (stability constants) for divalent and trivalent ions. The application of synthetic chelating agents and chelated micronutrients to the roots of some plants growing in alkaline soils can improve growth by supplying essential metals such as iron and zinc which are otherwise unavailable. The addition of such compounds to tissue culture media can help to make macro- and micro-nutrients more accessible to plant cells.

Iron chelates

A key property of iron is its capacity to be oxidized easily from the ferrous [Fe(II)] to the ferric [Fe(III)] state, and for ferric compounds to be readily reduced back to the ferrous form. In plants, iron is primarily used in the chloroplasts, mitochondria and peroxisomes of plants for effecting oxidation/reduction (redox) reactions. The element is required for the formation of amino laevulinic acid and protoporphyrinogen (which are respectively early and late precursors of chlorophyll) and deficiency leads to marked leaf chlorosis. Iron is also a component of ferredoxin proteins, which function as electron carriers in photosynthesis.

Iron is therefore an essential micronutrient for plant tissue culture media and can be provided from either ferrous or ferric salts. In early experiments, ferrous sulphate or ferric citrate or tartrate were used in media as a source of the element. Citric and tartaric acids can act as chelating agents for some divalent metals, but are not very efficient at keeping iron in solution. If Fe^{2+} and Fe^{3+} ions escape from the chelating agent, they are liable to be precipitated as iron phosphate. The iron may then not be available to plant cells, unless the pH of the medium falls sufficiently to bring free ions back into solution. The problem of precipitation is more severe in aerated media and where the pH of the medium drifts towards alkalinity. Under these conditions Fe^{2+} (ferrous) ions are oxidized to Fe^{3+} (ferric) ions and unchelated ferric ions may then also be converted to insoluble $Fe(OH)_3$. For plant hydroponic culture, the advantages of adding iron to nutrient solutions in the form of a chelate with EDTA was first recognized in the 1950's. Street *et al.*, (1952) soon found that iron in this form was less toxic and could be utilized by *in vitro* cultures of isolated tomato roots over a wider pH range than ferric citrate. Klein and Manos (1960) showed that callus cultures of several species grew more rapidly on White (1954) medium if Fe^{3+} ions from $Fe_2(SO_4)_3$ were chelated with EDTA, rather than added to the medium from the pure compound, and Doerschug and Miller (1967), that 0.036 mM Fe from NaFeEDTA was as effective as 0.067 mM Fe as ferric citrate, in promoting shoot bud initiation on lettuce cotyledons. Iron presented as ferric sulphate (0.025 mM Fe) was much less effective than either chelated form.

Skoog and co-workers began to use EDTA in media for tobacco callus cultures in 1956 and discussed their findings in the same paper that describes MS medium. The addition of an iron (Fe)-EDTA chelate once again greatly improved the availability of the element. Following this publication, (Fe)-EDTA complexes were rapidly recognized to give generally improved growth of all types of plant cultures. EDTA has now become almost a standard medium component and is generally preferred to other alternative chelating agents.

Preparation and use

(Fe)-EDTA chelates for tissue cultures are prepared in either of two ways.

1. A ferric or ferrous salt is dissolved in water with EDTA and the solution is heated;
2. A ready-prepared salt of iron salt of EDTA is dissolved and heated.

Heating can take place during the preparation of chelate stock solutions, or during the autoclaving of a medium.

The form of iron complexed is invariably Fe(III). If iron has been provided from ferrous salts, it is oxidized during heating in aerated solutions. The rate of oxidation of the ferrous ion is enhanced in some complexes and retarded in others. That of Fe^{2+}-EDTA is extremely rapid. Only a small proportion of Fe^{2+} is likely to remain: its chelate with EDTA is much less stable than the Fe(III) complex. Iron is however thought to be absorbed into plants in the ferrous form. Uptake of iron from EDTA probably occurs when molecules of Fe(III)-chelate bind to the outer plasma membrane (the plasmalemma) of the cytoplasm, where Fe(III) is reduced to Fe(II) and freed from the chelate.

In most recent plant tissue culture work, EDTA has been added to media at an equimolar concentration with iron, where it will theoretically form a chelate with all the iron in solution. However, it has been found in practice that the Fe(III)-EDTA chelate, although stable at pH 2-3, is liable to lose some of its bound iron in culture media at higher pH levels; the displaced iron may form insoluble ferric hydroxides and iron phosphate. If this occurs, free EDTA will tend to form chelates with other metal ions in solution. Some micronutrients complexed with EDTA may then not be available to the plant tissues. Re-complexing may also happen if the EDTA to Fe ratio is increased by decreasing the amount of iron added to the medium. It is not possible to add very much more than 0.1 mM EDTA to culture media because the chelating agent can become toxic to some plants.

Hill-Cottingham and Lloyd-Jones (1961) showed that tomato plants absorbed iron from FeEDTA more rapidly than they absorbed EDTA itself, but concluded that both Fe and Fe-chelate were probably taken up. They postulated that EDTA liberated by the absorption of Fe, would chelate other metals in the nutrient solution. Teasdale (1987) calculated that in many media, nearly all the copper and zinc, and some manganese ions might be secondarily chelated, but it is unclear whether micronutrients in this form are freely available to plant tissues. One presumes they are, for deficiency symptoms are not reported from *in vitro* cultures.

Ambiguous descriptions

In many early papers on plant tissue culture, the authors of scientific papers have failed to describe which form of EDTA was used in experiments, or have ascribed weights to EDTA, which should refer to its hydrated sodium salts. Singh and Krikorian (1980) drew attention to this lack of precision. They assumed that in papers where Na_2EDTA is described as a medium constituent, it indicates the use of the anhydrous salt (which would give 11 mol/l excess of EDTA to iron, with unknown consequences). However, the disodium salt of EDTA is generally made as the dihydrate and this is the form which will almost invariably have been used, Na_2EDTA merely being a shorthand way of indicating the hydrated salt without being intended as a precise chemical formula.

Further confusion has arisen through workers using ready-prepared iron-EDTA salts in media without specifying the weight or molar concentration of actual Fe used. Mono-, di-, tri-, and tetra-sodium salts of EDTA are possible, each with different (and sometimes alternative) hydrates, so that when a research report states only that a certain weight of 'FeEDTA' was used, it is impossible to calculate the concentration of iron that was employed with any certainty.

The compound 'monosodium ferric EDTA' with the formula NaFeEDTA (no water of hydration) exists, and is nowadays commonly selected as a source of chelated iron. However in some papers 'NaEDTA' has been used as an abbreviation for some other form of iron-EDTA salt. For example the paper of Eeuwens (1976) describing Y3 medium, says that to incorporate 0.05 mM iron, 32.5 mg/l 'sodium ferric EDTA' was used. The weight required using a compound with the strict molecular formula NaFeEDTA would be 18.35 mg/l. Hackett (1970) employed 'Na_4FeEDTA'. Gamborg and Shyluk (1981) and Gamborg (1982) said that to prepare B5 or MS medium with 0.1 mM Fe, 43 mg/l of 'ferric EDTA' or 'Fe-versenate' (EDTA) should be weighed. The compound recommended in these papers was probably the $Na_2FeEDTA.2H_2O$ chelate as was the 'FeEDTA' (13% iron) employed by Davis *et al.*, (1977). It should be noted that NaFeEDTA is the only source of Na in MS medium apart from the contamination in the gelling agents.

Alternatives to EDTA

A few other chelating agents have been used in culture media in place of EDTA. The B5 medium of Gamborg *et al.*, (1968) was originally formulated with 28 mg/l of the iron chelate 'Sequestrene 330 Fe'. According to Heberle-Bors (1980), 'Sequestrene 330 Fe' is FeDTPA, containing 10% iron. This means that the concentration of Fe in B5 medium was originally 0.05 mM. Gamborg and Shyluk (1981) have proposed more recently that the level of Fe should be increased to 0.1 mM. B5 medium is now often used with 0.1 mM FeEDTA, but some researchers still prefer FeDTPA, for example used it in place of FeEDTA in Lloyd and McCown (1981) WPM medium for shoot culture of several woody plants.

Growth regulatory effects of chelating agents

Although most iron, previously complexed to chelating agents such as EDTA, EDDHA and DTPA is absorbed as uncomplexed ions by plant roots, there is evidence that the chelating agents themselves can be taken up into plant tissues. Chelating compounds such as EDTA, in low concentrations, exert growth effects on plants, which are similar to those produced by auxins. The effects include elongation of oat coleoptiles, and etiolated lupin hypocotyls, the promotion of leaf epinasty and the inhibition of root growth. Hypotheses put forward to explain these observations have included that:

1. Chelating agents act as auxin synergists by sequestering Ca from the cell wall;
2. The biological properties of the natural auxin IAA may be related to an ability to chelate ions; other chelating agents therefore mimic its action.

Burstrom (1960) noted that EDTA inhibited root growth in darkness (not in light) but that the growth inhibition could be overcome by addition of Fe^{3+} or several other metal ions. He recognized that reversal of EDTA action by a metal does not mean that the metal is physiologically active but that it might only release another cation, which had previously been made unavailable to the tissue by chelation.

Effects in tissue culture

Growth and morphogenesis in tissue cultures have been noted on several occasions to be influenced by chelating agents other than EDTA.

It has not always been clear whether the observed effects were caused by the chelation of metal ions, or by the chelating agent *per se*.

The growth rate of potato shoot tips was increased by 0.01-0.3 mg/1 8-hydroxyquinoline (8-HQ) when cultured on a medium which also contained EDTA, and more callus cultures of a haploid tobacco variety formed shoots in the absence of growth regulators when DHPTA was added to Kasperbauer and Reinert (1967) medium which normally contains 22.4 mg/l EDTA. The DHPTA appears to have been used in addition to the EDTA, not as a replacement, and was not effective on callus of a diploid tobacco. In the same experimental system, Fe-DHPTA and Fe-EDDHA were more effective in promoting shoot formation from the haploid-derived tissue than Fe with CDTA, citric acid or tartaric acid.

The inclusion of EDTA into a liquid nutrient medium caused the small aquatic plant *Lemna perpusilla* to flower only in short day conditions whereas normally the plants were day-neutral. In the related species *Wolffia microscopica*, plants did not flower unless EDTA was present in the medium, and then did so in response to short days. When, however, Maheshwari and Seth (1966) substituted Fe-EDDHA for EDTA and ferric citrate, they found that plants not only flowered more freely under short days, but also did so under long days. The physiological effect of EDTA and EDDHA as chelating agents was thus clearly different. This was again shown by Chopra and Rashid (1969) who found that the moss *Anoectangium thomsonii* did not form buds as other mosses do, when grown on a simple medium containing ferric citrate or Fe-EDTA, but did so when 5-20 mg/1 Fe-EDDHA was added to the medium instead. An optimum concentration was between 5 and 8 mg/l. Rashid also discovered that haploid embryoids developed more freely from *in vitro* cultures of *Atropa belladonna* pollen microspores when Fe-EDDHA was incorporated into the medium, rather than Fe-EDTA. Heberle-Bors (1980) did not obtain the same result, and found that FeEDTA was superior to Fe-EDDHA for the production of pollen plants from anthers of this species and of two Nicotianas. In tobacco, the production of haploid plants was greatest with FeEDTA, next best with FeDTPA, FeEGTA, FeEDDHA, and poorest with Fe citrate. Each complex was tested at or about the same iron concentration. Heberle-Bors also showed that chelating agents are differentially absorbed by activated charcoal. In tissue culture of rose, citrus and red raspberry, it is advantageous to use FeEDDHA rather than FeEDTA.

Toxicity caused by chelating agents

Although low concentrations of EDTA markedly stimulate the growth of whole plants in hydroponic cultures by making iron more readily available, the compound begins to be toxic at higher levels. By comparisons with observations on animal tissues, Weinstein *et al.*, (1951) suggested that toxicity arose through competition between EDTA and enzymes (and other physiologically-active complexes) in the plant, for metals essential to their activity. This will occur if the avidity of the chelating agent is greater than the metal binding capacity of proteins on the surface of cells.

Toxicity can also occur in *in vitro* cultures. Legrand (1975) found that an optimum rate of adventitious shoot initiation occurred in endive leaf segments when only 7.5 mg/l EDTA (one fifth the concentration used in MS medium) was employed. In these circumstances, higher levels of EDTA were clearly inhibitory and more than 55 mg/l prevented shoot formation. Dalton *et al.*, (1983) found that 0.3 mM EDTA (compared to the 0.1 mM in MS medium) reduced the growth rate of *Ocimum* cell suspensions.

Flower buds of *Begonia franconis* died within a few days if cultured with a high level of FeEDTA (1-1.5 mM, i.e. 10-15 times the normal level) together with 0.4-1.6 mM $H_2PO_4^-$. Berghoef and Bruinsma (1979a) thought that Fe^{3+} released from the FeEDTA complex, had precipitated the phosphate. Necrosis was avoided by increasing $H_2PO_4^-$ concentration to 6.4 mM.

Tissues may be damaged by culture in media containing synthetic chelating agents where the pH approaches neutrality, because at these pH levels, EDTA and EGTA have been shown to remove calcium ions from the membranes of mitochondria and this inhibits NAD(P)H oxidation and respiration. Chelating agents have been found to inhibit the action of the growth substance ethylene and are thought to do so by sequestering Cu ions within plant tissues, thereby interfering with the synthesis or action of a Cu-containing enzyme responsible for ethylene metabolism. EDTA can also inhibit the activity of plant polyphenol oxidase enzymes *in vitro* and Smith (1968) thought that this might occur because EDTA made Cu ions less available for enzyme incorporation, when he found the chelating agent was able to prevent the blackening of freshly-isolated *Carex flacca* shoot tips. Several oxidative reactions are also biochemically catalyzed by ions such as Cu^{2+}, Co^{2+} and Zn^{2+}, and where this is the case chelating agents such as EDTA and CDTA are inhibitory.

ORGANIC SUPPLEMENTS

Growth and morphogenesis of plant tissue cultures can be improved by small amounts of some organic nutrients. These are mainly vitamins (including some substances that are not strictly animal vitamins), amino acids and certain undefined supplements. The amount of these substances required for successful culture varies with the species and genotype, and is probably a reflection of the synthetic capacity of the explant.

Vitamins

Vitamins are compounds required by animals in very small amounts as necessary ancillary food factors. Absence from the diet leads to abnormal growth and development and an unhealthy condition. Many of the same substances are also needed by plant cells as essential intermediates or metabolic catalysts, but intact plants, unlike animals, are able to produce their own requirements. Cultured plant cells and tissues can however become deficient in some factors; growth and survival is then improved by their addition to the culture medium.

In early work, the requirements of tissue cultures for trace amounts of certain organic substances were satisfied by "*undefined*" supplements such as fruit juices, coconut milk, yeast or malt extracts and hydrolysed casein. These supplements can contribute vitamins, amino acids and growth regulants to a culture medium. The use of undefined supplements has declined as the need for specific organic compounds has been defined, and these have become listed in catalogues as pure chemicals.

Development of Vitamin Mixtures

The vitamins most frequently used in plant tissue culture media are thiamine (Vit. B_1), nicotinic acid (niacin) and pyridoxine (Vit. B_6) and apart from these three compounds, and myo-inositol, there is little common agreement about which other vitamins are really essential.

The advantage of adding thiamine was discovered almost simultaneously by Bonner (1937, 1938), Robbins and Bartley (1937) and White (1937). Nicotinic acid and pyridoxine appear, in addition to thiamine, in media published by Bonner (1940), Gautheret (1942) and White (1943b); this was following the findings of Bonner and Devirian (1939) that nicotinic acid improved the growth of isolated roots of tomato, pea and radish; and the papers of Robbins and Schmidt (1939a,b) which indicated that pyridoxine was also required for tomato root culture. These four vitamins; myo-inositol, thiamine, nicotinic acid, and pyridoxine are ingredients of Murashige and Skoog (1962) medium and have been used in varying proportions for the culture of tissues of many plant species. However, unless there has been research on the requirements of a particular plant tissue or organ, it is not possible to conclude that all the vitamins which have been used in a particular experiment were essential.

The requirements of cells for added vitamins vary according to the nature of the plant and the type of culture. Welander (1977) found that Nitsch and Nitsch (1965) vitamins were not necessary, or were even inhibitory to direct shoot formation on petiole explants of *Begonia* x *hiemalis*. Roest and Bokelmann (1975) on the other hand, obtained increased shoot formation on *Chrysanthemum* pedicels when MS vitamins were present. Callus of *Pinus strobus* grew best when the level of inositol in MS medium was reduced to 50 mg/l whereas that of *P. echinata* proliferated most rapidly when no inositol was present.

Research workers often tend to adopt a '*belt and braces*' attitude to minor media components, and add unusual supplements just to ensure that there is no missing factor which will limit the success of their experiment. Sometimes complex mixtures of as many as nine or ten vitamins have been employed.

Experimentation often shows that some vitamins can be omitted from recommended media. Although four vitamins were used in MS medium, later work at Professor Skoog's laboratory showed that the optimum rate of growth of tobacco callus tissue on MS salts required the addition of only myo-inositol and thiamine. The level of thiamine was increased four-fold over that used by Murashige and Skoog (1962), but nicotinic acid, pyridoxine and glycine (amino acid) were unnecessary. A similar simplification of the MS vitamins was made by Earle and Torrey (1965) for the culture of *Convolvulus* callus.

Soczck and Hempel (1988) found that in the medium of Murashige *et al.* (1974) devised for the shoot culture of *Gerbera jamesonii*, thiamine, pyridoxine and inositol could be omitted without any reduction in the rate of shoot multiplication of their local cultivars. Ishihara and Katano (1982) found that *Malus* shoot cultures could be grown on MS salts alone, and that inositol and thiamine were largely unnecessary.

Specific Compounds

Myo-inositol

Myo-inositol (also sometimes described as *meso*-inositol or *i*-inositol) is the only one of the nine theoretical stereoisomers of inositol which has significant biological importance. Medically it has been classed as a member of the Vitamin B complex and is required for the growth of yeast and many mammalian cells in tissue culture. Rats and mice require it for hair growth and can develop dermatitis when it is not in the diet. *Myo*-inositol has been classed as a plant '*vitamin*', but note that some authors think that it should be regarded as a supplementary carbohydrate, although it does not contribute to carbohydrate utilization as an energy source or as an osmoticum.

Historical use in tissue cultures

Myo-inositol was first shown by Jacquiot (1951) to favour bud formation by elm cambial tissue when supplied at 20-1000 mg/l. Necrosis was retarded, though the proliferation of the callus was not promoted. *Myo*-inositol at 100 mg/l was also used by Morel and Wetmore (1951) in combination with six other vitamins for the culture of callus from the monocotyledon *Amorphophallus rivieri* (Araceae). Bud initials appeared on some cultures and both roots and buds on others according to the concentration of auxin employed. The vitamin was adopted by both Wood and Braun (1961) and Murashige and Skoog (1962) in combination with thiamine, nicotinic acid and pyridoxine in their preferred media fur the culture of *Catharanthus roseus* and *Nicotiana tabacum* respectively. Many other workers have since included it in culture media with favourable results on the rate of callus growth or the induction of morphogenesis. Letham (1966) found that *myo*-inositol interacted with cytokinin to promote cell division in carrot phloem explants.

Occurrence and biochemistry

Part of the growth promoting property of coconut milk is due to its myo-inositol content. Coconut milk also contains *scyllo*-inositol. This can also promote growth but to a smaller extent than the *myo*-isomer. Inositol is a constituent of yeast extract and small quantities may also be contained in commercial agar. *Myo*-inositol is a natural constituent of plants and much of it is often incorporated into phosphatidyl-inositol which may be an important factor in the functioning of membranes.

The phosphatidylinositol cycle controls various cellular responses in animal cells and yeasts, but evidence of it playing a similar role in plants is only just being accumulated. Enzymes which are thought to be involved in the cycle have been observed to have activities in plants and lithium chloride (which inhibits myo-inositol-l-phosphatase and decreases the cycle) inhibits callus formation in *Brassica oleracea*, and callus growth in *Amaranthus paniculatus*. In both plants the inhibition is reversed by *myo*-inositol.

As the *myo*-inositol molecule has six hydroxyl units, it can react with up to six acid molecules forming various esters. It appears that inositol phosphates act as second messengers to the primary action of auxin in plants: phytic acid (inositol hexa-phosphate) is one of these. Added to culture media it can promote tissue growth if it can serve as a source of inositol. In some species, auxin can be stored and may be transported as IAA-myo-inositol ester. *o*-Methyl-inositol is present in quite large quantities in legumes; inositol methyl ethers are known to occur in plants of several other families, although their function is unknown.

The stimulatory effect of myo-inositol in plant cultures probably arises partly from the participation of the compound in biosynthetic pathways leading to the formation of the pectin and hemicelluloses needed in cell walls and may have a role in the uptake and utilization of ions. In the experiments of Staudt (1984) mentioned below, when the PO_4^{3-} content of the medium was raised to 4.41 mM, the rate of callus growth of cv. 'Aris' was progressively enhanced as the *myo*-inositol in the medium was put up to 4000 mg/l. This result seems to stress the importance of inositol-containing phospholipids for growth.

Activity in tissue cultures

Cultured plant tissues vary in their capacity for myo-inositol biosynthesis. Intact shoots are usually able to produce their own requirements, but although many unorganized tissues are able to grow slowly without the vitamin being added to the medium the addition of a small quantity is frequently found to stimulate cell division. The compound has been discovered to be essential to some plants. In the opinion of Kaul and Sabharwal (1975) this includes all monocotyledons, the media for which, if they do not contain inositol, need to be complemented with coconut milk, or yeast extract.

Fraxinus pennsylvanica callus had an absolute requirement for 10 mg/l myo-inositol to achieve maximum growth; higher levels, up to 250 mg/l had no further effect on fresh or dry weight yields. The formation of shoot buds on callus of *Haworthia* spp was shown to be dependent on the availability of myo-inositol. In a revised Linsmaier and Skoog (1965) medium (containing 1.84 mM PO_4^{3-}), callus tissue of *Vitis vinifera* cv 'Muller-Thurgau' did not require myo-inositol for growth, but that of *Vitis vinifera* x *V. riparia* cv. 'Aris' was dependent on it and the rate of growth increased as the level of myo-inositol was increased up to 250 mg/l. Gupta *et al.* (1988) found that it was essential to add 5 g/l myo-inositol to Gupta and Durzan (1985) DCR-1 medium to induce embryogenesis (embryonal suspensor masses) from female gametophyte tissue of *Pseudotsuga menziesii* and *Pinus taeda*. The concentration necessary seems insufficient to have acted as an osmotic stimulus. *myo*-Inositol reduced the rate of proliferation in shoot cultures of *Euphorbia fulgens*.

Thiamine

Thiamine (Vit. B_1, aneurine) in the form of thiamine pyrophosphate, is an essential co-factor in carbohydrate metabolism and is directly involved in the biosynthesis of some amino acids. It has been added to plant culture media more frequently than any other vitamin. Tissues of most plants seem to require it for growth, the need becoming more apparent with consecutive passages, but some cultured cells are self sufficient. The maize suspension cultures of Polikarpochkina *et al.* (1979) showed much less growth in passage 2, and died in the third passage when thiamine was omitted from the medium.

MS medium contains 0.3 μM thiamine. That this may not be sufficient to obtain optimum results from some cultures is illustrated by the results of Barwale *et al.* (1986): increasing the concentration of thiamine-HCI in MS medium to 5 μM, increased the frequency with which zygotic embryos of *Glycine max* formed somatic embryos from 33% to 58%. Adding 30 μM nicotinic acid (normally 4 μM) improved the occurrence of embryogenesis even further to 76%. Thiamine was found to be essential for stimulating embryogenic callus induction in *Zoysia japonica*, a warm season turf grass from Japan. It has also been shown to stimulate adventitious rooting of *Taxus* spp.

There can be an interaction between thiamine and cytokinin growth regulators. Digby and Skoog (1966) discovered that normal callus cultures of tobacco produced an adequate level of thiamine to support growth providing a relatively high level of kinetin (ca. 1 mg/l) was added to the medium, but the tissue failed to grow when moved to a medium with less added kinetin unless thiamine was provided.

Sometimes a change from a thiamine-requiring to a thiamine-sufficient state occurs during culture. In rice callus, thiamine influenced morphogenesis in a way that depended on which state the cells were in. Presence of the vitamin in a pre-culture (Stage I) medium caused thiamine-sufficient callus to form root primordia on an induction (Stage II) medium, but suppressed the stimulating effect of kinetin on Stage II shoot formation in thiamine-requiring callus. It was essential to omit thiamine from the Stage I medium to induce thiamine-sufficient callus to produce shoots at Stage II.

Other Vitamins

Pantothenic acid

Pantothenic acid plays an important role in the growth of certain tissues. It favoured callus production by hawthorn stem fragments and stimulated tissue proliferation in willow and black henbane. However, pantothenic acid showed no effects with carrot, vine and Virginia creeper tissues which synthesize it in significant amounts (ca. 1 μg/ml).

Vitamin C

The effect of Vitamin C (L-ascorbic acid) as a component of culture media will be discussed elsewhere in this book. The compound is also used during explant isolation and to prevent blackening. Besides, its role as an antioxidant, ascorbic acid is involved in cell division and elongation, e.g., in tobacco cells. Ascorbic acid (4-8 $\times$ 10^{-4} M) also enhanced shoot formation in both young and old tobacco callus. It speeded up the shoot-forming process, and completely reversed the inhibition of shoot formation by gibberellic acid in young callus, but was less effective in old callus. Clearly its action here was not as a vitamin.

Vitamin D

Some vitamins in the D group, notably vitamin D_2 and D_3 can have a growth regulatory effect on plant tissue cultures.

Other vitamins

Evidence has been obtained that folic acid slows tissue proliferation in the dark, while enhancing it in the light. This is probably because it is hydrolysed in the light to *p*-aminobenzoic acid (PAB). In the presence of auxin, PAB has been shown to have a weak growth-stimulatory effect on cultured plant tissues.

Riboflavin which is a component of some vitamin mixtures, has been found to inhibit callus formation but it may improve the growth and quality of shoots. Suppression of callus growth can mean that the vitamin may either inhibit or stimulate root formation on cuttings. Riboflavin has been shown to stimulate adventitious rooting on shoots of *Carica papaya*, apple shoots and *Eucalyptus globulus*. It also enhances embryogenic callus induction in *Zoysia japonica* in association with cytokinins and thiamine.

Glycine is occasionally described as a vitamin in plant tissue cultures: its use has been described in the section on amino acids.

Adenine

Adenine (or adenine sulphate) has been widely used in tissue culture media, but because it mainly gives rise to effects which are similar to those produced by cytokinins.

Undefined Supplements

Many undefined supplements were employed in early tissue culture media. Their use has slowly declined as the balance between inorganic salts has been improved, and as the effect of amino acids and growth substances has become better understood. Nevertheless several supplements of uncertain and variable composition are still in common use.

The first successful cultures of plant tissue involved the use of yeast extract. Other undefined additions made to plant tissue culture media have been:

1. Meat, malt and yeast extracts and fibrin digest;
2. Juices, pulps and extracts from various fruits, including those from bananas and tomatoes;
3. The fluids which nourish immature zygotic embryos;
4. Extracts of seedlings or plant leaves;
5. The extract of boiled potatoes and corn steep liquor;
6. Plant sap or the extract of roots or rhizomes. Plant roots are thought to be the main site of cytokinin synthesis in plants;
7. Protein (usually casein) hydrolysates (containing a mixture of all the amino acids present in the original protein). Casein hydrolysates are sometimes termed casamino acids.

Many of these amendments can be a source of amino acids, peptides, fatty acids, carbohydrates, vitamins and plant growth substances in different concentrations.

Yeast Extract

Yeast extract (YE) is used less as an ingredient of plant media nowadays than in former times, when it was added as a source of amino acids and vitamins, especially inositol and thiamine (Vitamin B_1). In a medium consisting only of macro- and micro-nutrients, the provision of yeast extract was often found to be essential for tissue growth. The vitamin content of yeast extract distinguishes it from casein hydrolysate (CH) so that in such media CH or amino acids alone, could not be substituted for YE. It was soon found that amino acids such as glycine, lysine and arginine, and vitamins such as thiamine and nicotinic acid, could serve as replacements for YE, for example in the growth of tomato roots, or sugar cane cell suspensions.

The percentage of amino acids in a typical yeast extract is high (e.g. 7% amino nitrogen), but there is less glutamic acid than in casein or other protein hydrolysate. Malt extract contains little nitrogen (ca. 0.5% in total).

Yeast extract has been typically added to media in concentrations of 0.1-1 g/l; occasionally 5, 10 and even 20 g/l have been included. It normally only enhances growth in media containing relatively low concentrations of nitrogen, or where vitamins are lacking. Addition of 125-5000 mg/l YE to MS medium completely inhibited the growth of green callus of 5 different plants whereas small quantities added to Vasil and Hildebrandt (1966) THS medium (which contained 0.6 times the quantity of NO_3^- and NH_4^+ ions and unlike MS did not contain nicotinic acid or pyridoxine) gave more vigorous growth of carrot, endive and lettuce callus than occurred on MS. There was still no growth of parsley and tomato callus on THS medium: these tissues only grew well on unmodified MS.

Stage I media are sometimes fortified with yeast extract to reveal the presence of micro-organisms which may have escaped decontamination procedures: it is then omitted at later stages of culture.

Yeast extract has been shown to have some unusual properties which may relate to its amino acid content. It elicits phytoalexin accumulation in several plant species and in *Glycyrrhiza echinata* suspensions it stimulated chalcone synthase activity leading to the formation of narengin. It also stimulated furomocoumarin production in *Glehnia littoralis* cell suspensions. On Monnier (1976, 1978) medium 1 g/l yeast extract was found to inhibit the growth of immature zygotic embryos of *Linum*, an effect which, when 0.05 mg/l BAP and 400 mg/l glutamine were added, induced the direct formation of adventitious embryos.

Yeast extract is now purchased directly from chemical suppliers. In the 1930s and 1940s it was prepared in the laboratory. Brink *et al.* (1944) macerated yeast in water which was then boiled for 30 minutes and, after cooling, the starchy material was removed by centrifugation. However, Robbins and Bartley (1937) found that the active components of yeast could be extracted with 80% ethanol.

Potato Extract

Workers in China found that there was a sharp increase in the number of pollen plants produced from wheat anthers when they were cultured on an agar solidified medium containing only an extract of boiled potatoes, 0.1 mM FeEDTA, 9% sucrose and growth regulators. Potato extract alone or potato extract combined with components of conventional culture media has since been found to provide a useful medium for the anther culture of wheat and some other cereal plants. For example, the potato medium was found to be better for the anther culture of spring wheat than the synthetic (N6) medium. Sopory *et al.* (1978) obtained the initiation of embryogenesis from potato anthers on potato extract alone and Lichter (1981) found it beneficial to add 2.5 g/l Difco potato extract to a medium for *Brassica napus* anther culture, but it was omitted by Chuong and Beversdorf (1985) when they repeated this work. We are not aware of potato extract being added to media for micropropagation, apart from occasional reports of its use for orchid propagation. Sagawa and Kunisaki (1982) supplemented 1 litre of Vacin and Went (1949) medium with the extract from 100g potatoes boiled for 5 minutes, and Harvais (1982) added 5% of an extract from 200g potatoes boiled in 1 litre water to his orchid medium. Of interest was the finding that potato juice treatment enabled *in vitro* cultures of *Doritaenopsis* (Orchidaceae) to recover from hyperhydricity.

Malt Extract

Although no longer commonly used, malt extract seems to play a specific role in cultures of *Citrus*. Malt extract, mainly a source of carbohydrates, was shown to initiate embryogenesis in nucellar explants. Several recent studies showed a role for the extract in the multiplication of *Citrus sinensis* somatic embryos, and in other *Citrus* spp., in the promotion of plantlet formation from somatic embryos derived from styles of different *Citrus* cultivars, and in somatic embryogenesis and plantlet regeneration from pistil thin cell layers of *Citrus*. Malt extract also promoted germination of early cotyledonary stage embryos arising from the *in vitro* rescue of zygotic embryos of sour orange. The extract is commercially available and used at a level of 0.5–1 g/l.

Banana Homogenate

Homogenized banana fruit is sometimes added to media for the culture of orchids and is often reported to promote growth. The reason for its stimulatory effect has not been explained. One suggestion mentioned earlier is that it might help to stabilize the pH of the medium. Pierik *et al.* (1988) found that it was slightly inhibitory to the germination of *Paphiopedilum ciliolare* seedlings but promoted the growth of seedlings once germination had taken place.

Fluids which Nourish Embryos

The liquid which is present in the embryo sac of immature fruits of *Aesculus* (e.g. *A. woerlitzensis*) and *Juglans regia* has been found to have a strong growth-promoting effect on some plant tissues cultured on simple media, although growth inhibition has occasionally been reported. Fluid from the immature female gametophyte of *Ginkgo biloba* and extracts from the female gametophyte of *Pseudotsuga menziesii* and immature *Zea mays* grains (less than two weeks after pollination) can have a similar effect. The most readily obtained fluid with this kind of activity is coconut milk (water).

Coconut Milk/Water

When added to a medium containing auxin, the liquid endosperm of *Cocos nucifera* fruits can induce plant cells to divide and grow rapidly. The fluid is most commonly referred to as coconut milk, although Tulecke *et al.* (1961) maintained that the correct English term is '*coconut water*', because the term coconut milk also describes the white liquid obtained by grating the solid white coconut endosperm (the 'meat') in water and this is not generally used in tissue culture media. However, in this section, both terms are used.

Coconut milk was first used in tissue cultures by Van Overbeek *et al.* (1941, 1942) who found that its addition to a culture medium was necessary for the development of very young embryos of *Datura stramonium*. Gautheret (1942) found that coconut milk could be used to initiate and maintain growth in tissue cultures of several plants, and Caplin and Steward (1948) showed that callus derived from phloem tissue explants of *Daucus carota* roots grew much more rapidly when 15% coconut milk was added to a medium containing IAA. Unlike other undefined supplements to culture media (such as yeast extract, malt extract and casein hydrolysate) coconut milk has proved harder to replace by fully defined media. The liquid has been found to be beneficial for inducing growth of both callus and suspension cultures and for the induction of morphogenesis. Although commercial plant tissue culture laboratories (particularly those in temperate countries) would endeavour not to use this ingredient on account of its cost, it is still frequently employed for special purposes in research.

It is possible to get callus growth on coconut milk alone, but normally it is added to a recognized medium. Effective stimulation only occurs when relatively large quantities are added to a medium; the incorporation of 10-15 percent by volume is quite usual. For instance, Burnet and Ibrahim (1973) found that 20% coconut milk (i.e.

one-fifth of the final volume of the medium) was required for the initiation and continued growth of callus tissue of various *Citrus* species in MS medium; Rangan (1974) has obtained improved growth of *Panicum miliaceum* in MS medium using 2,4-D in the presence of 15% coconut milk. By contrast, Vasil and co-workers needed to add only 5% coconut milk to MS medium to obtain somatic embryogenesis from cereal callus and suspension cultures.

Many workers try to avoid having to use coconut milk in their protocols. It is an undefined supplement whose composition can vary considerably. However, adding coconut milk to media often provides a simple way to obtain satisfactory growth or morphogenesis without the need to work out a suitably defined formulation. Suggestions that coconut milk is essential for a particular purpose need to be treated with some caution. For instance, in the culture of embryogenic callus from root and petiole explants of *Daucus carota*, coconut milk could be replaced satisfactorily either by adenine or kinetin, showing that it did not contribute any unique substances required for embryogenesis.

Preparation

Ready prepared coconut water (milk) can be purchased from some chemical suppliers, but the liquid from fresh nuts (obtained from the greengrocer) is usually perfectly adequate. One nut will usually yield at least 100 ml. The water is most simply drained from dehusked coconuts by drilling holes through two of the micropyles. Only normal uncontaminated water should be used and so nuts should be extracted one by one, and the liquid endosperm from each examined to ascertain that it is unfermented before addition to a bulk supply. Water from green but mature coconuts may contain slightly different quantities of substances to that in the nuts purchased in the local market and has been said to be a more effective stimulant in plant media than that from ripe fruits, but Morel and Wetmore (1951) found to the contrary. Tulecke *et al.* (1961) discovered that the water from highly immature coconuts contained smaller quantities of the substances normally present in mature nuts.

Use of Coconut water

Coconut water is usually strained through cloth and deproteinized by being heated to 80-100°C for about 10 minutes while being stirred. It is then allowed to settle and the supernatant is separated from the coagulated proteins by filtration through paper. The liquid is stored frozen at -20°C. Borkird and Sink (1983) did not boil the water from fresh ripe coconuts, but having filtered it through several layers of cheesecloth, adjusted the pH to 10 with 2 N NaOH and then kept it overnight at 4°C. The following day the pH was re-adjusted to 7.0 with 5 N HCl, and the preparation was refiltered before being stored frozen at -20°C. Some workers autoclave media containing coconut milk; others filter-sterilize coconut milk and add it to a medium after autoclaving has been carried out. Morel and Wetmore (1951) used filter sterilization, but found that the milk lost its potency if stored sterile (but presumably unfrozen) for 3 months. Street (1977) advocated autoclaving coconut milk after it had been boiled and filtered; it was then stored at -20°C until required.

Active ingredients

The remarkable growth stimulating property of coconut milk has led to attempts to isolate and identify the active principles. This has proved to be difficult because the fractions into which coconut milk has been separated each possess only a small proportion of the total activity and the different components appear to act synergistically. Substances so far identified include amino acids, organic acids, nucleic acids, purines, sugars, sugar alcohols, vitamins, growth substances and minerals. The variable nature of the product is illustrated in the table by the analytical results obtained by different authors.

Auxin activity

The liquid has been found to have some auxin activity which is increased by autoclaving, probably because any such growth substances exist in a bound form and are released by hydrolysis. But although coconut milk can stimulate the growth of some *in vitro* cultures in the absence of exogenous auxin, it normally contains little of this kind of growth regulator and an additional exogenous supply is generally required. In modern media, where organic compounds are often added in defined amounts, the main benefit from using coconut milk is almost certainly due to its providing highly active natural cytokinin growth substances.

Cytokinin activity

Coconut milk was shown to have cytokinin activity by Kuraishi and Okumura (1961) and recognized natural cytokinin substances have since been isolated (9-β-D-ribo-furanosyl zeatin; zeatin and several unidentified ones; N, N'-diphenyl urea) but the levels of these compounds in various samples of coconut milk have not been published. An unusual cytokinin-like growth promoter, 2-(3-methylbut-2-enylamino)-purin-6-one was isolated by Letham (1982).

Because coconut milk contains natural cytokinins, adding it to media often has the same effect as adding a recognized cytokinin. This means that a beneficial effect on growth or morphogenesis is often dependent on the presence of an auxin. Steward and Caplin (1951) showed that there was a synergistic action between 2,4-D and coconut milk in stimulating the growth of potato tuber tissue. Lin and Staba (1961) similarly found that coconut milk gave significantly improved callus growth on seedling explants of peppermint and spearmint initiated by 2,4-D, but only slightly improved the growth initiated by the auxin 2- BTOA (2-benziothiazoleoxyacetic acid). The occurrence of gibberellin-like substances in coconut milk has also been reported.

Suboptimum stimulation and inhibition

In cases where optimal concentrations of growth adjuvants have been determined, it has been found that the level of the same or analogous substances in coconut milk may be suboptimal. La Motte (1960) noted that 150 mg/l of tyrosine most effectively induced morphogenesis in tobacco callus cultures, but coconut milk added at 15% would provide only 0.96 mg/l of this substance. Fresh and autoclaved coconut milk from mature nuts has proved inhibitory to growth or morphogenesis in some instances. It is not known which ingredients cause the inhibition but the growth of cultured embryos seems particularly liable to be prevented, suggesting that the compound responsible might be a natural dormancy-inducing factor such as abscisic acid. Van Overbeck *et al.* found that a factor was present in coconut milk which was essential for the growth of *Datura stramonium* embryos, but that heating the milk or allowing it to stand could lead to the release of toxic substances. These could be removed by shaking with alcohols or ether or lead acetate precipitation. Duhamet and Mentzer (1955) isolated nine fractions of coconut milk by chromatography, and found one of these to be inhibitory to cultured crown gall tissues of black salsify when more than 10-20% coconut milk was incorporated into the medium. Norstog (1965) showed that autoclaved coconut milk could inhibit the growth of barley embryos but that filter-sterilized milk was stimulatory. Coconut water inhibited somatic embryo induction in *Pinus taeda* and both autoclaved or filter-sterilized coconut milk inhibited the growth of wheat embryo-shoot apices.

ORGANIC ACIDS

Organic acids can have three roles in plant culture media:

1. They may act as chelating agents, improving the availability of some micronutrients,

2. They can buffer the medium against pH change,
3. They may act as nutrients.

A beneficial effect is largely restricted to the acids of the Krebs' cycle. Dougall *et al.* (1979) found that 20 mM succinate, malate or fumarate supported maximum growth of wild carrot cells when the medium was initially adjusted to pH 4.5. Although 1 mM glutarate, adipate, pimelate, suberate, azelate or phthalate controlled the pH of the medium, little or no cell growth took place.

Use as Buffers

The addition of organic acids to plant media is not a recent development. Various authors have found that some organic acids and their sodium or potassium salts stabilize the pH of hydroponic solutions or *in vitro media*, although it must be admitted that they are not as effective as synthetic biological buffers in this respect. Norstog and Smith (1963) discovered that 100 mg/l malic acid acted as an effective buffering agent in their medium for barley embryo culture and also appeared to enhance growth in the presence of glutamine and alanine. Malic acid, now at 1000 mg/l was retained in the improved Norstog (1973) Barley II medium. In the experiments of Schenk and Hildebrandt (1972) low levels of citrate and succinate ions did not impede callus growth of a wide variety of plants and appeared to be stimulatory in some species. The acids were also effective buffers between pH 5 and pH 6, but autoclaving a medium containing sodium citrate or citric acid caused a substantial pH increase.

Complexing with metals

Divalent organic acids such is citric, maleic, malic and malonic (depending on species) are found in the xylem sap of plants, where together with amino acids they can complex with metal ions and assist their transport. These acids can also be secreted from cultured cells and tissues into the growth medium and will contribute to the conditioning effect. Ojima and Ohira (1980) discovered that malic and citric acids, released into the medium by rice cells during the latter half of a passage, were able to make unchelated ferric iron available, so correcting an iron deficiency.

Nutritional role

As explained already, adding Krebs' cycle organic acids to the medium can enhance the metabolism of NH_4^+. Gamborg and Shyluk (1970) found that some organic acids could promote ammonium utilization and the incorporation of small quantities of sodium pyruvate, citric, malic and fumaric acids into the medium, was one factor which enabled Kao and Michayluk (1975) to culture *Vicia hajastana* cells at low density. Their mixture of organic acid ions has been copied into many other media designed for protoplast culture. Cultures may not tolerate the addition of a large quantity of a free acid which will acidify the medium. For example, *Triticale* anther callus grew well on Chu *et al.* (1975) N6 medium supplemented with 35 mg/l of a mixture of sodium pyruvate, malic acid, fumaric acid, citric acid, but not when 100 mg/l was added. When organic anions are added to the medium from the sodium or potassium salts of an acid there are metallic cations to counterbalance the organic anions, and it seems to be possible to add larger quantities without toxicity. Five mM (1240 mg/l .$3H_2O$) potassium succinate enhanced the growth of cultured peach embryos, and adding 15 mM (4052 mg/l .$4H_2O$) sodium succinate to MS medium (while also increasing the sucrose content from 3% to 6%) increased the cell volume and dry weight of *Brassica nigra* suspensions by 2.7 times.

Some plants seem to derive nutritional benefit from the presence of one particular organic acid. Murashige and Tucker (1969) showed that orange juice added to a medium containing MS salts promoted the growth of *Citrus* albedo callus. Malic and other Krebs' cycle acids also have a similar effect; of these, citric acid produces the most pronounced growth stimulation. A concentration of up to 10.4 mM can be effective. Succulent plants, in particular those in the family Crassulaccae, such as *Bryophyllum* and *Kalanchoe* fix relatively large amounts of carbon dioxide during darkness, converting it into organic acids, of which malic acid is particularly important. The organic acids are metabolized during daylight hours. In such plants, malic acid might be expected to prove especially efficient in enhancing growth if added to a culture medium. Lassocinski (1985) has shown this to be the case in chlorophyll-deficient cacti of three genera. The addition of L-malic acid to the medium of Savage *et al.* (1979) markedly improved the rate of survival and vigour of small cacti or areoles.

Organic acid (citrate, lactate, succinate, tartrate, and oxalate) pretreatment of alfalfa callus dramatically decreased the growth of callus, but increased the subsequent yield of somatic embryos and embryo development, as well as conversion to plantlets. They suggested that the acids may act in the physiological selection for embryogenic callus, by inducing preferential growth of slower-growing-compact cell aggregates compared to the faster growing friable callus.

Sugars - Nutritional and Regulatory Effects

Carbohydrates play an important role in *in vitro* cultures as an energy and carbon source, as well as an osmotic agent. In addition, carbohydrate-modulated gene expression in plants is known. Plant gene responses to changing carbohydrate status can vary markedly. Some genes are induced, some are repressed, and others minimally affected. As in microorganisms, sugar-sensitive plant genes are part of an ancient system of cellular adjustment to critical nutrient availability. However, there is no evidence that this role of carbohydrate is important in normal growth and organized development in cell cultures.

Sugars as Energy Sources

Carbohydrate autotrophy

Only a limited number of plant cell lines have been isolated which are autotrophic when cultured *in vitro.* Autotrophic cells are capable of fully supplying their own carbohydrate needs by carbon dioxide assimilation during photosynthesis. Many autotrophic cultures have only been capable of relatively slow growth, especially in the ambient atmosphere where the concentration of carbon dioxide is low. However, since these early trials, very good progress is being made with photoautrophic shoot cultures and photo-autotrophic micropropagation is now possible. Success is dependent on enriching the CO_2 concentrations in the vessels during the photoperiod, reducing or eliminating sugar from the medium, and optimizing the *in vitro* environment.

Nevertheless, for the normal culture of either cells, tissues or organs, it is necessary to incorporate a carbon source into the medium. Sucrose is almost universally used for micropropagation purposes as it is so generally utilizable by tissue cultures. Refined white domestic sugar is sufficiently pure for most practical purposes. The presence of sucrose in tissue culture media specifically inhibits chlorophyll formation and photosynthesis making autotrophic growth less feasible.

Alternatives to Sucrose

Other Sugars

The selection of sucrose as the most suitable energy source for cultures follows many comparisons between possible alternatives. Some of the first work of this kind on the carbohydrate nutrition of plant tissue was done by Gautheret (1945) using normal carrot tissue. Sucrose was found to be the best source of carbon followed by glucose, maltose and raffinose; fructose was less effective and mannose and lactose were the least suitable. Sucrose has almost invariably been found to be the best carbohydrate; glucose is generally found to support growth equally well, and in a few plants it may result in better *in vitro* growth than sucrose, or promote organogenesis where sucrose will not; but being more expensive than sucrose, glucose will only be preferred for micropropagation where it produces clearly advantageous results.

Multiplication of *Alnus crispa*, *A. cordata* and *A. rubra* shoot cultures was best on glucose, while that of *A. glutinosa* was best on sucrose. Direct shoot formation from *Capsicum annum* leaf discs in a 16 h day required the presence of glucose. Glucose is required for the culture of isolated roots of wheat and some other monocotyledons.

Some other monosaccharides such as arabinose and xylose; disaccharides such as cellobiose, maltose and trehalose; and some polysaccharides; all of which are capable of being broken down to glucose and fructose, can also sometimes be used as partial replacements for sucrose. In *Phaseolus* callus, Jeffs and Northcote (1967) found that sucrose could be replaced by maltose and trehalose (all three sugars have an alpha-glucosyl radical at the non-reducing end), but not by glucose or fructose alone or in combination, or by several other different sugars. Galactose has been said to be toxic to most plant tissues; it inhibits the growth of orchids and other plants in concentrations as low as 0.01% (0.9 mM). However, cells can become adapted and grown on galactose, e.g., sugar cane cells. The key was the induction of the enzyme galactose kinase, which converts galactose to galactose-1-phosphate. More recently, other reports on galactose use have appeared. It promoted callus growth in rugosa rose, but inhibited somatic embryogenesis. Galactose promoted early somatic embryo maturation stages in European silver fir. When used instead of sucrose, it improved rooting of *Annona squamosa* microshoots. In addition, galactose has been found to reduce or overcome hyperhydricity in shoot cultures. Fructose has also been reported to be effective in preventing hyperhydricity.

There are some situations where fructose supports growth just as well as sucrose or glucose and occasionally it gives better results. Some orchid species have been reported to grow better on fructose than glucose. Fructose was the best sugar for the production of adventitious shoots from *Glycine max* cotyledonary nodes, especially if the concentration of nutrient salts supplied was inadequate. Shoot and leaf growth and axillary shoot formation in *Castanea* shoot cultures was stimulated when sucrose was replaced by 30 g/l fructose. The growth of basal callus was reduced and it was possible to propagate from mature explants of *C. crenata*, although this was not possible on the same medium supplemented with sucrose. However, fructose was reported to be toxic to carrot tissue if, as the sole source of carbon, it was autoclaved with White (1943a) A medium. When filter sterilized, fructose supported the growth of callus cultures which had a final weight 70% of those grown on sucrose.

Sucrose in culture media is usually hydrolysed totally, or partially, into the component monosaccharides glucose and fructose and so it is logical to compare the efficacy of combinations of these two sugars with that of sucrose. Kromer and Kukulczanka (1985) found that meristem tips of *Canna indica* survived better on a mixture of 25 g/l glucose plus 5 g/l fructose, than on 30 g/l sucrose. Germination of *Paphiopedilum* orchid seeds was best on a medium containing 5 g/l fructose plus 5 g/l glucose; a mixture of 7.5 g/l of each sugar was optimal for further growth of the seedlings. In spite of its rapid hydrolysis to glucose and fructose, sucrose appears to have a specific stimulatory effect on embryo development in Douglas fir, that was not observed when it was replaced by the monosaccharides.

The general superiority of sucrose over glucose for the culture of organized plant tissues such as isolated roots may be on account of the more effective translocation of sucrose to apical meristems. In addition, there could be an osmotic effect, because, from an equal weight of compound, a solution of glucose has almost twice the molarity of a sucrose solution, and will thus, in the absence of inversion of the disaccharide, induce a more negative water potential.

Maltose

Plant species vary in their ability to utilize unusual sugars. For instance, although Gautheret (1945) could grow carrot callus on maltose, Mathes *et al.* (1973) obtained only minimal growth of *Acer* tissue on media supplemented with this sugar. Similarly, growth of soybean tissue on maltose is normally very slow, but variant strains of cells have been selected which can utilize it, perhaps because the new genotypes possessed an improved capacity for its active transport. Later studies have given a more prominent role to maltose as a component of tissue culture media. Maltose serves as both a carbon source and as an osmoticum. Compared to sucrose there is a slower rate of extracellular hydrolysis, it is taken up more slowly, and hydrolysed intracellularly more slowly. Maltose led to a substantial increase in somatic embryos from *Petunia* anthers. It also led to an increase in callus induction and plantlet regeneration during *in vitro* androgenesis of hexaploid winter triticale and wheat. Maltose also increased callus induction in rice microspore culture, with an acceleration of initial cell divisions. For barley microspore culture, the inclusion of maltose led to a higher frequency of green plants. Maltose has been reported to equal or surpass sucrose in supporting embryogenesis in a number of species, including carrot, alfalfa, wild cherry, *Malus*, *Abies* and loblolly pine. The number of plants regenerated from *indica*, and *japonica* rice varieties was also greater when protoplasts were cultured with maltose rather than sucrose. Transfer from a medium containing sucrose or glucose to one supplemented with maltose has been used by Stuart *et al.* (1986) and Redenbaugh *et al.* (1987) to enhance the conversion of alfalfa embryos. Similarly, maltose led to a much higher germination rate from asparagus somatic embryos than sucrose.

Lactose

The disaccharide lactose has been detected in only a few plants. When added to tissue culture media it has been found to induce the activity of β-galactosidase enzyme which can be secreted into the medium. The hydrolysis of lactose to galactose and glucose then permits the growth of *Nemesia strumosa* and *Petunia hybrida* callus, cucumber suspensions, cotton callus and cell suspensions, and Japanese morning glory callus. The key to lactose utilization in Japanese morning glory was not only the extracellular hydrolysis of this disaccharide, but the induction of galactose kinase, which prevented the accumulation of toxic galactose. Rodriguez and Lorenzo Martin (1987) found that adding 30 g/l lactose to MS medium instead of sucrose increased the number of shoots produced by a *Musa accuminata* shoot culture, but no new shoots were produced on subsequent subculture, although they were when sucrose was present.

In addition to lactose, plant cells have been shown to become adapted and then to grow on other galactose-containing oligosaccharides, including melibiose, raffinose, and stachyose.

Corn syrups

Kinnersley and Henderson (1988) have shown that certain corn syrups can be used as carbon sources in plant culture media and that they may induce morphogenesis which is not provoked by supplementing with sucrose. Embryogenesis was induced in a 10-year old non-embryogenic cell line of *Daucus carota* and plantlets were obtained from *Nicotiana tabacum* anthers by using syrups. Those used contained a mixture of glucose, maltose, maltotriose and higher polysaccharides. Their stimulatory effect was reproduced by mixtures of maltose and glucose.

Sugar alcohols

Sugar alcohols were thought not usually to be metabolized by plant tissues and therefore unavailable as carbon sources. For this reason, mannitol and sorbitol have been frequently employed as osmotica to modify the water potential of a culture medium. In these circumstances, sufficient sucrose must also be present to supply the energy requirement of the tissues. Adding either mannitol or sorbitol to the medium may make boron unavailable.

Mannitol was found to be metabolized by *Fraxinus* tissues. Later, studies with carrot and tobacco suspensions and cotyledon cultures of radiata pine showed that although mannitol was taken up very slowly, it was readily metabolized. Thus, this sugar alcohol is only of value as a short-term osmotic agent. In contrast, sorbitol is readily taken up and metabolized in some species. It has been found to support the growth of apple callus and that of other rosaceous plants, occasionally giving rise to more vigorous growth than can be obtained on sucrose. The ability of Rosaceae to use sorbitol as a carbon source is reported to be variety dependent. Albrecht (1986) found that shoot cultures of one crabapple variety required sorbitol for growth and would not grow on sucrose; another benefited from being grown on a mixture of sorbitol and sucrose and the growth of a third suffered if any sucrose was replaced by sorbitol. The apple rootstock 'Ottawa 3' produced abnormal shoots on sorbitol. Evidence is accumulating to show that sugar alcohols generally exhibit non-osmotic roles in regulating morphogenesis and metabolism in plants that do not produce polyols as primary photosynthetic products. In addition to being metabolized to varying degrees in heterotrophic cultures, such as tobacco, maize, rice, citrus and chichory, sugar alcohols stimulate specific molecular and physiological responses, where they apparently act as chemical signals.

The cyclic hexahydric alcohol myo-inositol does not seem to provide a source of energy and its beneficial effect on the growth of cultured tissues when used as a supplementary nutrient must depend on its participation in biosynthetic pathways.

Starch

Cultured cells of a few plants are able to utilize starch in the growth medium and appear to do so by release of extracellular amylases. Growth rates of these cultures are increased by the addition of gibberellic acid, probably because it increases the synthesis or secretion of amylase enzymes.

Hydrolysis of Sucrose

The remainder of this section on sugars is devoted to the apparent effects of sucrose concentration on cell differentiation and morphogenesis. Reports on the subject should be tempered with the knowledge that some or all of the sucrose in the medium is liable to be broken down into its constituent hexose sugars, and that such inversion will also occur within plant tissues, where reducing sugar levels of at least 0.5 per cent are likely to occur.

Autoclaving

A partial hydrolysis of sucrose takes place during the autoclaving of media the extent being greater when the compound is dissolved together with other medium constituents than when it is autoclaved in pure aqueous solution. In a fungal medium, not dissimilar to a plant culture medium, Bretzloff (1954) found that sucrose inversion during autoclaving (15 min at 15 lbs/in^2) was dependent on pH in the following way:

pH 3.0	100%
pH 3.4	75%
pH 3.8	40%
pH 4.2	25%
pH 4.7	12.5%
pH 5.0	10%
pH 6.0	0%

These results suggest that the proportion of sucrose hydrolysed by autoclaving media at conventional pH levels (5.5-5.8) should be negligible. Most evidence suggests that this is not the case and that 10-15% sucrose can be converted into glucose and fructose. Cultures of some plants grow better in media containing autoclaved (rather than filter-sterilized) sucrose suggesting that the cells benefit from the availability of glucose and/or fructose. However, Nitsch and Nitsch (1956) noted that glucose only supported the growth of *Helianthus* callus if it had been autoclaved, and Romberger and Tabor (1971) that the growth of *Picea* shoot apices was less when the medium contained sucrose autoclaved separately in water (or sucrose autoclaved with only the organic constituents of the medium) than when all the constituents had been autoclaved together. They suggested that a stimulatory substance might be released when sugars are autoclaved with agar.

In other species there has been no difference between the growth of cultures supplied with autoclaved or filter sterilized sucrose or growth has been less on a medium containing autoclaved instead of filter sterilized sucrose.

Enzymatic breakdown

Sucrose in the medium is also inverted into monosaccharides during the *in vitro* culture of plant material. This occurs by the action of invertase located in the plant cell walls or by the release of extracellular enzyme. In most cultures, inversion of sucrose into glucose and fructose takes place in the medium; but, because the secretion of invertase enzymes varies, the degree to which it occurs differs from one kind of plant to another. After 28 days on a medium containing either 3 or 4% sucrose, single explants of *Hemerocallis* and *Delphinium* had used 20-30% of the sugar. Of that which remained in the medium which had supported *Hemerocallis*, about 45% was sucrose, while only 5% was sucrose in the media in which *Delphinium* had been grown. In both cases the rest of the sucrose had been inverted.

The sucrose-inverting capacity of tomato root cultures was greatest in media of pH 3.6-4.7. Activity sharply declined in less acid media. Helgeson *et al.* (1972) found that omission of IAA auxin from the medium in which tobacco callus was cultured, caused there to be a marked rise in reducing sugar due to the progressive hydrolysis of sucrose, both in the medium and in the tissue. A temporary increase in reducing sugars also occurred at the end of the lag phase when newly transferred callus pieces started to grow rapidly. It is interesting to note that cell wall invertase possessed catalytic activity *in situ,* whether or not tobacco tissue was grown on sucrose.

In cultures of some species, uptake of sugar may depend on the prior extracellular hydrolysis of sugar. This is the case in sugar cane; and possibly also in *Dendrobium* orchids and carrot. Nearly all the sucrose in suspension cultures of sugar cane and sugar beet was hydrolysed in 3 days and *Daucus carota* suspensions have been reported to hydrolyse all of the sucrose in the medium into the constituent hexoses within 3 days or within 24 h. However, most species can take up sucrose directly as was shown through studies with asymetrically labeled ^{14}C-sucrose, and metabolize it intracellularly. Within the cell, soluble invertase, sucrose phosphate synthetase and sucrose synthetase serve to hydrolyse sucrose. Thus, in *Acer* the soluble invertase activity paralleled growth rate, while in tobacco and Japanese morning glory sucrose synthetase was more important. In the last species, the change in activity of sucrose synthetase was greater than that of sucrose phosphate synthetase, an enzyme not extensively examined in cultured cells.

The growth of shoots from non-dormant buds of mulberry is not promoted by sucrose, only by maltose, glucose or fructose. Even though mulberry tissue hydrolysed sucrose into component monosaccharides, shoots did not develop. In the presence of 3% fructose, sucrose was actually inhibitory to shoot development at concentrations as low as 0.2%.

Uptake

The uptake of sugar molecules into plant tissues appears to be partly through passive permeation and partly through active transport. The extent of the two mechanisms may vary. Active uptake is associated with the withdrawal of protons (H^+) from the medium. Charge compensation is effected by the excretion of a cation (H^+ or K^+). Glucose was taken up preferentially by carrot suspensions during the first 7 days of a 14 day passage; fructose uptake followed during days 7-9. At concentrations below 200 mM, glucose was taken up more rapidly into strawberry fruit discs and protoplasts than either sucrose or fructose.

Effective Concentrations

In most of the comparisons between the nutritional capabilities of sugars discussed above, the criterion of excellence has been the most rapid growth of unorganized callus or suspension-cultured cells. For this purpose 2-4% sucrose w/v is usually optimal. Similar concentrations are also used in media employed for micropropagation, but laboratories probably pay insufficient attention to the effects of sucrose on morphogenesis and plantlet development. Sucrose levels in culture media which result in good callus growth may not be optimal for morphogenesis, and either lower or higher levels may be more effective.

The optimum concentration of sucrose to induce morphogenesis or growth differs between different genotypes, sometimes even between those which are closely related. For instance, Damiano *et al.* (1987) found that the concentration of sucrose necessary to produce the best rate of shoot proliferation in *Eucalyptus gunnii* shoot cultures varied between clones. The influence of sucrose concentration on direct shoot formation from *Chrysanthemum* explants varied with plant cultivar.

Experiments of Molnar (1988) have shown that the optimum level of sucrose may depend upon the other amendments added to a culture medium. The most rapid growth of *Brassica nigra* suspensions on one containing MS salts (but less iron and B5 vitamins) occurred when 2% sucrose was added. However, when it was supplemented with 1-4 g/l casein hydrolysate or a mixture of 3 defined amino acids, growth was increased on up to 6% sucrose. A similar result was obtained if, instead of the amino acids, 15 mM sodium succinate was added. There was an extended growth period and the harvested dry weight of the culture was 2.8 times that on the original medium with 2% sucrose.

The level of sucrose in the medium may have a direct effect on the type of morphogenesis. Thus, sucrose (87 mM) favored organogenesis, while a higher level (350 mM) favoured somatic embryogenesis from immature zygotic embryos of sunflower. *In vitro* minicrowns of asparagus developed short, thickened storage roots at high frequencies when the sucrose concentration in the medium was increased to 6%. Lower sucrose concentrations, even with the addition of non- or poorly metabolized carbohydrates, such as cellobiose, maltose, mannose, melibiose and sorbitol produced thin fibrous roots, indicating that the additional sucrose was nutritional rather than osmotic.

The respiration rate of cultured plant tissues rises as the concentration of added sucrose or glucose is increased. In wheat callus, it was found to reach a maximum when 90 g/l (0.263 M) was added to the medium, even though 20 g/l produced the highest rate of growth and number of adventitious shoots. The uptake of inorganic ions can be dependent on sugar concentration and the benefit of adding increased quantities of nutrients to a medium may not be apparent unless the amount of sugar is increased at the same time.

Cell differentiation

Formation of vascular elements

Although sugars are clearly involved in the differentiation of xylem and phloem elements in cultured cells, it is still uncertain whether they have a regulatory role apart from providing a carbon energy source necessary for active cell metabolism. Sucrose is generally required to be present in addition to IAA before tracheid elements are differentiated in tissue cultures. The number of both sieve and xylem elements formed (and possibly the proportion of each kind) depends on sucrose concentration. In *Helianthus tuberosus* tuber slices, although sucrose, glucose and trehalose were best for supporting cell division and tracheid formation, maltose was only a moderately effective carbon source. Shininger (1979) has concluded that only carbohydrates which enable significant cell division are capable of promoting tracheary element formation. The occurrence of lignin in cultured cells is not invariably associated with thickened cell walls. Sycamore suspension cultures produced large amounts of lignin when grown on a medium with abnormally high sucrose (more than 6%) and 2,4-D levels. It was deposited within the cells and released into the medium.

Chlorophyll formation

Levels of sucrose normally used to support the growth of tissue cultures are often inhibitory to chlorophyll synthesis but the degree of inhibition does vary according to the species of plant from which the tissue was derived. In experiments of Hildebrandt *et al.* (1963) and Fukami and Hildebrandt (1967) for example, carrot and rose callus had a high chlorophyll content on Hildebrandt *et al.* (1946) tobacco medium with 2-8% sucrose; but tissue of endive, lettuce and spinach only produced large amounts of the pigment on a medium with no added sucrose (although a small amount of sugar was probably supplied by 15-16% coconut milk).

Cymbidium protocorms contain high chlorophyll levels only if they are cultured on media containing 0.2-0.5% sucrose. Their degree of greening declines rapidly when they are grown on sucrose concentrations higher than this. Likewise, orchid protocorm-like bodies will not become green and cannot develop into plantlets if sucrose is present in the medium beyond the stage of their differentiation from the explants. Where added sucrose does reduce chlorophyll formation, it is thought that the synthesis of 5- aminolaevulinic acid (ALA - a precursor of the porphyrin molecules of which chlorophyll is composed) is reduced due to an inhibition of the activity of the enzyme ALA synthase. Sugars apart from sucrose are not inhibitory. It has been said that cells grown on sucrose for prolonged periods may permanently lose the ability to synthesize chlorophyll. Plastids become converted to amyloplasts packed with starch and this may change the expression of plastid DNA or result in a reduction of plastid RNA.

A small amount of photosynthesis may be carried out by cultured shoots providing they are not maintained on media containing a high concentration of sucrose. Photosynthesis increased in *Rosa* shoot cultures when they were grown initially on 20 or 40 g/l sucrose which was decreased to 10 g/l in successive subcultures. An increase in photosynthesis occurs when sucrose is omitted from the medium in which rooted plantlets are growing, but these treatments are not successful in ensuring a greater survival of plantlets when they are transferred *extra vitrum*. A more recent study also showed the relative contribution of autotrophic and heterotrophic carbon metabolism in cultured potato plants. With 8% sucrose in the medium 90% of the tissue carbon was of heterotrophic origin in light-grown plants; while on 3% sucrose, only 50% was of heterotrophic origin.

Starch Accumulation and Morphogenesis

Starch deposition preceeding morphogenesis

Cells of callus and suspension cultures commonly accumulate starch in their plastids and it is particularly prevalent in cells at the stationary phase. Starch in cells of rice suspensions had different chemical properties to that in the endosperm of seeds. In searching for features which might be related to later morphogenetic events in *Nicotiana* callus, Murashige and co-workers noticed that starch accumulated preferentially in cells sited where shoot primordia ultimately formed. The starch is produced from sucrose supplied in the culture medium. As a result of this work, it was suggested that starch accumulation might be a prerequisite of morphogenesis. In tobacco, starch presumably acts as a direct cellular reserve of the energy required for morphogenesis, because it disappears rapidly as meristenoids and shoot primordia are formed. Morphogenesis is an energy-demanding process and callus maintained on a shoot-inducing medium does have a greater respiration rate than similar tissue kept on a non-inductive medium. Organ-forming callus of tobacco has been found to accumulate starch prior to shoot or root formation, whereas callus not capable of morphogenesis did not do so.

Morphogenesis without starch deposition

Cells of other plants which become committed to initiate organs do not necessarily accumulate starch as a preliminary to morphogenesis and it seems likely that the occurrence of this phenomenon is species-related. The deposition of starch was observed as an early manifestation of organogenesis in *Pinus coulteri* embryos, but not in those of *Picea abies*. Although zygotic embryos of the latter species immediately began to accumulate starch (particularly in the chloroplasts of cells in the cortex) when they were placed on a medium containing sucrose. It was never observed in meristematic cells from which adventitious buds developed. However, if a major role for the accumulation of starch prior to the initiation of organized development is for energy production, this role would be satisfied by the lipid reserves in zygotic embryos of conifers. Indeed, the rapid and nearly linear degradation of triglycerides during the period of high respiration during shoot initiation in excised cotyledons of radiata pine would support this view.

Meristematic centres in bulb scales of *Nerine bowdenii* can be detected as groups of cells from which starch is absent. Starch was not accumulated in caulogenic callus of *Rosa persica* x *R. xanthina* initiated from recently-initiated shoot cultures, but cells did accumulate starch when the shoot forming capacity of the callus was lost after more than three passages. Callus derived from barley embryos was noted to accumulate starch very rapidly and this was accompanied by a reduction in osmotic pressure within the cells. Gibberellic acid, which in this plant could be used to induce shoot formation, brought about an increase in cell osmolarity.

There are some further examples where a diminution of the amount of stored carbohydrate in cultured tissues has restored or improved their organogenetic capacity. A salt-tolerant line of alfalfa cells which showed no ability for shoot regeneration after three and a half years in culture on 3% sucrose was induced to form shoots and plantlets by being cultured for one passage of 24 days on 1% sucrose, before being returned to a medium containing 3% sucrose and a high 2,4-D level. Cells which in 3% sucrose were full of starch became starch-depleted during culture on a lower sucrose level. The number of somatic embryos formed by embryogenic 'Shamouti' orange callus, was increased when sucrose was omitted from the medium for one passage, before being returned to Murashige and Tucker (1969) medium with 5-6% sucrose.

Unusual sugars

In some plants, unusual sugars are able to regulate morphogenesis and differentiation. Galactose stimulates embryogenesis in *Citrus* cultures and can enhance the maturation of alfalfa embryos. Callus of *Cucumis sativus* grew most rapidly on raffinose and was capable of forming roots when grown on this sugar; somatic embryos were only differentiated when the callus was cultured on sucrose (88-175 mM), but if a small amount of stachyose (0.3 mM) was added to 88 mM sucrose the callus produced adventitious shoots instead. Stachyose is the major translocated carbohydrate in cucurbits.

Osmotic Effect of Media Ingredients

Besides having a purely nutritive effect, solutions of inorganic salts and sugars, which compose tissue culture media, influence plant cell growth through their osmotic properties. A discussion is most conveniently accommodated at this point, as many of the papers published on the subject stress the osmotic effects of added sugars.

Osmotic and Water Potentials

Water movement into and out of a plant cell is governed by the relative concentrations of dissolved substances in the external and internal solutions, and by the pressure exerted by its restraining cell wall. The manner of defining the respective forces has changed in recent years, and as both old and new terminology are found in the tissue culture literature, the following brief description may assist the reader. More detailed explanations can be found in many text books on plant physiology.

In the older concept, cells were considered to take up water by suction induced by the osmotically active concentration of dissolved substances within the cell. The suction force or *suction pressure* (SP) was defined as that resulting from the osmotic pressure of the cell sap (OP_{cs}) minus the osmotic pressure of the external solution (OP_{ext}), and the pressure exerted on, and stretching the cell wall, *turgor pressure* (TP) - so called because it is at a maximum when the cells are turgid. This may be represented by the equations:

$$SP = (OP_{cs} - OP_{ext}) - TP \text{ or}$$
$$SP = OP_{cs} - (OP_{ext} + TP).$$

These definitions devised by botanists, are not satisfactory thermodynamically because water should be considered to move down an energy gradient, losing energy as it does so. The term water potential (Ψ - Greek capital letter psi) is now used, and solutions of compounds in water are said to exert an osmotic potential. As the potential of pure water is defined as being zero, and dissolved substances cause it to be reduced, solutions have osmotic potentials, Ψs (or $\Psi\pi$ - Greek small letter pi) which are negative in value. Water is said to move from a region of high potential (having a less negative value) to one that is lower (having a more negative value). Both osmotic pressure and osmotic potential are used as terms in tissue culture literature and are equivalent except that they are opposite in sign (i.e. an OP of +6 bar equals a Ψs of -6 bar). Thermodynamically, the cell's turgor pressure is defined as a positive pressure potential (Ψp). The water potential of a cell (Ψ cell) is then equal to the sum of its osmotic and pressure potentials plus the force holding water in microcapillaries or bound to the cell wall matrix (Ψm):

$$\Psi \text{ cell} = \Psi s + \Psi p + \Psi m$$

Modern statements of the older suction pressure concept are therefore that the difference in water potential (i.e. the direction of water movement) between a cell and solution outside is given by:

$$\Delta T = (\Psi s_{\text{inside cell}} - \Psi s_{\text{outside cell}}) - \Psi p \text{ or}$$

$$\Delta T = \Psi \text{ cell} - \Psi \pi_{\text{outside}}$$

and when $\Delta\Psi = 0$ (at equilibrium)

$$\Psi \pi_{\text{outside}} = \Psi \text{ cell}$$

The force of water movement between two cells, 'A' and 'B', of different water potential, is given by:

$$\Delta\Psi = \Psi \text{ cellA} - \Psi \text{ cellB}$$

the direction of movement being towards the more negative water potential.

The osmotic potential (pressure) of solutions is determined by their molar concentration and by temperature. The water potential of a plant tissue culture medium (Ψtcm) is equivalent to the osmotic potential of the dissolved compounds (Ψs). There is no pressure potential but, if they are added, substances such as agar and Gelrite contribute a matric potential (Ψm):

$$\Psi tcm = \Psi s + \Psi m$$

Osmotic pressure and water potential are measured in standard pressure units thus:

$$1 \text{ bar} = 0.987 \text{ atm}$$
$$= 10^6 \text{ dynes cm}^{-2}$$
$$= 10^5 \text{ Pa } (0.1 \text{ MPa} = 1 \text{ bar}).$$

Whereas molarity is defined as number of gram moles of a substance in one litre of a solution (i.e. one litre of solution requires less than one litre of solvent), molality is the number of gram moles of solute per kilogram of solvent, and thus, unlike osmotic potential (osmolality, measured in pressure units) is independent of temperature. It is therefore more convenient to give measurements of osmotic pressure in osmolality units. The osmole (Osm) is defined as:

The unit of the osmolality of a solution exerting an osmotic pressure equal to that of an ideal non-dissociating substance which has a concentration of one mole of solute per kilogram of solvent.

The osmolality of a very dilute solution of a substance which does not dissociate into ions, will be the same as its molality (i.e. g moles per kilogram of solvent). The osmolality of a weak solution of a salt, or salts, which has completely dissociated into ions, will equal that of the total molality of the ions.

The osmotic potential of dilute solutions approximates to Van't Hoff's equation:

$$\Psi = -cRT$$

where

c = concentration of solutes in mol/litre;

R = the gas constant and

T = temperature in °K

From the above equation, at 0°C one litre of a solution containing 1 mole of an undissociated compound, or 1 mole of ions, could be expected to have an osmolality of 1 Osm/kg, and an osmotic (water) potential of

$$\Psi = -1 \text{ (mole)} \times 0.082054 \text{ (atm /mole/°C)} \times 273.16 \text{ (°K)} = -22.414 \text{ atm}$$

Thus, although in practice the Van't Hoff equation must be corrected by the osmotic coefficient, Φ:

$$\Psi = -\Phi \text{ cRT}$$

it is possible to give approximate figures for converting osmolality into osmotic potential pressure units. This table can also be used to estimate the osmotic potential of non-dissociating molecules such as sugars or mannitol. Thus at 25°C, 30 g/l sucrose (molecular wt. 342.3) should exert an osmotic potential of

$$-2.4789 \times \frac{30}{342.3} = -0.217 \text{ MPa}$$

The observed potential is -0.223 Mpa.

A definition

The osmotic properties of solutions can be difficult to describe without confusion. In this book, the addition of solutes to a solvent (which makes the osmotic or water potential more negative, but makes the osmolality of the solution increase to a larger positive value), has been said to reduce (decrease) the osmotic potential of solutions. MS medium containing 3% sucrose (osmolality, ca. 186 mOsm/kg; ca. -461 kPa at 25°C) is thus described as having a lower water potential than the medium of White (1954) supplemented with 2% sucrose (osmolality, ca. 78 mOsm/kg; ca. –193 kPa at 25°C).

Osmotic Potential of Tissue Culture Media

Total osmotic potential of solutes

The approximate total osmotic potential of a medium due to dissolved substances, can be estimated from: $\Psi s = \Psi s_{macronutrients} + \Psi s_{sugars}$.

When 3% w/v sucrose is added to Murashige and Skoog (1962) medium, the osmolality of a filter sterilized preparation rises from

0.096 to 0.186 Osm/kg and at 25°C, the osmotic potential of the medium decreases from –0.237 to –0.460 MPa. Sugars are thus responsible for much of the osmotic potential of normal plant culture media. Even without any inversion to monosaccharides, the addition of 3% w/v sucrose is responsible for over four fifths of the total osmotic potential of White (1954) medium, 60% of that of Schenk and Hildebrandt (1972), and for just under one half that of MS.

Contribution of the gelling agent

The water potential of media solidified with gels is more negative than that of a liquid medium, due to their matric potential, but this component is probably relatively small. In the following sections, the matric potential of semi-solid media containing ca. 6 g/l agar has been assumed to be –0.01 MPa at 25°C, but as adding extra agar to media helps to prevent hyperhydricity, it is possible that this is an underestimate.

Contribution of nutrient salts

Inorganic salts dissociate into ions when they are dissolved in water, so that the water potential of their solutions (especially weak solutions) does not depend on the molality (or molarity) of undissociated compounds, but on the molality (or molarity) of their ions. Thus a solution of KC1 with a molality of 0.1, will have a theoretical osmolality of 0.2, because in solution it dissociates into 0.1 mole K^+ and 0.1 mole Cl^-. Osmolality of a solution of mixed salts is dependent on the total molality of ions in solution.

Dissociation may not be complete, especially when several different compounds are dissolved together as in plant culture media, which is a further reason why calculated predictions of water potential may be imprecise. In practice, osmotic potentials should be determined by actual measurement with an osmometer. Clearly though, osmotic potential of a culture medium is related to the concentration of solutes, particularly that of the macronutrients and sugar.

Of the inorganic salts in nutrient media, the macronutrients contribute most to the final osmotic (water) potential because of their greater concentration. The osmolality of these relatively dilute solutions is very similar to the total osmolarity of the constituent ions at 0°C, and can therefore be estimated from the total molarity of the macronutrient ions. Thus based on its macronutrient composition, a liquid Murashige and Skoog (1962) medium (without sugar) with a total macronutrient ion concentration of 95.75 mM, will have an osmolality of ca. 0.0958 osmoles (Osm) per kilogram of water solvent

(95.8 mOsm/kg), at 25°C, an osmotic potential of ca. –0.237 MPa (237 kPa). Estimates of osmolality derived in this way agree closely to actual measurements of osmolality or osmolarity for named media given in the papers of Yoshida *et al.* (1973), Kavi Kishor and Reddy (1986) and Lazzeri *et al.* (1988).

Contribution of sugars

Those of mannitol and sorbitol solutions of equivalent molarities will be approximately comparable. At concentrations up to 3% w/v, the osmolality of sucrose is close to molarity. It will be seen that the osmotic potential (in MPa) of sucrose solutions at 25°C can be roughly estimated by multiplying the % weight/volume concentration by –0.075: that of the monosaccharides fructose, glucose, mannitol and sorbitol (which have a molecular weight approximately 0.52 times that of sucrose), by multiplying by –0.14.

If any of the sucrose in a medium becomes hydrolysed into monosaccharides, the osmotic potential of the combined sugar components (sucrose + glucose + fructose), (Ψs_{sugars}), will be lower (more negative) than would be estimated. From this data it seems that 40-50% of the sucrose added to MS medium by Lazzeri *et al.* (1988) was broken down into monosaccharides during autoclaving. Hydrolysis of sucrose by plant-derived invertase enzymes will also have a similar effect on osmotic potential. In some suspension cultures, all sucrose remaining in the medium is inverted within 24 h. A fully inverted sucrose solution would have almost double the negative potential of the original solution, but as the appearance of glucose and fructose by enzymatic hydrolysis usually occurs concurrently with the uptake of sugars by the tissues, it will tend to have a stabilizing effect on osmotic potential during the passage of a culture. Reported measurements of the osmolality of MS medium containing 3% sucrose, after autoclaving, are:

230 mOsm/kg

240 ± 20 mOsm/kg (0.6-0.925% agar)

230 ± 50 mOsm/kg (0.2-0.4%)

Decreasing osmotic potential with other osmotica

By adding soluble substances in place of some of the sugar in a medium, it can be shown that sugars not only act as a carbohydrate source, but also as osmoregulants. Osmotica employed for the deliberate modification of osmotic potential, should be largely lacking other biological effects. Those most frequently selected are the sugar alcohols mannitol and sorbitol. It is assumed that plants that do not have a native pathway for sugar alcohol biosynthesis are also deficient in pathways to assimilate them. Sugar alcohols, though, are usually translocated, and may be metabolized and utilized to various degrees. Polyethylene glycol may be more helpful as an inert nonpenetrating osmolyte although it may contain toxic contaminants. Mannitol can easily penetrate cell walls, but the plasmalemma is considered to be relatively impermeable to it, whereas high-molecular-weight polyethylene glycol 4000 is too large to penetrate cell walls. Thus, a nonpenetrating osmolyte cannot penetrate into the plant cells, but inhibits water uptake. Sodium sulphate and sodium chloride have also been used in some experiments.

Effects and Uses of Osmolytes in Tissue Culture Media

Protoplast isolation and culture

The osmotic potential of a plant cell is counter-balanced by the pressure potential exerted by the cell wall. To safely remove the cell wall during protoplast isolation without damaging the plasma membrane, it has been found necessary to plasmolyse cells before wall-degrading enzymes are used. This is done by placing the cells in a solution of lower water potential than that of the cell.

Glucose, sucrose and especially mannitol and sorbitol, are usually added to protoplast isolation media for this purpose, either singly or in combination, at a total concentration of 0.35-0.7 M. These addenda are then retained in the subsequent protoplast culture medium, their concentration being progressively reduced as cell colonies start to grow. Tobacco cell suspensions take up only a very small amount of mannitol from solution and its effect as an osmotic agent appears to be exerted outside the cell. When protoplasts were isolated from *Pseudotsuga* and *Pinus* suspensions, which required a high concentration of inositol to induce embryogenesis, it was found to be essential to add 60 g/l (0.33 M) myo-inositol (plus 30 g/l sucrose, 20 g/l glucose and 10 g/l sorbitol) to the isolation and culture media. Mannitol and further amounts of sorbitol could not serve as substitutes.

Osmotic effects on growth

Solutions of different concentrations partly exert their effect on growth and morphogenesis by their nutritional value, and partly through their varying osmotic potential. Lapefla *et al* (1988) estimated that three quarters of the sucrose necessary to promote the optimum rate of direct adventitious shoot formation from *Digitalis obscura* hypocotyls, was required to supply energy, while the surplus regulated morphogenesis osmotically.

How osmotic potential influences cellular processes is still far from clear. Cells maintained in an environment with low (highly negative) osmotic potential, lose water and in consequence the water potential of the cell decreases. This brings about changes in metabolism and cells accumulate high levels of proline. The activity of the main respiratory pathway of cells (the cytochrome pathway) is reduced in conditions of osmotic stress, in favour of an alternative oxidase system. The increase of the concentration of osmolytes, may also result in high levels of the plant hormone abscisic acid, both *extra vitrum* and *in vitro*.

Equilibrium between the water potential of the medium and that of *Echinopsis* callus, only occurred when the callus was dead. Normally the water potential of the medium was greater so that water flowed into the callus. Clearly this situation could not occur in media which were too concentrated, and Cleland (1977) proposed that a critical water potential needs to be established within a cell before cell expansion and cell division, can occur. The osmotic concentration of culture media could therefore be expected to influence the rate of cell division or the success of morphogenesis of the cells or tissues they support. Both the inorganic and organic components will be contributory. The cells of many plants which are natives of sea-shores or deserts (e.g. many cacti) characteristically have a low water potential (Ψ cell) and in consequence may need to be cultured in media of relatively low (highly negative) osmotic potentials. Sucrose concentrations of 4.5-6% have sometimes been found to be beneficial for such plants.

Above normal sucrose concentrations can often be beneficial in media for anther culture, and for the culture of immature embryos.

If the osmotic potential of the medium does indeed influence the growth of tissue cultures, one might expect the sucrose concentration, which is optimal for growth, to vary from one medium to another, more sucrose being required in dilute media than in more concentrated ones. Evans *et al.* (1976) found this was so with cultures of soybean tissue. Maximum rates of callus growth were obtained in media containing either:

1. 50-75% of MS basal salts + 3-4% sucrose, or,
2. 75-100% of MS basal salts + 2% sucrose.

Similarly Yoshida *et al.* (1973) obtained equally good growth rates of *Nicotiana glutinosa* callus with nutrient media in which

1. The salts exerted –0.274 MPa and sucrose –0.223 MPa, or
2. The salts exerted –0.365 MPa and sucrose –0.091 MPa.

It was essential to add 60 g/l sucrose (i.e. $\Psi s_{sucrose}$ = –0.461 MPa at 25°C) to the 'MEDIUM' salts of de Fossard *et al.* (1974) (Ψ = –0.135 MPa) to obtain germination and seedling growth from immature zygotic embryos of tomato. However, if the 'HIGH' salts (Ψs_{medium} = –0.260 MPa) were used, 60 g/l sucrose gave only slightly better growth than 2.1 g/l ($\Psi s_{sucrose}$ = –0.156 MPa).

A detailed examination of osmotic effects of culture media on callus cultures was conducted by Kimball *et al.* (1975). Various organic substances were added to a modified Miller (1961) medium (which included 2% sucrose), to decrease the osmotic potential (Ψs) from –0.290 MPa. Surprisingly, the greatest callus growth was said to occur at –1.290 to –1.490 MPa (unusually low potentials which normally inhibit growth) in the presence of mannitol or sorbitol, and between –1.090 to –1.290 MPa when extra sucrose or glucose were added. On the standard medium, many cells of the callus were irregularly shaped; as the osmotic potential of the solution was decreased there were fewer irregularities and at about Ψs = –1.090 MPa all the cells were spherical. The percentage dry matter of cultures also increased as Ψs was decreased.

Doley and Leyton (1970) found that decreasing the water (osmotic) potential of half White (1963) medium by –0.100 or –0.200 MPa through adding more sucrose (and/or polyethylene glycol), caused the rate of callus growth from the cut ends of *Fraxinus* stem sections to be lower than on a standard medium. At the reduced water potential, callus had suberized surfaces and grew through the activity of a vascular cambium. It also contained more lignified xylem and sclereids. At each potential there was an optimal IAA concentration for xylem differentiation.

When the concentration of sucrose in a high salt medium such as MS is increased above 4-5 per cent, there begins to be a progressive inhibition of cell growth in many types of culture. This appears to be an osmotic effect because addition of other osmotically-active substances (such as mannitol and polyethyleneglycol) to the medium causes a similar response. Usually, high concentrations of sucrose are not toxic, at least not in the short term, and cell growth resumes when tissues or organs are transferred to media containing normal levels of sugar. Increase of Ψs is one method of extending the shelf life of cultures. Pech and Romani (1979) found that the addition of 0.4 M mannitol to MS medium (modified organics), was able to prevent the rapid cell lysis and death which occurred when 2,4-D was withdrawn from pear suspension cultures.

Decreasing the osmotic potential (usually by adding mannitol) together with lowering of the temperature has been used to reduce the growth rate for preservation of valuable genotypes *in vitro.* This has been reported, among others, for potato, *Dioscorea alata* and enset.

Tissue water content

The water content of cultured tissues decreases as the level of sucrose in the medium is increased. Isolated embryos of barley were grown by Dunwell (1981) on MS medium in sucrose concentrations up to 12 per cent. Dry weight increased as the sucrose concentration was raised to 6 or 9 per cent and shoot length of some varieties was also greater than on 3 per cent sucrose. Water content of the developing plantlets was inversely proportional to the sucrose level.

The dry weight of *Lilium auratum* bulbs and roots on MS medium increased as the sucrose concentration was augmented to 90 g/l but decreased dramatically with 150 g/l because growth was inhibited. The fresh weight/dry weight ratio of both kinds of tissue once again declined progressively as sucrose was added to the medium (0 - 150 g/l). In a medium containing 30 mg/l sucrose, the number and fresh weight of bulblets increased as MS salt strength was raised from one eighth to two times its normal concentration. An interaction between salt and sucrose concentrations was demonstrated in that optimum dry weight of bulbs could be obtained in either single strength MS + 120 g/l sucrose (osmolality, without sucrose inversion = ca. 492 mOsm/kg; Ψs = ca. −1.22 MPa), or double strength MS + 60 g/l sucrose (osmolality, without sucrose inversion = ca. 378 mOsm/kg; Ψs = ca. -0.94 MPa).

Morphogenesis

The osmotic effect of sucrose in culture solutions was well demonstrated by a series of experiments on tobacco callus by Brown *et al.,* (1979): rates of callus growth and shoot regeneration which were optimal on culture media containing 3 per cent sucrose, could be maintained when the sugar was replaced partially by osmotically equivalent levels of mannitol. The optimal (medium plus sucrose) here was between −0.4 and −0.6 MPa, and increasing sucrose levels above 3 per cent brought a progressive decrease in shoot regeneration. Similar results were obtained by Barg and Umiel (1977), but when they kept the osmotic potential of the culture solution roughly constant by additions of mannitol, the sucrose concentrations optimal for tobacco callus growth or morphogenesis were not the same.

Brown and Thorpe (1980) subsequently found that callus of *Nicotiana* capable of forming shoots, had a water potential (Ψ) of -0.8 MPa, while non-shoot-forming callus had a Ψ of –0.4 MPa. The two relationships were:

Shoot forming callus

$$\Psi cell = \Psi s + \Psi p + \Psi m$$
$$-0.8 = -1 + 0.4 + 0 \text{ MPa}$$

Non-shoot-forming callus

$$\Psi cell = \Psi s + \Psi p + \Psi m$$
$$-0.4 = -0 + 0.3 + 0 \text{ MPa}$$

Correct water potential

An optimum rate of growth and adventitious shoot formation of wheat callus occurred on MS medium containing 2% sucrose. Only a small number of shoots were produced on the medium supplemented with 1% sucrose, but if mannitol was added so that the total Ψcell = Ψs was the same as when 2% sucrose was present, the formation of adventitious shoots was stimulated. Very similar results were obtained by Laperia *et al.* (1988). A small number of shoot buds were produced from *Digitalis obscura* hypocotyls on MS medium containing 1% sucrose: more than twice as many if the medium contained 2% sucrose (total Ψs given as –0.336 MPa), or 1% sucrose plus mannitol to again give a total Ψs equal to –0.336 MPa.

Water potential can modify commitment

Morphogenesis can also be regulated by altering the water potential of media. Shepard and Totten (1977) found that very small (ca. 1-2 mm) calluses formed from potato mesophyll protoplasts were unable to survive in 1 or 2% sucrose, and the base of larger (5-10 mm) ones turned brown, while the upper portions turned green but formed roots and no shoots. The calli became fully green only on 0.2-0.5% sucrose. At these levels shoots were formed in the presence of 0.2-0.3 M mannitol. When the level of mannitol was reduced to 0.05 M, the proportion of calli differentiating shoots fell from 61% to 2%. The possibility of an osmotic affect was suggested because equimolar concentrations of myo-inositol were just as effective in promoting shoot regeneration.

Another way to modify morphogenesis is to increase the ionic concentration of the medium. Pith phloem callus of tobacco proliferates on Zapata *et al.* (1983) MY1 medium supplemented with 10^{-5} M IAA and 2.5 × 10^{-6} M kinetin, but forms shoots on Murashige *et al.* (1972) medium containing 10^{-5} M IAA and 10^{-5} M kinetin. These two media contain very similar macronutrients (total ionic concentrations, respectively 96 and 101 mM), yet adding 0.5-1.0% sodium sulphate (additional osmolality 89-130 mOsm/kg) decreased the shoot formation of callus grown on the shoot-forming medium, but increased it on the medium which previously only supported callus proliferation. Callus cultured in the presence of sodium sulphate retained its shoot-producing capacity over a long period, although the effect was not permanent. In these experiments shoot formation was also enhanced by sodium chloride and mannitol.

Increasing the level of sucrose from 1 to 3 per cent in MS medium containing 0.3 mg/l IAA, induced tobacco callus to form shoots, while further increasing it to 6 per cent resulted in root differentiation. The formation of adventitious shoots from *Nicotiana tabacum* pith callus is inhibited on a medium with MS salts if 10-15% sucrose is added. Preferential zones of cell division and meristemoids produced in 3% sucrose then become disorganized into parenchymatous tissue.

It should be noted that auxin which has a very important influence on the growth and morphogenesis of cultured plant cells, causes their osmotic potential to be altered. When tobacco callus is grown on a medium which promotes shoot regeneration, the cells have a greater osmotic pressure (or more negative water potential) than callus grown on a non-inductive medium.

Apogamous buds on ferns

Whittier and Steeves (1960) found a very clear effect of glucose concentration on the formation of apogamous buds on prothalli of the fern *Pteridium*. (Apogamous buds give rise to the leafy and spore-producing generation of the plant which has the haploid genetic constitution of the prothallus). Bud formation was greatest between 2-3% glucose (optimum 2.5%). Results which confirm this observation were obtained by Menon and Lal (1972) in the moss *Physcomitrium pyriforme*. Here apogamous sporophytes were formed most freely in low sucrose concentrations (0.5-2%) and low light conditions (50-100 lux), and were not produced at all when prothalli were cultured in 6% sucrose or high light (5000-6000 lux). Whittier and Steeves (loc. cit.) noted that they could not obtain the same rate of apogamous bud production by using 0.25% glucose plus mannitol or polyethyleneglycol; on the other hand, adding these osmotica to 2.5% glucose (so that the osmotic potential of the solution was equivalent to one with 8% sucrose) did reduce bud formation. It therefore appeared that the stimulatory effect of glucose on morphogenesis was mainly due to its action as a respiratory substrate, but that inhibition might be caused by an excessively depressed osmotic potential.

Differentiation of floral buds

Pieces of cold-stored chicory root were found by Margara and Rancillac (1966) to require more sucrose (up to 68 g/l; 199 mM) to form floral shoots than to produce vegetative shoots (as little as 17 g/l; 50 mM). Tran Thanh Van and co-workers have similarly shown that the specific formation of vegetative buds, flower buds, callus or roots by thin cell layers excised from tobacco stems, could be controlled by selecting appropriate concentrations of sugars and of auxin and cytokinin growth regulators.

Root formation and root growth

Media of small osmotic potential are usually employed for the induction and growth of roots on micropropagated shoots. High salt levels are frequently inhibitory to root initiation. Where such levels have been used for Stage II of shoot cultures, it is common to select a low salts medium (e.g. 1/4 or 1/2 MS), when detached shoots are required to be rooted at Stage III. By testing four concentrations of MS salts (quarter, half, three quarters and full strength) against four levels of sucrose (1, 2, 3 and 4%), Harris and Stevenson (1979) found that correct salt concentration (1/2 or 1/4 MS) was more important than sucrose concentration for root induction on grapevine cuttings *in vitro*. The benefit of low salt levels for root initiation may be due more to the need for a low nitrogen level, than for an increased osmotic potential. Dunstan (1982) showed that microcuttings of several tree-fruit rootstocks rooted best on MS salts, but that in media of these concentrations, the amount of added sugar was not critical, although it was essential for there to be some present. For the best rooting of *Castanea*, it was important to place shoots in Lloyd and McCown (1981) WPM medium containing 4% sucrose (Serres, 1988).

There are reports that an excessive sugar concentration can inhibit root formation. Green cotyledons of *Sinapis alba* and *Raphanus sativus* were found by Lovell *et al.* (1972) to form roots in 2% sucrose in the dark, but not in light of 5500 lux luminous intensity. In the light, rooting did occur if the explants were kept in water, or (to a lesser extent) if they were treated with DCMU (a chemical inhibitor of photosynthesis) before culture in 2% sucrose. The authors of this paper suggested that sugars were produced within the plant tissues during photosynthesis, which, added to the sucrose absorbed from the medium, provided too great a total sugar concentration for rooting. Rahman and Blake (1988) reached the same conclusion in experiments on *Artocarpus heterophyllus*. When shoots of this plant were kept on a rooting medium in the dark, the number and weight of roots formed on shoots, increased with the inclusion of up to 80 g/l sucrose. The optimum sucrose concentration was 40 g/l if the shoots were grown in the light.

Root formation on avocado cuttings in 0.3 MS salts was satisfactory with 1.5, 3 or 6% sucrose, and only reduced when 9% sucrose was added.

Although 4% (and occasionally 8%) sucrose has been used in media for isolated root culture, 2% has been used in the great majority of cases. In an investigation into the effects of sucrose concentration on the growth of tomato roots, Street and McGregor (1952) found that although sucrose concentrations of between 1.5 and 2.5% caused the same rate of increase of root fresh weight, 1.5% sucrose was optimal. It produced the best rate of growth of the main root axis, and the greatest number and total length of lateral roots.

Somatic embryogenesis

The osmotic potential of a medium can influence whether somatic embryogenesis can occur and can regulate the proper development of embryos. As will be shown below, a low osmotic potential is often favourable, but this is not always the case. For instance immature cotyledons of *Glycine max* produced somatic embryos on Phillips and Collins (1979) L2 medium containing less than 2% sucrose, but not if the concentration of sugar was increased above this level.

Placing tissues in solutions with high osmotic potential will cause cells to become plasmolyzed, leading to the breaking of cytoplasmic interconnections between adjacent cells (*plasmodesmata*). Wetherell (1984) has suggested that when cells and cell groups of higher plants are isolated by this process, they become enabled to develop independently, and express their totipotency. He pointed out that the isolation of cells of lower plants induces regeneration, and plasmolysis has long been known to initiate regeneration in multicellular algae, the leaves of mosses, fern prothallia and the gemmae of liverworts. Carrot cell cultures pre-plasmolysed for 45 min in 0.5-1.0 M sucrose or 1.0 M sorbitol gave rise to many more somatic embryos when incubated in Wetherell (1969) medium with 0.5 mg/l 2,4-D than if they had not been pre-treated in this way. Moreover embryo formation was more closely synchronized. Ikeda-Iwai *et al.* (2003) found that in *Arabidopsis* a 6-12 hour treatment with 0.7 M sucrose, sorbitol or mannitol resulted in somatic embryogenesis.

Callus derived from hypocotyls of *Albizia richardiana*, produced the greatest numbers of adventitious shoots on B5 medium containing 4% sucrose, but somatic embryos grew most readily when 2% sucrose was added. At least 1% sucrose was necessary for any kind of morphogenesis to take place. A similar result was obtained by Culafic *et al.* (1987) with callus from axillary buds of *Rumex acetosella*: adventitious shoots were produced on a medium containing MS salts and 2% sucrose (Ψs = ca. –0.39 MPa at 25°C), but embryogenesis occurred when the sucrose concentration was increased to 6% ($\Psi s_{sucrose}$ = –0.46 MPa, Ψs = ca. –0.70 MPa at 25°C) or if the medium was supplemented with 2% sucrose plus 21.3 g/l mannitol or sorbitol (which together have the same osmolality as 6% sucrose).

A low (highly negative) osmotic potential helps to induce somatic embryogenesis in some other plants. Adding 10-30 g/l sorbitol to Kumar *et al.* (1988) L-6 medium (total macronutient ions 64.26 mM; 20 g/l sucrose), caused there to be a high level of embryogenesis in *Vigna aconitifolia* suspensions and the capacity for embryogenesis to be retained in long-term cultures. The formation of somatic embryos in ovary callus of *Fuchsia hybrida* was accelerated by adding 5% sucrose to B5 medium, and the induction of embryogenic callus of *Euphorbia longan* required the culture of young leaflets on B5 medium with 6% sucrose.

There are exceptions, particularly with regard to embryo growth. The proportion of *Ipomoea batatas* somatic embryos forming shoots was greatest when a medium containing MS inorganics contained 1.6%, rather than 3% sucrose. Protocorm proliferation of orchids is most rapid when tissue is cultured in high concentrations of sucrose, but for plantlet growth, the level of sucrose must be reduced.

The induction of embryogenic callus from immature seed embryos of *Zea mays* was best on MS medium with 12% sucrose, and Ho and Vasil (1983) used 6-10% sucrose in MS medium to promote the formation of pro-embryoids from young leaves of *Saccharum officinarum*. However, in the experiments of Ahloowalia and Maretski (1983), somatic embryo formation from callus of this plant was best on MS medium with 3% sucrose, but growth of the embryos into complete plantlets required that the embryos should be cultured first on MS medium with 6% sucrose and then on MS with 3% sucrose.

Polyethylene glycol 4000 (PEG 4000) improves root and shoot emergence without limiting embryo histodifferentiation in soybean somatic embryos. Likewise in spruce, it was reported that polyethylene glycol might improve the quality of somatic embryos by promoting normal differentiation of the embryonic shoot and root. Non-penetrating osmotica like polyethylene glycol cannot enter plant cells, but restrict water uptake and provide a simulated drought stress during embryo development. A combination of ABA and an osmoticum prevents precocious germination in white spruce and allows embryo development to proceed. Advantageous effects of polyethylene glycol and ABA have been reported in a number of species (*Hevea brasiliensis*, *Picia abies*, white spruce, *Corydalis yanhusuo*, and *Panax ginseng*).

When cultured on Sears and Deckard (1982) medium, embryogenesis in callus initiated from immature embryos of 'Chinese Spring' and some other varieties of wheat was incomplete, because shoot apices germinated and grew before embryos had properly formed. More typical somatic embryos could be obtained by adding 40 mM sodium or potassium chloride to the medium. The salts had to be removed to allow plantlets to develop normally. Transferring somatic embryos to a medium of lower (more negative) water potential is often necessary to ensure their further growth and/or germination. High sucrose levels are often required in media for the culture of zygotic embryos if they are isolated when immature. The use of 50-120 g/l sucrose in media is then reported, the higher concentrations usually being added to very weak salt mixtures. Embryos which are more fully developed when excised, grow satisfactorily in a medium with 10-30 g/l sugar.

Storage organ formation

At high concentration, sucrose promotes the formation of tubers, bulbs and corms. This promotion might be mediated by ABA since osmotic stress induces ABA synthesis and ABA promotes bulb and tuber formation. The situation is, however, more complex. Suttle and Hultstrand (1994) did not find a reduction of tuber formation in potato by adding fluridone, an ABA-synthesis inhibitor, and Xu *et al.* (1998) did not observe an increased ABA-level at high sucrose concentration. So in potato, the effect of sucrose is not likely to be mediated by ABA. Exogenous ABA does not promote *Gladiolus* corm formation but bulb formation of lily was completely inhibited by fluridone and restored by simultaneous addition of ABA. To establish the effect of osmoticum directly, experiments have been carried out with addition of mannitol instead of sucrose. Mannitol did not promote corm production of

Gladiolus or bulb production in onion, but data for lily suggested that, although it had a toxic effect, in this plant mannitol did stimulate bulb formation.

Anther culture

The use of high concentrations of sucrose is commonly reported in papers on anther culture where the addition of 5-20% sucrose to the culture medium is found to assist the development of somatic embryos from pollen microspores. This appears to be due to an osmotic regulation of morphogenesis, for once embryoid development has commenced, such high levels of sucrose are no longer required, or may be inhibitory. A high concentration of mannitol has been used for pretreatment before the culture of barley anthers and pollen; tobacco anthers and pollen; and before wheat microspore culture. A high concentration of mannitol has also been used to induce osmotic stress in microspore derived embryos of *Brassica napus* and before anther culture of *Brassica campestris*. Isolated microspores of *Brassica napus* cultured on a high concentration of mannitol and at a low concentration of sucrose (0.08-0.1%) yield no embryos whereas on high polyethyleneglycol 4000 the embryo yield is comparable to that of the sucrose control. These results demonstrate that in microspore embryogenesis of *Brassica napus* the level of metabolizable carbohydrate required for microspore embryo induction and formation may be very low and that an appropriate osmoticum (polyethylene glycol 4000 or sucrose) is required.

The temporary presence of high sucrose concentrations is said to prevent the proliferation of callus from diploid cells of the anther that would otherwise swamp the growth of the pollen-derived embryoids. The concentration of macronutrient ions generally used in anther culture media is not especially low. In a sample of reports it was found to be 68.7 mM, and so the total osmotic potential, Ψ_s, (salts plus sucrose) of many anther culture media is in the range –0.55 to –1.15 MPa.

Relative humidity

The vapour pressure of water is reduced by dissolving substances in it. This means that the relative humidity of the air within closed culture vessels is dependent on the water potential of the medium according to the equation:

$$\Psi = \frac{1000RT}{W_0} \ln_c \left(\frac{p}{p_0} \right) \cdot \left(p - c\frac{dp}{dc} \right)$$

where: Ψ, R and T are as in the Van't Hoff equation, W_0 is the molecular weight of water, c is the concentration of the solution in moles per litre, and p_0 and p are respectively the vapour pressures of water and the solution.

If the change in vapour pressure dependent on the density of the solution, $p - c\,(dp/dc)$, is treated as unity (legitimate perhaps for very dilute solutions, or when the equation is expressed in molality, rather than molarity), it is possible to estimate relative humidity ($100 \times p/p_0$) above tissue culture media of known water potentials, from:

$$\Psi = \frac{1000RT}{18.016} \ln_c \left(\frac{p}{p_0} \right),$$

The relative humidity above most plant tissue cultures in closed vessels is thus calculated to be in the range 99.25-99.75%, the osmolality of some typical media being

1. White (1963), liquid, 2% sucrose, 106 mOsm/kg
2. MS, agar, 3% sucrose, 230 mOsm/kg
3. MS, agar, 6% sucrose, 359 mOsm/kg
4. MS, agar, 12% sucrose (unusual), 659 mOsm/kg

(in these cases, 50% hydrolysis of sucrose into monosaccharides is assumed to have taken place during autoclaving).

Relative humidity can be reduced below the levels indicated above by, for example, covering vessels with gas-permeable closures while using non-gelatinous support systems.

pH OF TISSUE CULTURE MEDIA

The relative acidity or alkalinity of a solution is assessed by its pH. This is a measure of the hydrogen ion concentration; the greater the concentration of H^+ ions (actually H_3O^+ ions), the more acid the solution. As pH is defined as the negative logarithm of hydrogen ion concentration, acid solutions have low pH values (0-7) and alkaline solutions, high values (7-14). Solutions of pH 4 (the concentration of H^+ is 10^{-4} mol.l^{-1}) are therefore more acid than those of pH 5 (where the concentration of H^+ is 10^{-5} mol.l^{-1}); solutions of pH 9 are more alkaline than those of pH 8. Pure water, without any dissolved gases such as CO_2, has a neutral pH of 7. To judge the effect of medium pH, it is essential to discriminate between the various sites where the pH might have an effect: (1) in the explant, (2) in the medium and (3) at the interface between explant and medium.

The pH of a culture medium must be such that it does not disrupt the plant tissue. Within the acceptable limits the pH also:

1. Governs whether salts will remain in a soluble form;
2. Influences the uptake of medium ingredients and plant growth regulator additives;
3. Has an effect on chemical reactions (especially those catalyzed by enzymes); and
4. Affects the gelling efficiency of agar.

This means that the effective range of pH for media is restricted. As will be explained, medium pH is altered during culture, but as a rule of thumb, the initial pH is set at 5.5 – 6.0. In culture media, detrimental effects of an adverse pH are generally related to ion availability and nutrient uptake rather than cell damage.

pH of Media

Buffering

The components of common tissue culture media have only little buffering capacity. Vacin and Went (1949) investigated the effect on pH of each compound in their medium. The chemicals which seemed to be most instrumental in changing pH were $FeSO_4.7H_2O$ and $Ca(NO_3)_2.4H_2O$. Replacing the former with ferric tartrate at a weight which maintained the original molar concentration of iron, and substituting $Ca_3(PO_4)_2$ and KNO_3 for $Ca(NO_3)_2.4H_2O$, they found that the solution was more effectively buffered. While amino acids also showed promise as buffering agents, KH_2PO_4 was ineffective unless it was at high concentration.

In some early experiments attempts were made to stabilize pH by incorporating a mixture of KH_2PO_4 and K_2HPO_4 into a medium, but Street and Henshaw (1966) found that significant buffering was only achieved by soluble phosphates at levels inhibitory to plant growth. For this reason Sheat *et al.* (1959) proposed the buffering of plant root culture media with sparingly-soluble calcium phosphates, but unfortunately if these compounds are autoclaved with other medium constituents, they absorb micronutrients which then become unavailable.

A buffer is a compound which can poise the pH level at a selected level: effective buffers should maintain the pH with little change as culture proceeds. As noted before, plant tissue culture media are normally poorly buffered. However, pH is stabilized to a certain extent when tissues are cultured in media containing both nitrate and ammonium ions. Agar and Gelrite gelling agents may have a slight buffering capacity.

Organic acids

Many organic acids can act as buffers in plant culture media. By stabilizing pH at ca. pH 5.5, they can facilitate the uptake of NH_4^+ when this is the only source of nitrogen, and by their own metabolism, assist the conversion of NH_4^+ into amino acids. There can be improvement to growth from adding organic acids to media containing both NH_4^+ and NO_3^-, but this is not always the case. Norstog and Smith (1963) noted that 0.75 mM malic acid was an effective buffer and appeared to enhance the effect of the glutamine and alanine which they added to their medium. Vyskot and Bezdek (1984) found that the buffering capacity of MS medium was increased by adding either 1.25 mM sodium citrate or 1.97 mM citric acid plus 6.07 mM dibasic sodium phosphate. Citric acid and some other organic acids have been noted to enhance the growth of *Citrus* callus when added to the medium (presumably that of Murashige and Tucker, 1969). For the propagation of various cacti from axillary buds, Vyskot and Jara (1984) added sodium citrate to MS medium to increase its buffering capacity.

Recognized biological buffers

Unlike organic acids, conventional buffers are not metabolized by the plant, but can poise pH levels very effectively. Compounds which have been used in plant culture media for critical purposes such as protoplast isolation and culture, and culture of cells at very low inoculation densities, include:

1. TRIS, Tris(hydroxymethyl)aminomethane;
2. Tricine, N-tris(hydroxymethyl)methylglycine;
3. MES, 2-(N-morpholino)ethanesulphonic acid;
4. HEPES, 4-(2-hydroxyethyl)-1-piperazine(2-ethanesulphonic acid); and
5. CAPS, 3-cyclohexylamino-1-propanesulphonic acid.

Such compounds may have biological effects which are unrelated to their buffering capacity. Depending on the plant species, they have been known to kill protoplasts or greatly increase the rate of cell division and/or plant regeneration. The 'biological' buffers MES and HEPES have been developed for biological research.

MES

MES is one of the few highly effective and commercially available buffers with significant buffering capacity in the pH range 5-6 to which plant culture media are usually adjusted and has only a low capacity to complex with micronutrients. It is not toxic to most plants, although there are some which are sensitive. Ramage and Williams (2002) report that shoot regeneration from tobacco leaf discs was not affected by MES when increasing the concentration up to 100 mM. De Klerk et al (2007), though, observed a decrease of rooting from apple stem slices with increasing MES concentration. This effect of MES was not understood. It only occurred during the first days of the rooting process and was not observed during the outgrowth phase after the meristems had been formed.

During the culture of thin cell layers of *Nicotiana*, Tiburcio *et al.* (1989) found that the pH of LS medium could be kept close to 5.8 for 28 days by adding 50 mM MES, whereas without the buffer pH gradually decreased to 5.25. Regulating pH with MES alters the type of morphogenesis which occur in this (and other) tissues.

Parfitt *et al.* (1988) found 10 mM MES to be an effective buffer in four different media used for tobacco, carrot and tomato callus cultures, and peach and carnation shoot cultures, although stabilizing pH did not result in superior growth. The tobacco, peach and carnation cultures were damaged by 50 mM MES. Tris was toxic at all concentrations tested, although Klein and Manos (1960) had found that the addition of only 0.5 mM Tris effectively increased the fresh weight of callus which could be grown on White (1954) medium when iron was chelated with EDTA.

MES has also been used successfully to buffer many cultures initiated from single protoplasts and its inclusion in the culture medium can be essential for the survival of individual cells and their division to form callus colonies (e.g. those of *Datura innoxia*). MES was found to be somewhat toxic to single protoplasts of *Brassica napus*, but 'Polybuffer 74' (PB-74, a mixture of polyaminosulphonates), allowed excellent microcolony growth in the pH range 5.5-7.0. Used at 1/100 of the commercially available solution, it has a buffering capacity of a 1.3 mM buffer at its pK value.

Banana homogenate is widely used in orchid micropropagation media. Ernst (1974) noted that it appeared to buffer the medium in which slipper orchid seedlings were being grown.

Uptake of ions and molecules

The pH of the medium has an effect on the availability of many minerals. In general, the uptake of negatively charged ions (anions) is favoured at acid pH, while that of cations (positively charged) is best when the pH is increased. As mentioned before, the relative uptake of nutrient cations and anions will alter the pH of the medium. The release of hydroxyl ions from the plant in exchange for nitrate ions results in media becoming more alkaline; when ammonium ions are taken up in exchange for protons, media become more acid.

Nitrate and ammonium ions

The uptake of ammonium and nitrate ions is markedly affected by pH. Excised plant roots can be grown with NH_4^+ as the sole source of nitrogen providing the pH is maintained within the range 6.8 to 7.2 and iron is available in a chelated form. At pH levels below 6.4 the roots grow slowly on ammonium alone and have an abnormal appearance. This accords with experiments on intact plants where ammonium as the only source of nitrogen is found to be poorly taken up at low pH. It was most effectively utilized in *Asparagus* in a medium buffered at pH 5.5. At low pH (ca. 4 or less) the ability of the roots of most plants to take up ions of any kind may be impaired, there may be a loss of soluble cell constituents and growth of both the main axis and that of laterals is depressed.

Nitrate on the other hand, is not readily absorbed by plant cells at neutral pH or above. Growth of tumour callus of *Rumex acetosa* on a nitrate-containing medium was greater at pH 3.5 than pH 5.0, while Chevre *et al.* (1983) reported that axillary bud multiplication in shoot cultures of *Castanea* was most satisfactory when the pH of MS was reduced to 4 and the Ca^{2+} and Mg^{2+} concentrations were doubled.

pH stabilization

One of the chief advantages of having both NO_3^- and NH_4^+ ions in the medium is that uptake of one provides a better pH environment for the uptake of the other. The pH of the medium is thereby stabilized. Uptake of nitrate ions by plant cells leads to a drift towards an alkaline pH, while NH_4^+ uptake results in a more rapid shift towards acidity. For each equivalent of ammonium incorporated into organic matter, about 0.8-1 H^+ (proton) equivalents are released into the external medium; for each equivalent of nitrate assimilated, 1-1.2 proton equivalents are removed from the medium. Raven (1986) calculated that there should be no change of pH resulting from NO_3^- or NH_4^+ uptake, when the ratio of the two is 2 to 1.

The pH shifts caused by uptake of nitrate and ammonium during culture (see above) can lead to a situation of nitrogen deficiency if either is used as the sole nitrogen source without the addition of a buffer. The uptake of NH_4^+, when this is the sole source of nitrogen is only efficient when a buffer or an organic acid is also present in the medium. In media containing both NO_3^- and NH_4^+ with an initial pH of 5-6, preferential uptake of NH_4^+ causes the pH to drop during the early growth of the culture. This results in increased NO_3^- utilization and a gradual pH rise. The final pH of the medium depends on the relative proportions of NO_3^- and NH_4^+ which are provided. After 7 days of root culture, White (1943a) medium (containing only nitrate), adjusted to pH 4.8-4.9, had a pH of 5.8-6.0, but Sheat *et al.* (1959) could stabilize the medium at pH 5.8 by having one fiftieth of the total amount of nitrogen as ammonium ion, the rest as nitrate. Changes in the pH of a medium do however vary from one kind of plant to another. Ramage and Williams (2002) observed a decrease in pH when tobacco leaf discs were cultured with only NH_4^+ nitrogen whereas no such decrease was observed on medium with both NH_4^+ and NO_3^-. No organogenesis occurred when the medium with only NH_4^+ was unbuffered but the inclusion of MES resulted in the formation of meristems (but no shoots).

Media differing in total nitrogen levels (but all having the same ratio of nitrate to ammonium as MS medium), had a final pH of ca. 4.5 after being used for Stage III root initiation on rose shoots, whereas those containing ammonium alone had a final pH of ca. 4.1. A similar observation with MS medium itself was made by Delfel and Smith (1980). No matter what the starting value in the range 4.5-8.0, the final pH after culture of *Cephalotaxus* callus was always 4.2. The medium of De Jong *et al.* (1974) always had a pH of 4.8-5.0 after *Begonia* buds had been cultured, whatever the starting pH in the range 4.0-6.5. This is not to say that the initial pH was unimportant, because there was an optimum for growth and development.

Other compounds

The availability and uptake of other inorganic ions and organic molecules is also affected by pH. As explained above, the uptake of phosphate is most efficient from acid solutions. Petunia cells took up phosphate most rapidly at pH 4 and its uptake declined as the pH was raised. Vacin and Went (1949) noted the formation of iron phosphate complexes in their medium by pH changes. Insoluble iron phosphates can also be formed in MS medium at pH 6.2 or above unless the proportion of EDTA to iron is increased.

The uptake of Cl^- ions into barley roots is favoured by low pH, but Jacobson *et al.* (1971) noticed that it was only notably less at high pH in solutions strongly stirred by high aeration. They therefore suggested that H^+ ions secreted from plant roots as a result of the uptake of anions, can maintain a zone of reduced pH in the Nernst layer, a stationary film of water immediately adjacent to cultured plant tissue. In plant tissue culture, uptake of ions and molecules may therefore more liable to be affected by adverse pH in agitated liquid media, than in media solidified by agar.

Lysine uptake into tobacco cells was found by Harrington and Henke (1981) to be stimulated by low pH. The effect of pH on uptake is especially relevant for auxins. Depending on the pH and their pKa, auxins are either present as an undissociated molecule or as an anion. For influx, the undissociated auxin molecule may pass through the membrane by diffusion whereas the anion is taken up by a carrier. Dissociation dependent upon the pH. In the apoplast, the pH is low, ca. 5. When taken up, the auxin enters the cytoplasm with pH 7. At this pH, most auxin is present as anion and cannot diffuse out. Efflux of the anion is brought about by an efflux carrier system. Thus, the net uptake into cells of plant growth regulators which are weak lipophilic acids (such as IAA, NAA, 2,4-D and abscisic acid) will be greater, the more acid the medium and the greater the difference between its pH and that of the cell cytoplasm. Shvetsov and Gamburg (1981) did in fact find that the rate of 2,4-D uptake into cultured corn cells increased as the pH of the medium fell from 5.5 to 4. Increased uptake of auxin at low pH was also found in apple microcuttings, both for IBA and IAA. As previously mentioned, auxins can themselves modify intra- and extra-cellular pH. Adding 2,4-D to a medium increased the uptake of nicotine into culture of *Acer pseudoplatanus* cells.

Choosing the pH of culture media: Starting pH

Many plant cells and tissues *in vitro,* will tolerate pH in the range of about 4.0-7.2; those inoculated into media adjusted to pH 2.5-3.0 or 8.0 will probably die. Best results are usually obtained in slightly acid conditions. In a random sample of papers on micropropagation, the average initial pH adopted for several different media was found to be 5.6 (mode 5.7) but adjustments to as low as 3.5 and as high as 7.1 had been made.

The pH for most plant cultures is thus lower than that which is optimum for hydroponic cultures, where intact plants with their roots in aerated solution usually grow most rapidly when the pH of the solution is in the range 6.0-7.3.

Suspension cultured cells of *Ipomoea*, grew satisfactorily in Rose and Martin (1975) P2 medium adjusted initially to pH 5.6 or pH 6.3, but the yield of dry cells was less at two extremes (pH 4.8 and pH 7.1). Martin and Rose (1976) supposed that a low yield of cells from a culture started at pH 7.1 was due to the inability of the culture to utilize NO_3^-, but the cause of a reduction in growth at pH 4.8 was less obvious. It might have resulted from the plant having to expend energy to maintain an appropriate physiological pH internally.

Kartha (1981) found that pH 5.6-5.8 supported the growth of most meristem tips in culture and that cassava meristems did not grow for a prolonged period on a medium adjusted to pH 4.8. The optimum pH (before autoclaving) for the growth of carnation shoot tips on Linsmaier and Skoog (1965) medium, was 5.5-6.5. When the medium was supplemented with 4 mg/l adenine sulphate and 2 g/l casein hydrolysate, the optimum pH was 5, and on media adjusted to 6.0 and 6.5, there was significantly less growth. Shoot proliferation in *Camellia sasanqua* shoot cultures was best when the pH of a medium with MS salts was adjusted to 5-5.5. Only in these flasks was the capacity of juvenile explants to produce more shoots than adult ones really pronounced.

Norstog and Smith (1963) suggested that the pH of the medium used for the culture of isolated zygotic embryos, should not be greater than 5.2.

pH adjustment

Because there is then no need to take special aseptic precautions, and it is impractical to adjust pH once medium has been dispensed into small lots, the pH of a medium is usually adjusted with acid or alkali before autoclaving. According to Krieg and Gerhardt (1981), agar is partially hydrolysed if autoclaved at pH 6 or less and will not solidify so effectively when cooled. They recommend that agar media for bacteriological purposes should be sterilized at a pH greater than 6 and, if necessary, should be adjusted to acid conditions with sterile acid after autoclaving, when they have cooled to 45-50°C. The degree of hydrolysis in plant culture media autoclaved at pH 5.6-5.7, is presumably small.

Effect of autoclaving

Autoclaving changes the pH of media. In media without sugars, the change is usually small unless the phosphate concentration is low, when more significant fluctuations occur. Media autoclaved with sucrose generally have a slightly lower pH than those autoclaved without it, but if maltose, glucose, or fructose have been added instead of sucrose, the post-autoclave pH is significantly reduced.

The pH of liquid media containing MS salts containing 3-3.4% sucrose, has been found to fall during autoclaving from an adjusted level of 5.7, to pH 5.17, to pH 5.5, or to pH 4.6. The drop in pH may vary according to the pH to which the medium was initially adjusted. In the experiments of Skirvin *et al.*, the pH of a medium adjusted to 5.0, fell to 4.2; one adjusted to 6.4, fell to 5.1; that set at pH 8.5, fell to 8.1.

Most agars cause the pH of media to increase immediately they are dissolved. Knudson (1946) noticed that the pH of his medium shifted from 4.6-4.7 to 5.4-5.5 once agar had been added and dissolved; and Singha (1982) discovered that the unadjusted pH of MS medium rose from pH 4 to pH 5.2, depending on the amount of agar added. However if a medium containing agar was adjusted to pH 5.7, and then autoclaved, the medium became more acid than if agar had not been added, the fall in pH being generally in proportion to the amount of agar present.

Effect of storage

The pH of autoclaved plant media tends to fall if they are stored. Vacin and Went (1949) noted that autoclaving just accelerated a drop in the pH of Knudson (1946) C medium, as solutions left to stand showed similar changes. Sterilization by filtration was not a satisfactory alternative, as it too effected pH changes. The compounds particularly responsible were thought to be unchelated ferrous sulphate (when the pH had initially been set between 3 and 6), and calcium nitrate (when the original pH was 6 to 9). Complex iron phosphates, unavailable to plants, were produced from the ferrous sulphate, but if iron was added as ferric citrate or ferric tartrate (chelates), no significant pH changes resulted from autoclaving, and plants showed no iron-deficiency symptoms.

Skirvin *et al.* (1986) found that both with and without agar, autoclaved MS medium tended to become more acid after 6 weeks storage, for example:

Time	MS medium	
	Liquid	With 0.6% Difco Bacto Agar
Starting pH	5.7	5.7
After autoclaving	4.6	4.6
6 weeks later	4.1	4.4

To minimize a change in the pH of stored media, it is suggested that they are kept in the dark: Owen *et al.*, (1991) reported that the pH of MS containing 0.1M sucrose or 0.8% Phytagar, remained relatively stable after autoclaving if kept in the dark at 4°C, but fell by up to 0.8 units if stored in the light at 25°C. De Klerk *et al.* (2007) using BBL agar, also observed a shift of pH, but this was negligible when 10 mM MES was added.

Hydrolysis

Some organic components of culture media are liable to be hydrolysed by autoclaving in acid media. The degree of hydrolysis of different brands of agar may be one factor influencing the incidence of hyperhydricity in plant cultures. Agar media may not solidify satisfactorily when the initial pH has been adjusted to 4-4.5. The reduction in pH which occurs in most media during autoclaving may also cause unsatisfactory gelling of agar which has been added in low concentrations. Part of the sucrose added to plant culture media adjusted to pH 5.5 is also hydrolysed during autoclaving: the proportion degraded increases if the pH of the medium is much less than this. Hydrolysis of sucrose is, however, not necessarily detrimental.

Activated charcoal

The presence of activated charcoal can alter the pH of a medium. As in the production of activated charcoal it is, at one stage washed with HCl, the pH of a medium can be lowered by acid residues.

pH changes during culture

Due to the differential uptake of anions and cations into plant tissues, the pH of culture media does not usually stay constant, but changes as ions and compounds are absorbed by the plant. It is usual for media containing nitrate and ammonium ions to decline slowly in pH during a passage, after being adjusted initially to pH 5.4-5.7, Sometimes after a preliminary decrease, the pH may begin to rise and return to a value close to, or even well above that at which the culture was initiated. The pH of White (1954) medium, which contains only NO_3^- nitrogen, drifted from an initial 5.0 to 5.5 towards neutrality as callus was cultured on it.

With some cultures, the initial pH of the medium may have little effect. Cell suspension cultures of *Dioscorea deltoides* in the medium of Kaul and Staba (1968) [containing MS salts], adjusted to a pH either 3.5, 4.3, 5.8 or 6.3, all had a pH of 4.6-4.7 within 10 h of inoculation. There was a further fall to pH 4.0-4.2 in the next 2-3 days, but during the following 15-17 days the pH was 4.7-5.0, finally rising to pH 6.0-6.3 on about day 19. Similarly, Skirvin *et al.* (1986) found that MS medium adjusted to pH 3.33, 5.11, 6.63, or 7.98 before autoclaving, and then used for the culture of *Cucumis melo* callus, had a pH after 48 h in the range 4.6-5.0. Visseur (1987) also reported that although the pH of his medium (similar macronutrients to MS but more Ca^{2+} and PO_4^{3-}) decreased if it was solidified with agar, but on a 2-phase medium, the final pH was 6.9 ± 0.4 irrespective of whether it was initially 4.8, 5.5 or 6.2.

Despite the above remarks, it should be noted that the nature of the pH drift which occurs in any one medium, differs widely, according to the species of plant grown upon it. The pH of the medium supporting shoot cultures of *Disanthus cercidifolius* changed from 5.5 to 6.5 over a 6 week period, necessitating frequent subculturing to prevent the onset of senescence, whereas in the medium in which shoots of the calcifuge *Lapageria rosea* were grown, the pH, initially set to 3.5-5.0, only changed to 3.8-4.1. Note also that the reversion of media to a homeostatic pH, may be due to the presence of both NO_3^- and NH_4^+ ions. Adjustment of media containing only one of these nitrogen sources to a range of pH levels, would be expected to result in a more variable set of final values. As the pH of media deviate from the original titration level, simple unmonitored cultures may not provide the most favourable pH for different phases of growth and differentiation. In *Rosa* 'Paul's Scarlet' suspensions, the optimum pH for the cell division phase was 5.2-5.4: pH 5.5-6.0 was best for the cell expansion phase. The maximum growth rate of *Daucus carota* habituated callus on White (1954) medium with iron as Fe-EDTA, occurred at pH 6.0.

It has been suggested that acidification of media is partly due to the accumulation of carbon dioxide in tightly closed culture flasks, but the decrease in pH associated with incubating anther cultures with 5% CO_2 was found by Johansson and Eriksson (1984) to be only ca. 0.1 units. Removal of CO_2 from an aerated cell suspension culture of *Poinsettia* resulted in an increase of about 0.2 pH-units.

Auxin plant growth regulators promote cell growth by inducing the efflux of H^+ ions through the cell wall. Hydrogen ion efflux from the cell is accompanied by potassium ion influx. When cultures are incubated in a medium containing an auxin, the medium will therefore become more acid while the pH of the cell sap will rise. The extent of these changes was found by Kurkdjian *et al.* (1982) to be proportional to auxin (2,4-D) concentration.

pH Control within the Plant

The various compartments of cells have a different pH and this pH is maintained. In the symplasm, the pH of the cytoplasm is ca. 7

and of the vacuole ca. 5. The apoplasm has a pH of ca. 5. Plant cells typically generate an excess of acidic compounds during metabolism which have to be neutralized. One of the most important ways by which this is accomplished is for H^+, or K^+ to be pumped out of the cell, in exchange for anions (e.g. OH^-), thereby decreasing the extracellular pH. Plant cells also compensate for an excess of H^+ by the degradation of organic acids. Synthesis of organic acids, such as malate, from neutral precursors is used to increase H^+ concentration when the cytoplasmic pH rises, for instance if plants are grown in alkaline soils. In intact plants, there is usually a downwards gradient from the low pH external to the cell, to higher pH levels in more mature parts, and this enables the upwards transport of non-electrolytic compounds such as sugars and amino acids.

Altering the pH of the external solution surrounding roots or cells can alter the pH of the cell. Because of necessary controls, the pH of the cytoplasm may be only slightly altered, that of vacuoles may show a more marked change. Changing the pH of the medium in which photo-autotrophic *Chenopodium rubrum* suspensions were cultured from 4.5 to 6.3, caused the pH of the cytoplasm to rise from 7.4 to 7.6 and that of the cell vacuoles to increase from 5.3 to 6.6. The increase in cytoplasmic pH caused there to be a marked diversion of carbon metabolism, away from sugar and starch, into the production of lipids, amino acids and proteins. The maintenance of the pH is also illustrated in an experiment with detached leaves of *Vicia faba*. When the leaves were placed in a 10 mM MES-TRIS buffer and transferred to buffer with another pH, changes in the pH of the apoplasm were small: with an initial buffer pH = 4.1 and transfer to buffer pH = 6.8, the apoplastic pH of substomatal cavities increased from 4.71 to 5.13 and in the reverse transfer decreased from 5.13 to 4.70. This indicates that the pH of the apoplasm is not strongly influenced by the medium but stays close to the 'natural' pH of ca. 5.0. No exact details are given but in this experiment the distance between the site of the pH measurements and the MES-TRIS solution in which the leaves had been placed was probably large. The symplasm has a much larger capacity to buffer so that its pH will be even less influenced by the medium pH. Thus, within the explant the pH of both apoplasm and symplasm will be affected only little by the medium pH. The situation may be different at the interface between explant and medium. The influence of medium pH will extend towards inner tissues of the explant as the buffering capacity of the medium is increased (and thus overcomes buffering by the tissue). Inside the explant, the pH will also greatly influence movement through membranes, i.e. uptake in cells, as this often depends on the dissociation of compounds which is pH dependent.

Changing the pH of the medium can thus have a regulatory role on plant cultures which is similar to that of plant growth regulating chemicals, one of the actions of which is to modify intracellular pH and the quantity of free calcium ions. Auxins can modify cytoplasmic pH by triggering the release of H^+ from cells. In plant tissue culture these ions can acidify the medium. Proton release is thought to be the first step in acid-triggered and turgor-triggered growth. It should be noted that pH changes themselves may act as a signal.

Effect of pH on Cultures

Initiating cultures at low pH

Plants of the family Ericaceae which only grow well on acid soils (e.g. rhododendrons and blueberries), have been said to grow best on media such as Anderson (1975), Anderson (1978; 1980) and Lloyd and McCown (1981) WPM, when the pH is first set to ca. 4.5, but for highly calcifuge species such as *Magnolia soulangiana*, a starting pH of 3.5 can result in the highest rate of shoot proliferation in shoot cultures. Chevre *et al.* (1983) state that chestnut shoot cultures grew and proliferated best at pH 4 provided MS medium was modified by doubling the usual levels of calcium and magnesium.

De Jong *et al.* (1974), using a specially developed medium, found that a low pH value favoured the growth of floral organs. A similar result was seen by Berghoef and Bruinsma (1979c) with *Begonia* buds. Growth was greatest when the pH of the medium was initially adjusted to acid, 4.5-5.0 being optimal. At pH 4.0 the buds became glassy.

Before the discovery of effective chelating agents for plant cultures, root cultures were grown on media with a low pH. Tomato roots were, for instance, unable to grow on media similar to those of White (1943a), when the pH rose to 5.2. Boll and Street (1951) were able to show that the depression of growth at high pH was due to the loss of Fe from the medium and that it could be overcome by adding a chelated form of iron. Using FeEDTA, Torrey (1956) discovered that isolated pea roots grew optimally at pH 6.0-6.4 but were clearly inhibited at pH 7.0 or greater; and Street (1969) reported that growth of tomato roots could be obtained between pH 4.0 and pH 7.2, if EDTA was present in the medium. Because agar does not gel properly when the initial pH of the medium is adjusted to 4, it is necessary to use liquid media for low pH cultures; or employ another gelling agent, or a mechanical support.

Some other cultures may also be beneficially started at low pH, which may indicate that the tissues have an initial requirement for NO_3^-. Embryogenesis of *Pelargonium* was induced more effectively if MS, or other media, were adjusted to pH 4.5-5.0 before autoclaving (rather than pH 5.5 and above).

Differentiation and morphogenesis

Differentiation and morphogenesis are frequently found to be pH-dependent. Xylogenesis depends on the medium pH. The growth of callus and the formation of adventitious organs from thin cell layers excised from superficial tissues of the inflorescence rachis of *Nicotiana*, depended on the initial pH of Linsmaier and Skoog (1965) medium containing 0.5 1,1M IBA and 3 1.1M kinetin. Pasqua *et al.* (2002) reported many quantitative effects of pH during regeneration from tobacco thin cell layers. The types of callus produced from the plumules of *Hevea* seedlings differed according to the pH of the medium devised by Chua (1966). Soft and spongy callus formed at acid (5.4) or alkaline (8.0) pH. A compact callus was obtained between pH 6.2 and 6.8.

Adventitious root formation

There are several reports in the literature which show that the pH of the medium can influence root formation of some plants *in vitro*. A slightly acid pH seems to be preferred by most species. Zatkyo and Molnar (1986), who showed a close correlation between the acidity of the medium (pH 7.0 to 3.0) and the rooting of *Vitis*, *Ribes nigrum* and *Aronia melancarpa* shoots, suggested that this was because acidity is necessary for auxin action.

Sharma *et al.* (1981) found it advantageous to reduce the pH of the medium to 4.5 to induce rooting of *Bougainvillea* shoots and a reduction of the pH of MS medium to 4.0 (accompanied by incubation in the dark) was required for reliable root formation of two *Santalum* species and *Correa decumbens* and *Prostanthera striatifolia*. Other Australian woody species rooted satisfactorily at pH 5.5 and pH 4 was inhibitory. Shoots from carnation meristem tips rooted more readily at pH 5.5 than pH 6.0, and rooting of excised potato buds was best at pH 5.7, root formation being inhibited at pH 4.8 and at pH 6.2 or above.

Direct root formation on *Nautilocalyx leaf* segments was retarded if a modified MS medium containing IAA was adjusted initially to an acid pH (3.5 or 4.0) or a neutral pH (6.5). Good and rapid root formation occurred when the medium was adjusted to between pH 5.0 to 6.3. De Klerk *et al* (2007), working with apple stem slices, found only a small effect of pH on rooting: when the pH was set before autoclaving at 4.5 (after autoclaving the pH was 4.54), the number of roots was 4.5, and with the pH set at 8.0 (after autoclaving the pH had dropped to 5.65), the number of roots per slice increased to 7. In medium buffered with MES, the maximum number of roots was formed at pH 4.4 (measured after auto-claving). In these experiments, the dose-response curve for root number did not correspond with the effect of pH on IAA uptake. Such discrepancy between the effects of the pH on uptake and root number, was also reported by Harbage, Stimart and Auer (1998). Direct formation of roots from *Lilium auratum* bulb scales occurred when MS medium was adjusted within the range 4-7 but was optimal at pH 6. The pH range for adventitious bulblet formation in this plant was from 4 to 8, but most bulblets were produced when the initial pH was between 5 and 7. Substrates which are to acid or too alkaline can adversely affect rooting *ex vitro*.

Embryogenesis

Smith and Krikorian (1989) discovered that pre-globular stage pro-embryos (PGSP) of carrot could be made to proliferate from tissues capable of direct embryo formation, with no auxin, on a medium containing 1-5 mM NH_4^+ (and no nitrate). Somatic embryos were formed when this tissue was moved to MS medium. The pH of the 'ammonium-nitrogen' medium fell from 5.5 to 4.0 in each subculture period, and it was later found that culture on a medium of low pH was essential for the maintenance of PGSP cultures. Sustained culture at a pH equal or greater than 5.7, with no auxin, allowed somatic embryo development. A similar observation to that of Smith and Krikorian had been made by. Although the pH of a suspension culture of alfalfa 'Regen-S' cells in a modified Schenk and Hildebrandt (1972) medium with 15 mM NH_4^+ was adjusted to 5.8, it quickly fell back to pH 4.4-5.0 in a few hours. The pH then gradually increased as somatic embryos were produced, until at day 14 it was 5.0. In certain suspension cultures, the pH was titrated daily to 5.5, but on each occasion soon returned to nearly the same pH as that in flasks which were untouched. Even so, the pH-adjusted suspensions produced more embryos than the controls.

The ammonium ion has been found to be essential for embryogenesis. Is one of its functions to reduce the pH of the medium through rapid uptake and metabolism, thereby facilitating the uptake of nitrate, upon which embryogenesis is really dependent?

Embryogenesis from leaf explants of *Ostericum koreanum*, was found to depend strongly on pH. As the explants were cultured continuously with NAA, it is possible that the observed relationship was caused by differential NAA uptake. This is also suggested by the slower rate of embryo development seen at low pH, because this would be expected where there is a high internal NAA concentration.

Liquid Media and Support Systems

The nutritional requirements of plant cultures can be supplied by liquid media but growth in liquid medium may be retarded and development affected by oxygen deprivation and hyperhydration. The oxygen concentration of liquid media is often insufficient to meet the respiratory requirements of submerged cells and tissues. It can be increased either by raising the oxygen concentration of the medium or placing cells or tissues in direct contact with air. If the water potential of the medium is greater (less negative) than that of a cell, water flows into the cell and the vacuole becomes distended.

Cells and tissues affected in this way are described as hyperhydric. Shoots often show physiological disturbances with symptoms that can be recognized visually. The term hyperhydric is preferable to the previously used term 'vitrified' for reasons explained by Debergh *et al.* (1992). Water potential is determined by osmotic potential of the solutes in the medium and, in the case of a gelled medium, by the matric potential of the gel. Reductions in hyperhydricity can be achieved by increasing the concentrations of the solutes and the gel. Hyperhydricity may also be reduced through evaporation of water from tissues if they are placed in contact with air.

Contact of cultured tissues with air, to alleviate problems of hyperhydricity and hypoxia, can be achieved by the use of either porous or semi-solid (gelled) supports. The advantage of supports, as opposed to thin layers of liquid medium, is that tissues can be placed in a sufficient volume of medium to prevent depletion of nutrients and allow for the dispersion of any toxins that might be produced by the plant tissues. The relative advantages of liquid medium, solid supports or gelled media, varies with the type of material being cultured.

Liquid Media

Liquid medium, without supporting structures, is used for the culture of protoplasts, cells or root systems for the production of secondary metabolites, and the propagation of somatic embryos, meristematic nodules, microtubers and shoot clusters. In liquid medium, these cultures often give faster growth rates than on agar-solidifed medium. Cultures may be fully or only partially immersed in the medium.

Aeration of liquid medium in stationary Petri dishes is sometimes adequate for the culture of protoplasts and cells because of the shallow depth of the medium, but may still be suboptimal. Anthony *et al.* (1995) cultured protoplasts of cassava, in liquid medium in Petri dishes with an underlying layer of agarose in which glass rods were embedded vertically. Sustained protoplast division was observed in the cultures with glass rods but not in the controls without glass rods. The authors suggested that the glass rods extended the liquid meniscus, where the cell colonies were clustered thus causing gaseous exchange between the liquid and the atmosphere above to be facilitated.

Anthony *et al.* (1997) cultured protoplasts of *Passiflora* and *Petunia* in 30 ml glass bottles containing protoplast suspensions in 2 ml aliquots, either alone or in the presence of the oxygen carriers Erythrogen or oxygenated Perfluorodecalin. Cell division in each of the two species was stimulated by both oxygen carriers.

Laboratory-scale experimentation on immersed cultures of cells, tissues and organs, may be carried out in 125 ml or 250 ml Erlenmeyer flasks. Large-scale cultures are usually carried out in bioreactors with a capacity of 1 litre or more. The concentration of oxygen in the medium is raised by oxygen in the gas phase above and air bubbles inside the liquid. Increasing the oxygen concentration and circulation of the medium is facilitated in flasks by the use of gyratory shakers and in bioreactors by stirring and/or bubbling air through the medium. The use of bioreactors often involves the automated adjustment of the culture medium. The design of bioreactors for plant cells and organs was reviewed by Doran (1993) and the use of shake-flasks and bioreactors for the scale-up of embryogenic plant suspension cultures has been reviewed by Tautorus and Dunstan (1995). The importance of oxygen concentration in bioreactors can be illustrated by an investigation into the growth of hairy roots of *Atropa belladonna*. They found that no growth occurred at oxygen tensions of 50% air saturation but between 70% and 100% air saturation, total root length and the number of root tips increased exponentially. Hyperhydricity in liquid cultures may be avoided by adding to the medium osmoregulators, such as mannitol, maltose and sorbitol, and inhibitors of gibberellin biosynthesis including ancymidol and paclobutrazol.

Plantlets and microtubers can be cultured by partial submersion in liquid medium. One method of aerating tissues is by the automated flooding and evacuation of tissues by liquid medium. This method has been used to produce microtubers of potato from single node cuttings. An alternative approach to aeration is to apply the liquid medium over the plant tissues as a nutrient mist. For example, Kurata *et al.* (1991) found that nodes of potato grew better in nutrient mist than on agar-based cultures.

Support by Semi-solid Matrices

Gelled media provide semi-solid, supporting matrices that are widely used for protoplast, cell, tissue and organ culture. Agar, agarose, gellan gums and various other products have been used as gelling agents.

Agar

Agar is very widely employed for the preparation of semi-solid culture media. It has the advantages that have made it so widely used for the culture of bacteria, namely:

1. It forms gels with water that melt at approx. 100°C and solidify at approx. 45°C, and are thus stable at all feasible incubation temperatures;
2. Gels are not digested by plant enzymes;
3. Agar does not strongly react with media constituents.

To ensure adequate contact between tissue and medium, a lower concentration of agar is generally used for plant cultures than for the culture of bacteria. Plant media are not firmly gelled, but only rendered semi-solid. Depending on brand, concentrations of between 0.5-1.0% agar are generally used for this purpose. Agar is thought to be composed of a complex mixture of related polysaccharides built up from galactose. These range from an uncharged neutral polymer fraction, agarose, that has the capacity to form strong gels, to highly charged anionic polysaccharides, sometimes called agaropectins, which give agar its viscosity. Agar is extracted from species of *Gelidium* and other red algae, collected from the sea in several different countries. It varies in nature according to country of origin, the year of collection and the way in which it has been extracted and processed. The proportion of agarose to total polysaccharides can vary from 50 to 90%. Agars contain small amounts of macro- and micro-elements; particularly calcium, sodium, potassium, and phosphate, carbohydrates, traces of amino acids and vitamins that affect the osmotic and nutrient characteristics of a gel. They also contain phenolic substances and less pure grades may contain long chain fatty acids, inhibitory to the growth of some bacteria.

As agar can be the most expensive component of plant media, there is interest in minimizing its concentration. Concentrations of agar can be considered inadequate if they do not support explants or lead to hyperhydricity. Hyperhydricity decreases as the agar concentration is raised but there may be an accompanying reduction in the rate of growth. For example, Debergh *et al.* (1981) found that shoot cultures of *Cynara scolymus* were hyperhydric on medium containing 0.6% Difco 'Bacto' agar. No hyperhydricity occurred on medium containing 1.1% agar but shoot proliferation was reduced. Likewise, Hakkaart and Versluijs (1983) found that shoots of carnation were hyperhydric on medium containing 0.6% agar whereas growth was severely reduced on medium containing 1.2%. During studies to optimize the production of morphogenic callus from leaf discs of sugarbeet, Owens and Wozniak (1991) found large differences in the numbers of somatic embryos and shoots according to the gelling agent employed. They found that water availability, determined by gel matric potential, was the dominant factor involved. When they adjusted the concentrations of the gelling agents to give media of equal gel matric potential, somatic embryos and shoots were found in similar numbers on Bacto agar (0.7%), HGT agarose (0.46%), Phytagar (0.62%) and Gelrite (0.12%).

Various brands and grades of agar are available commercially. These differ in the amounts of impurities they contain and their gelling capabilities. The gelling capacity of Difco brands of agar increased with increasing purity i.e. 'Noble' > 'Purified' > 'Bacto'. After impurities of agar were removed by sealing agar in a semipermeable bags and washing in deionized water, the water potentials of gels of three brands were substantially lower than unwashed gels. A ten-fold difference in the regeneration rate of sugar cane was observed by Anders *et al.* (1988) on media gelled with the best and least effective of seven brands of agar. Scholten and Pierik (1998) investigated the different growth characteristics of seven different agar brands on the growth of axillary shoots, adventitious shoots and adventitious roots of rose, lily and cactus. They concluded that no single bioassay could identify 'good or bad' agars for a large group of plant species but Merk 1614, Daishin, MC29, and BD Purified gave the best results in most experiments. 'Daishin' showed no batch-to-batch variations. They found no relationship between price and quality of the brands of agar.

Agarose

Agarose is the high gel strength moiety of agar. It consists of β-D(1→3) galactopyranose and 3,6-anhydro-α-L(1 → 4) galactopyranose polymer chains of 20-160 monosaccharide units alternatively linked to form double helices. There are also several different agaro-pectin products available, which have been extracted from agar and treated to remove most of the residual sulphate side groupings. Because preparation involves additional processes, agarose is much more expensive than agar and its use is only warranted for valuable cultures, including protoplast and anther culture. Brands may differ widely in their suitability for these applications. Concentrations of 0.4-1.0% are used. Agarose derivatives are available which melt and gel at temperatures below 30°C, making them especially suitable for testing media ingredients that are heat-labile, or for embedding protoplasts. Low melting-point agarose is prepared by introducing hydroxyethyl groups into the agarose molecules. Another advantage of agarose over agar lies in the removal of toxic components of agar during its preparation. Bolandi *et al.* (1999) preferred to embed protoplasts in agarose, rather than use liquid medium, because in agarose the semi-solid matrix applies a direct pressure on the plasma membrane of the protoplasts. They mixed protoplasts of sunflower with 0.5% agarose, pipetted 50 μl droplets of the mixture into Petri dishes and covered them with a thin layer of culture medium. A similar method was used to culture protoplasts of *Dioscorea* by Tor *et al.* (1999). The use of droplets has the advantage that a high plating density can be achieved in the droplets, while exposing the protoplasts to a larger reservoir of culture medium in the liquid phase. Bishoi *et al.* (2000) initially cultured anthers of Basmati rice on liquid medium, then used 1.0% agarose to culture calli derived from microspores.

Gellan gum

Gellan gum is a widely used gelling agent in plant tissue culture, that is marketed under various trade names including Gelrite, Phytagel and Kelcogel. It is an exopolysaccharide that encapsulates cells of the bacterium *Sphingomonas paucimobilis* (= *Auromonas elodea* = *Pseudomonas elodea*), from which it is obtained by industrial fermentation. The structure, physico-chemical properties and the rheology of solutions of gellan gum and related polysaccharides has been reviewed by Banik *et al.* (2000). Gellan gum consists of a linear repeating tetrasaccharide of D-glucose, D-glucuronic acid, D-glucose and L-rhamnose. Heating solutions of gellan gum in solutions that contain cations, such as K^+, Ca^{2+}, Mg^{2+}, causes the polysaccharide to form a gel in which the polymers form a half-staggered parallel double helix. The commercial deacetylated and purified polysaccharide forms a firm non-elastic gel. The gel sets rapidly at a temperature, determined by the concentrations of the polysaccharide and the cation, which varies between 35-50°C. The commercial product contains significant quantities of potassium, sodium, calcium and magnesium but is said to be free of the organic impurities found in agar. It is unclear whether or not these cations remain fully available to plant cultures. Some researchers add an extra 750 mg l^{-1} $MgCl_2$ to a medium containing MS salts to aid the gelling of 1.6% Gelrite. In most other reports on the use of Gelrite, cations in the medium have been sufficient to produce a gel.

Beruto *et al.* (1995) found that 0.12 % Gelrite and 0.7% Bacto agar have equivalent matric potential and support equivalent adventitious regeneration in leaf discs of sugarbeet. As gellan gum is used in lower concentration than agar, the cost per litre of medium is less. It produces a clear gel in which plant tissues can be more easily seen and microbial contamination more easily detected than in agar gels. It has proved to be a suitable gelling agent for tissue cultures of many herbaceous plants and there are reports of its successful use for callus culture, the direct and indirect formation of adventitious organs and somatic embryos, shoot culture of herbaceous and semi-woody species and the rooting of plantlets. In most cases the results have been as good as, and sometimes superior to, those obtainable on agar-solidified media. Anders *et al.* (1988) described the regeneration of greater numbers of plants from of sugar cane on Gelrite than on the most productive brand of agar, and Koetje *et al.* (1989) obtained more somatic embryos from callus cultures of rice on media solidified with Gelrite than with Bacto-agar.

Shoot cultures, particularly of some woody species, are liable to become hyperhydric if Gelrite, like agar, is used at too low a concentration. There are sharp differences in the response of different species to concentration of Gelrite. Turner and Singha (1990) found the highest rate of shoot proliferation in *Geum* occurred on 0.2% and in *Malus* on 0.4%. Garin *et al.* (2000) obtained more mature somatic embryos of *Pinus strobus* on gellan gum at 1% concentration of than at 0.6%. Pasqualetto *et al.* (1986a,b) used mixtures of Gelrite (0.1-0.15%) and Sigma @ agar (0.2-0.3%) to prevent the hyperhydricity that occurred in shoot cultures of *Malus domestica* 'Gala' on media solidified with Gelrite alone. Nairn (1988) used a mixture of Gelrite

(0.194%) and Difco 'Bacto' agar (0.024%) to prevent the hyperhydricity that occurred in shoot cultures of *Pinus radiata* on medium gelled with 0.2% Gelrite alone.

Alginates

Alginic acid is a binary linear heteropolymer 1,4-β-D-mannuronic acid and 1,4-α-L-guluronic acid. It is extracted from various species of brown algae. When the sodium salt is exposed to calcium ions, gelation occurs. Alginates are widely used for protoplast culture and to encapsulate artificial seed.

Protoplasts embedded in beads or thin films of alginate can plated densely while yet exposed to a large pool of medium that dilutes inhibitors and toxic substances. Embedded protoplasts can be surrounded by nurse cells, either free in the surrounding medium or separated by filters or membranes. Alginate has the advantages over agarose that protoplasts do not have to be exposed to elevated temperatures when they are mixed with the gelling agent and the gel can be liquified by adding sodium citrate, releasing protoplasts or cell colonies for transfer to other media. The method of embedding protoplasts in beads as employed by Larkin *et al.* (1988) involved mixing the protoplast suspensions with an aqueous solution of sodium alginate and dropping it, through a needle, into a solution of calcium chloride. Beads containing the protoplasts were formed when the alginate made contact with the calcium ions. Beads with a final concentration of 1.5% sodium alginate were washed and cultured in liquid medium on an orbital shaker. Protoplasts may also be captured in thin layers of alginate.

Synthetic seeds (synseeds, somatic seeds) encapsulated in alginate, can be prepared from somatic embryos, shoot tips, apical and axillary buds, single nodes, and cell aggregates from hairy roots. The uses of synthetic seeds include direct planting into soil, storage of tissues and transfer of materials between laboratories under sterile conditions. The methods of encapsulation of somatic embryos of carrot were described by Timbert *et al.* (1995). Torpedo-shaped embryos were mixed with a 1% sodium alginate solution. The mixture was dropped into a solution 100 mM calcium chloride in 10% sucrose. The beads (3-3.5 mm in diameter) were then rinsed in a 10% sucrose solution.

Starch

Sorvari (1986a,c) found that plantlets formed in higher frequencies in anther cultures of barley on a medium solidified with 5% corn starch or barley starch rather than with agar. Corn starch only formed a weak gel and it was necessary to place a polyester net on its surface to prevent the explants from sinking. Sorvari (1986b) found that it took 5-14 weeks for adventitious shoots to form on potato discs on agar-solidified medium but only 3 weeks on medium containing barley starch. Henderson and Kinnersley (1988) found that embryogenic carrot callus cultures grew slightly better on media gelled with 12% corn starch than on 0.9% Difco 'Bacto' agar.

'Kappa'-carrageenan

Carrageenan is a product of sea weeds of the genus *Euchema* and the kappa form has strong gelling properties. Like gellan gum, kappa-carrageenan requires the presence of cations for gelation. In Linsmaier and Skoog (1965) medium at 0.6% w/v, the gel strength was slightly less than that of 0.2% Gelrite or 0.8% of extra pure agar. Chauvin *et al.* (1999) found that regeneration from cultures of tulip, gladiolus and tobacco shoots was possible in the presence of 200 mg l^{-1} kanamycin, whereas in several other gelling agents a concentration of 100 mg l^{-1} inhibited regeneration.

Pectins

A mixture of pectin and agar can be a less expensive substitute for agar. For example, a semi-solid medium consisting of 0.2% agar plus 0.8-1.0% pectin, was employed for shoot culture of strawberry and some other plants.

Other gelling agents

Battachary *et al.* (1994) found sago (from *Metroxylon sagu*) and isubgol (from *Plantago ovata*) were satisfactory substitutes for agar at, respectively, one eighth and one tenth of the cost of Sigma purified agar A7921.

Porous Supports

Aeration of the tissues on a porous substrate is usually better than it would be in agar or static liquid.

Chin *et al.* (1988) used a buoyant polypropylene membrane floated on top of a liquid medium to culture cells and protoplasts of *Asparagus*. The membrane has a pore size of 0.04 mm and is autoclavable. Conner and Meredith (1984) found that cells grew more rapidly on filter papers laid over polyurethane foam pads saturated with medium than on agar. Young *et al.* (1991) supported shoots of tomato over liquid culture medium on a floating microporous polypropylene membrane and entrained the growing shoots through polypropylene netting. They reported opportunities for the development of this method for mechanization by mass handling.

Cheng and Voqui (1977) and Cheng (1978) used polyester fleece to support cultures of Douglas fir that were irrigated with liquid media in Petri dishes. Plantlets that were regenerated from cotyledon explants were cultured on 3 mm-thick fabric. When a protoplast suspension was dispersed over 0.5 mm-thick fabric, numerous colonies were produced in 12 days, whereas in the absence of the support, cell colonies failed to proliferate beyond the 20 cell stage. A major advantage in using this type of fabric support is that media can be changed simply and quickly without disturbing the tissues. The system has also been used for protoplast culture of other plants.

Heller and Gautheret (1949) found that tissues could be satisfactorily cultured on pieces of ashless filter paper moistened by contact with liquid medium. Very small explants, such as meristem tips, that might be lost if placed in a rotated or agitated liquid medium, can be successfully cultured if placed on an M-shaped strip of filter paper (sometimes called a 'Heller' support). When the folded paper is placed in a tube of liquid medium, the side arms act as wicks. This method of support ensures excellent tissue aeration but the extra time required for preparation and insertion has meant that paper wicks are only used for special purposes such as the initial cultural stages of single small explants which are otherwise difficult to establish. Whether explants grow best on agar or on filter paper supports, varies from one species of plant to another. Davis *et al.* (1977) found that carnation shoot tips grew less well on filter paper bridges than on 0.6% agar but axillary bud explants of *Leucospermum* survived on bridges but not on agar.

Paper was also used in the construction of plugs known as Sorbarods. These are cylindrical (20 mm in length and 18 mm in diameter) and consist of cold-crimped cellulose, longitudinally folded, wrapped in cellulose paper. The plug has a porosity (total volume - volume of cellulose) of 94.2% and high capillarity, so that the culture medium is efficiently drawn up into the plug, leaving the sides of the plug in direct contact with air. Roots permeate the plugs and are protected by the cellulose during transfer to soil. Plantlets of chrysanthemum in Sorbarods formed longer stems, larger leaves, more roots, and developed greater fresh mass, dry weights and fresh to dry mass ratios than plantlets in agar-solidified medium. The greater fresh to dry mass ratio indicates that contact with liquid medium led to greater hydration of tissues. This was subsequently controlled by the inclusion of a growth retardant, paclobutrazol (1 mg l^{-1}), in the culture medium. Other porous materials that have been used to support plant growth include rockwool, polyurethane foam, vermiculite, a mixture of vermiculite and Gelrite.

Afreen-Zobayed *et al.* (2000) cultured sweet potato, on sugar-free medium in autotrophic conditions, on mixtures of paper pulp and vermiculite in various proportions. Optimal growth was obtained on a mixture containing 70% paper pulp. On this mixture, the fresh mass of plantlets was greater by a factor of 2.7 than on agar-solidified medium. Mixtures of paper pulp and vermiculite, in unspecified proportions, are prepared in a commercial product known as Florialite. Afreen-Zobayed *et al.* (1999) found that growth rates of plantlets of sweet potato grown autotrophically on Florialite were greater, in ascending order, on agar, gellan gum, vermiculite, Sorbarods and Florialite (best). The dry mass of leaves and roots were greater by factors of 2.9 and 2.8, respectively, on Florialite than on an agar matrix. These authors observed that roots spread better in Florialite than in Sorbarods. They attributed this to the net-like orientation of fibers in Florialite that contrasted with the vertical orientation of fibres in Sorbarods. Ichimura and Oda (1995) found three substances that stimulated plant growth in extracts of paper pulp. Each was characterized by low molecular weight and high polarity. It is possible that these substances contribute to the superior growth observed on substrates containing paper pulp.

Opportunities for improved ventilation and photoautotrophy

When plantlets are cultured in vessels containing air at a *relative humidity* (RH) of less than 100%, transpiration occurs which is an important factor in reducing hyperhydricity. Relative humidity in culture vessels can be reduced through ventilation, but gelled substrates are then unsuitable because the absorbance of water by the roots of a transpiring plant is impeded by the gel's low hydraulic conductivity and this increases as the gel dries. Thus a common feature of studies using ventilated vessels has been the use of liquid medium supported by porous materials.

For example, when plantlets of chrysanthemum were grown in Sorbarods in a culture vessel with air 94% RH, a reduction in the tissue hydration was indicated by a significantly lower fresh to dry mass ratio than at 100% RH and increases in stem length and leaf area. In this investigation, the RH was reduced to 94% by gaseous diffusion through a gas-permeable membrane that covered holes drilled in the lid of the culture vessel. The use of such ventilated culture vessels can significantly improve plant growth by reducing hyperhydration and facilitating the movement of solutes to the leaves in the transpiration stream. It also provides opportunities for photoautotrophic growth in sugar-free media. When plantlets are cultured in closed vessels, carbon dioxide concentrations fall to low levels in the light period, as Kozai and Sekimoto (1988) demonstrated in cultures of strawberry plants. Photosynthesis requires an adequate supply of carbon dioxide and suitable lighting. Adequate concentrations of carbon dioxide for photoautotrophy can be maintained in ventilated culture vessels with or without elevated levels of carbon dioxide in the atmosphere outside the culture vessel. Adequate lighting can be achieved under lights delivering a photosynthetic photon flux of 150 μmol m^{-2} s^{-1} in a culture room or in day-light in a greenhouse.

Immobilised Cells

Yields of secondary metabolites are usually greater in differentiated, slow-growing cells than in fast growing, undifferentiated cells. By immobilizing cells in a suitable matrix, their rate of growth can be slowed and the production of secondary products enhanced. Several ingenious methods of immobilization have been employed. Examples include immobilization in spirally wound cotton, glass fibre fabric reinforced with a gelling solution of hybrid SiO_2 precursors, loofa sponge and polyurethane foam and alginate beads.

6

Nuclear Import and Export of Proteins

In this Chapter, we will focus primarily on protein import into the nucleus of plants. As in other eukaryotes the partitioning of genetic information into the nucleus necessitates the import and export of macromolecules such as proteins, nucleic acids and protein/nucleic acid complexes across the nuclear envelope. These transport processes are essential and are subject to stringent regulatory controls. In plants, it is clear that in addition to the maintenance of basic cellular processes, the regulated import of proteins plays a vital role in development. As we will discuss, the nuclear import of proteins in a selective manner is essential in responses of plants to light and results in the dramatic morphological changes that occur as plants switch from growth in the dark to growth in the light. Protein translocation across the nuclear envelope is also a process that is utilized by pathogenic viruses and tumor-inducing bacteria in the genus *Agrobacterium* to transport protein/nucleic acid complexes into the nucleus for replication and even incorporation of pathogen DNA into the host genome.

The import pathway for proteins, themselves key factors regulating nuclear transport processes, is the best understood of the nuclear transport processes in plants. Numerous import signals have been characterized, and we are beginning to identify and understand the major components of the import machinery. As expected many of these factors are conserved between plants, animals and fungi, but there are surprising results indicating subtle differences in preferences for *nuclear localization signals* (NLSs), the mechanics of import receptor function, and potential plant-specific import factors. As we move forward plants are contributing new knowledge such as potential mechanisms for the targeting of proteins to the nuclear envelope and *nuclear pore complex* (NPC). Recent efforts have begun to focus on other nuclear transport processes in plants such as the export of proteins and nucleic acids.

Protein Import in Animals and Yeast

Our knowledge of nuclear transport in vertebrates and yeast is more advanced than our understanding in plants. Among the reasons for this historically are: (i) the ease of recovering nuclei from *Xenopus* oocytes that are intact for morphological studies, (ii) the development of in vitro systems in *Xenopus* for reconstituting the assembly of nuclei and NPCs, (iii) the reconstitution of cytosol-dependent nuclear import in permeabilized mammalian cells, (iv) the rapid genetics possible in yeast, (v) the large biomedical research community and potential importance of nuclear processes for medicine. However as we are discovering nuclear translocation is a critical aspect of plant growth and development and thus has broad implications for agriculture in terms of disease and stress resistance and other crop improvements, which are the keys to feeding an expanding worldwide population. To place our knowledge of plant nuclear import in perspective, we will overview processes in animals and yeast throughout the chapter highlighting important advances and unique aspects in plants. There are also excellent recent reviews that are focused on the nuclear/ cytoplasmic transport of proteins, nucleic acids and their complexes as well as the structure and function of the NPCs in vertebrates and yeast.

Nuclear Translocation in Plants

Nuclear Pore Complex

The channels through which all transport substrates must pass are the NPCs which are macromolecular complexes embedded in the double-membrane nuclear envelope. The NPCs are estimated to have a mass of 125 MDa in higher eukaryotes and to be composed of 50 to 100 different proteins collectively known as nucleoporins. Morphologically, an NPC is composed of a nucleoplasmic and a cytoplasmic ring. Eight spokes are found within the rings that extend toward a central channel resulting in eight 9 nm channels thought to function in the diffusion of small molecules across the nuclear envelope. A basket-like structure extends from the nucleoplasmic face and fibrils have been observed extending from the cytoplasmic face. Thus far, fewer than 20 nucleoporins have been purified from vertebrates. The NPCs in yeast are less complex (mass about 66 MDa) and are not dissociated and reassembled during mitosis as in higher eukaryotes; nevertheless the development of methods to purify intact complexes has permitted the identification and sequencing of all 30 of its nucleoporins. Even with such information available understanding the assembly and function of the NPC will be a daunting task.

A number of vertebrate NPC proteins have been implicated in nuclear import. They include Nup which is located on the cytoplasmic filaments. Nup 358 has multiple Ran binding sites and binds to the protein transporter importin β via FXFG (single amino acid code where X represents an amino acid with a small or polar side chain) repeats which are characteristic of many nucleoporins. Other nucleoporins reported to bind to importin β include Nup153, Nup214, Nup116p, Nup100p, and p62. The p62 protein was one of the first nucleoporins to be purified, and like many nucleoporins from vertebrates it is modified by single *O*-linked *N*-acetylglucosamine (O-GlcNAc) residues. While the O-GlcNAc is probably not essential for import, the binding of the lectin *wheat germ agglutinin* (WGA) inhibits nuclear import in vertebrates and has been used for the identification and purification of nucleoporins from higher eukaryotes.

From electron microscopy studies beginning in the 1970s we know that nuclear pores in plants are morphologically similar to those of other organisms. However there have been few reports in which plant nucleoporins have been purified. Scofield et al localized a protein at the NPC and identified a 100 kDa polypeptide in a nuclear matrix fraction from carrots using an antibody to the yeast nucleoporin NSP; however successful purification was not reported. Other studies indicate that nuclear envelope fractions from maize and tobacco nuclei contain a subset of proteins in the NPC fraction that can bind NLSs specifically. The binding site is at the NPC indicating a role for plant nucleoporins in protein import. This finding is consistent with the unusually tight association of at least one plant importin α NLS receptor with the nuclear envelope in purified nuclei and intact cells.

Using WGA as a probe it is clear that GlcNAc-modifications are present at the periphery of the nucleus, and electron microscopy has shown that some of these modifications are present at the NPC. In fact biochemical characterization of tobacco nuclear fractions has shown that the glycans are attached to proteins via an *O*-linkage and the moities are longer than five sugar residues in length ending with a terminal GlcNAc residue. This is a novel modification not found in vertebrates, which contain only a single O-GlcNAc residue. Although the functional significance of O-GlcNAc modifications is not clear the modification can be used to advantage for purifying plant NPC proteins. Using lectin affinity chromatography four O-GlcNAc proteins were purified from nuclear envelope fractions of cultured tobacco cells. Peptide sequence was obtained from a protein of 40 kDa (gp40) that shares about 30% identity with aldose-1-epimerases (also known as *mutarotases*) which are involved in aldose sugar metabolism in bacteria. Their role in higher eukaryotes is not well defined but they do share structural similarities to the glucose carrier from erythrocytes. Interestingly, it has been reported that glycosylated proteins can be imported in a sugar-specific and NLS-independent manner in mammalian cells in vitro. Thus one speculation is that gp40 may be involved in such an import system in plants by binding to glycosylated proteins destined for import. Regardless of the role eventually assigned to gp40 a procedure to purify NPC proteins from plants should permit additional proteins to be characterized. The availability of *Arabidopsis* genome sequence should also contribute to the identification of plant nucleoporins, although it should be noted that beyond several very short repeat motifs such as FXFG there are few similarities even between vertebrate and yeast nucleoporins. While interesting in itself, it suggests that the identification of plant nucleoporins based on protein identity across kingdoms will be only partially successful.

Import Signals in Plants

Several types of transport can occur through NPCs. Ions and small proteins that are typically less than 20 to 30 kDa can pass by simple diffusion through the 9 nm channel. However even small macromolecules such as histones (14 kDa) or tRNAs cross the NPC via active processes permitting their translocation to be under cellular control. There is evidence that even calcium ions may be subject to selective concentration within the nucleus; this may relate to the suggestion that calcium plays a role in the regulation of active import. Nuclear transport processes are mediated by specific import receptors that recognize signals located within their respective substrates. In the case of protein import, NLSs interact with the import machinery that facilitates translocation through the NCP. Unlike most signals for organelles including the chloroplasts, mitochondria, vacuoles, peroxisomes and endoplasmic reticulum, NLSs are not proteolytically removed following import. This permits NLS-containing proteins to shuttle in and out of the nucleus and to be re-imported following post-mitotic nuclear assembly. Many transcription factors and cell cycle-regulatory proteins are able to exert their activities upon the cell based on their relative abundances in the cytosol compared to the nuceloplasm. Most NLSs are classified as either monopartite or bipartite. The classic monopartite NLS is from the SV40 large T-antigen and is composed of a single short region enriched in the basic residues arginine and lysine, whereas the nucleoplasmin NLS, which first defined the bipartite class, is composed of two basic domains separated by a spacer of variable length and composition. There are also NLSs that are less typical. One unusual NLS class is typified by the signal within the Matα2 protein which requires both basic and hydrophobic residues.

As in other organisms nuclear localization signals in plants cannot be defined by a strict consensus sequence and regions of basic amino acids are common within proteins, particularly regions involved in DNA binding. Thus, putative NLSs must be examined for activity in vivo to confirm their function. A number of NLSs have been carefully defined in such a manner in plants. These signals include at least one member for each of the three NLS classes. As examples, the transcriptional activitor R from maize possesses three functional NLSs, two SV40-like and one Matα2-like signal, each of which are sufficient to target a reporter protein to the nucleus in vivo. Another transcription factor from maize, Opaque-2 possesses two signals, one SV40-like and one bipartite signal, that are sufficient to target a reporter protein to the nucleus. For R and Opaque2, mutations within the intact proteins indicate that multiple NLSs are necessary for efficient targeting in vivo suggesting cooperativity among NLSs for import which is also true in vertebrates.

It is generally accepted that most NLSs can function across kingdoms pointing to a high degree of functional conservation. For example the SV40 large T-antigen NLS functions in plantsand the single bipartite NLS from the VirD2 protein of the plant pathogen *Agrobacterium* functions in plant, *Xenopus*, *Drosophila*, mammalian, and yeast cells. *Agrobacterium* is an opportunistic pathogen that infects a wide variety of plant species. In the coarse of pathogenesis *Agrobacterium* interacts with the host cell and transfers pathogen DNA (T-DNA) into the host cell nucleus through the NPC via the cooperative action of the VirD2 and VirE2 proteins. The T-DNA integrates into the plant genome and utilizes host factors to transcribe pathogen sequences.

Plant import is not strictly conserved when compared to import in other kingdoms however. The yeast Matα2 NLS targets a β-glucuronidase (GUS) reporter protein to the nucleus in onion epidermal cells which is consistent with the specific association of this class of NLSs with an import receptor from *Arabidopsis*. Interestingly, the Matα2 signal does not function in mammals although other yeast NLSs are known to function in vertebrates. Another exception is from *Agrobacterium*. As mentioned the NLS from VirD2 is broadly functional across kingdoms. Fascinatingly, VirE2 which contains two functional bipartite signals does not function in any of the non-plant organisms described for VirD2. However a single amino acid change which alters one NLS to conform to the animal bipartite consensus permits the NLS to function in *Xenopus* and *Drosophila*. Overall these results indicate that there may be subsets of import receptors or other components in plants that are not present in animals and fungi.

Components and Mechanisms of Protein Import

The key components of the classical NLS protein import pathway have been identified within the past decade and are the NLS-receptor importin α, the broad specificity transporter importin β, the GTPase Ran, and the Ran-interacting factor NTF2. In the first event of the protein import pathway, NLSs within nuclear proteins are recognized and bound by the NLS receptor importin β in the cytoplasm. Another factor, importin β interacts with importin α (via an importin β binding domain within importin α) completing the trimeric import complex. It is importin β that then interacts with specific proteins of the NPC facilitating translocation through the NPC. Thus for protein import importin β functions as an adapter that recognizes the NLS and associates with importin β, whereas importin β functions as the actual transporter.

The directionality of import is determined by the binding of the small *ras*-related GTPase Ran to importin β. Following import of the trimeric complex, the GTP-bound form of Ran (Ran-GTP) binds to importin β in the nucleoplasm resulting in the release of importin α and the NLS-containing cargo from the complex. The importin β/ Ran-GTP heterodimer is then exported to the cytoplasm where Ran-GTP is hydrolyzed to Ran-GDP leading to the release of importin β for subsequent rounds of import via reassociation with importin α and NLS-containing cargo. Since monomeric importin α does not typically interact directly with the NPC it is exported back to the cytoplasm via its own export receptor, CAS, for subsequent rounds of import.

The energy and directionality of import are hypothesized to depend on the enrichment of Ran-GTP in the nucleoplasm compared to the cytoplasm which contains mostly Ran-GDP. This is accomplished through the action of the nucleotide exchange factor RCC1 in the nucleus that enhances nucleotide exchange favoring Ran-GTP and the GTPase-activating protein RanGAP1 that favors conversion to Ran-GDP in the cytoplasm. The GTPase activity is further stimulated by Ran-BP1. Import is effectively a process of facilitated diffusion utilizing a gradient of Ran-GTP; a loose analogy would be to envision the import apparatus as an antiporter that causes accumulation of protein against a gradient of Ran-GTP. In the cytoplasm, the factor NTF2 interacts with Ran-GDP and NPC proteins and functions as a receptor/transporter for the re-import of Ran preventing its depletion from the nucleoplasm. It should be noted that Ran plays essential roles beyond nuclear trafficking such as regulation of cell cycle progression.

The selectivity and range of cargoes shuttled across the NPC is determined by different isoforms of importin α and importin β within the cell. The importin α receptor for NLS-mediated protein import has been found either as a single gene (SRP1) in yeast or as a small gene family in other organisms. In vertebrates at least six genes encoding importin α isoforms has been reported. There is also evidence of distinct but overlapping preferences for different NLSs and possible cell-specific roles for the different importin αs. Conversely, the range of importin β-like proteins, many of which are only distantly related, is far more diverse. This is due to the fact that members of the importin β family participate not only in the import of NLS-containing proteins but also function directly as import (importins) and export (exportins) receptor/transporters for other essential cargoes via their interaction with the NPC. Some examples include transportin 1 (import of hnRNP proteins), CRM1 (exportin 1; export of proteins containing a nuclear export signal), exportin-t (export of tRNAs), importin 7 (import of ribosomal proteins), and snurportin 1 (import of U snRNPs). The recent understanding of this receptor/transporter family has provided mechanistic details about the transport of diverse cargoes and highlighted potential ways in which the translocation of essential macromolecules is coordinated during the cell cycle.

In higher plants we have only recently begun to identify components of the import apparatus. Hicks et al identified a homologue of the importin α receptor, At-IMP α from *Arabidopsis*. Immunologically-related proteins are found in all organs examined including roots, stems, leaves, and flowers as expected for an essential factor. There is also evidence that At-IMP α is phosphylated in vitro in the presence of cytosolic extracts suggesting a potential mechanism for controlling NLS binding or interaction with other proteins. At the cellular level, At-IMP α is localized in the cytoplasm and nucleoplasm and at the nuclear envelope as expected for a receptor that shuttles between compartments. One unusual aspect of At-IMP α is its tight association with the nuclear envelope even in plant cells that have been treated to permeabilize the plasma membrane and deplete cytosolic contents; in animal cells endogenous importin α is mostly cytosolic and easily depleted from permeabilized cells. The plant receptor binds specifically in vitro to monopartite, bipartite and Matα2-like NLSs indicating broad NLS selectivity. Unlike yeast and mammalian orthologs that require importin β for high-affinity binding to NLSs, At-IMP α binds with high affinity (K_d of 5 to 10 nM) in the absence of an importin β subunit, although At-IMP α is capable of binding to mouse importin β. This is consistent with the finding that At-IMP α can mediate association of import substrate at the nuclear envelope in permeabilized animal cells, whereas mouse importin α absolutely requires importin β. In fact, At-IMP α can mediate nuclear protein import in permeabilized animal cells in the absence of exogenous importin β indicating that At-IMP α shares some properties with importin β. This is a surprising result given that At-IMPα shares significant homology with other importin αs which require importin β and indicates the possibility of an importin β-independent pathway in plants.

Although At-IMP α has some unusual properties, other plant importin α homologues are more typical of their animal and yeast counterparts. Using the *Agrobacterium* protein VirD2 as bait in a yeast two-hybrid screen of an *Arabidopsis* library, Ballas and Citovsky identified an importin α homologue (AtKAP α) that has high identity with At-IMP α except for a 64 amino acid extension at the amino-terminus. AtKAP α interacts specifically with the carboxy-terminal NLS of VirD2 both in vitro and in vivo in the two-hybrid assay, and the AtKAP α gene was found to complement a temperature-sensitive yeast mutant *srp1-31* (yeast importin α). Cytosolic extract containing AtKAP α protein was able to restore import to cells of *srp1-31* in an in vitro import system using permeabilized yeast cells in which import is dependent upon exogenous cytosol. Interestingly, VirE2 does not interact with AtKAP α suggesting an alternative pathway for its import.

A two-hybrid screen has identified a candidate for a VirE2 import protein (VIP1) that is related to basic leucine zipper proteins rather than to importin α. One possibility is that VirE2 is imported via a "*piggy-back*" mechanism through association with VIP1 which is a nuclear protein likely having functions other than in import. It is likely that importin α in *Arabidopsis* is encoded by a small gene family as at least 4 members have been reported. Another member of this family has been found in a two-hybrid screen using a WD40 type regulatory protein as a bait. PRL1 when disrupted by the insertion of a T-DNA tag results in a pleiotropic phenotype conferring hypersensitivity to glucose, sucrose and hormones. PRL1 was found to interact in vivo and in vitro with ATHKAP2 which has high identity with At-IMP α, but ATHKAP2 has a truncated carboxy terminal end being about 60 amino acids shorter than AtIMP α. As with the other characterized importin αs it possesses motifs required for interaction with importin β as well as the characteristic eight conserved armadillo repeats presumably for protein-protein interactions.

An importin α homologue (importin α1) from rice has been characterized that is 76% identical to At-IMP α. Expression of the gene in rice is suppressed by light in both etiolated seedlings and in leaves but not in roots or calli which display constitutive expression. Binding to NLSs was examined in vitro, and importin α1 was found to bind to the SV40 large T antigen NLS and the bipartite signal from Opaque . However no association was observed between importin α1 and either the yeast Matα2 signal or the Matα2-like signal from the R protein. This again points to some selectivity in NLS recognition in plants. Rice importin α binds to importin β from mouse, although with an affinity much lower than that of mouse importin α. Nevertheless in vitro import in permeabilized HeLa cells could be made dependent upon the presence exogenous rice importin α1 and mouse importin β indicating that rice importin α1 can function in a manner similar to that of the animal and yeast homologues.

Importin β in plants has been less studied than importin α, although there should be a significant number of genes for importin βs by analogy with vertebrates. There are a few plant importin βs that are characterized, and the results indicate a high degree of functional conservation between kingdoms. The first importin β homologues to be reported are from rice, and these are designated rice importins β1 and β2. Recombinant importin β1 can interact in vitro with rice importin α1 and a second rice importin α homologue (rice importin α2). This was studied by examining mobility shifts of protein complexes on native polyacrylamide gels. Using this approach it is argued that importin β1 specifically interacts with Ran-GTP and not Ran-GDP which is consistent with the ability of Ran-GTP to dissociate the importin α/importin β import complex in animals and yeast. In permeabilized HeLa cells, exogenous rice importin β1 and importin β1 can support import in the presence of mouse Ran. This result plus the finding that exogenous importin β1 can bind directly to the nuclear envelope in permeabilized tobacco BY2 cells argues that importin α and β functions are conserved. It will be interesting to examine the importin βs and other components for tissue specific or light regulation as has been observed for rice importin α1 because regulation of development by light is an essential and unique feature of plants.

The Ran GTPase has been characterized in plants and homologues have been reported in *Arabidopsis*, tomato and tobacco among other species. While a direct role for Ran in nuclear import in plants has not been established, the protein is localized to the nucleus and appears to be encoded by a small gene family of at least 3 members that are expressed throughout the plant including tissues that are not actively growing. Furthermore, it has been demonstrated that homologues from tomato or tobacco when expressed in *Schizosaccharomyces pombe* can suppress the phenotype of the *pim46-1* mutant which is defective in cell cycle progression. Ran is known to be involved in cell cycle control suggesting that the tomato and tobacco Ran homologues have functions analogous to those in other organisms.

Both RanGAP1 and RanBP1 can stimulate Ran GTPase activity in the cytoplasm favoring Ran-GDP in that compartment. RanBP1 has been identified functionally from *Arabidopsis* by using the Ran homologue AtRan1 as bait in a two-hybrid screen in yeast. Haizel et al found interaction with two RanBP1-like proteins (At-RanBP1a and At-RanBP1b) having 60% identity with vertebrate Ran BP1 and possessing conserved domains for Ran binding. Further study indicates that both At-RanBP1s are capable of associating in vivo in yeast with the GTP-bound form of each of the three Ran homologues from *Arabidopsis* (AtRan1, AtRan2, AtRan3). Neither the intracellular location nor the ability of At-RanBP1a or At-RanBP1b to stimulate GTPase activity has been reported. Likewise, neither RanGAP1 nor the nuclear exchange factor RCC1 has been characterized functionally in plants. Two putative RanGAPs have been identified from *Arabidopsis*, and they appear to have a unique domain not present in Ran GAPs from animals and yeast. The motif which has been named a WPP domain is shared with MAF1. MAF1 is a recently discovered protein that is composed mostly of WPP repeats and appears to be localized to the nuclear envelope via interaction with another envelope protein, MFP1. The WPP domain appears to be specific to plants and is hypothesized to be involved in protein-protein interaction. If true, RanGAP may associate with the nuclear envelope and NPCs in plants through this interaction.

It is increasingly apparent that there is functional conservation of the basic nuclear import pathway in plants, vertebrates and yeast. However many of the interesting biological questions will no doubt reside in the ways that plants have adapted the basic nuclear import pathways to fit their sessile photosynthetic life style. Exceptions have been noted already (importin β independence of At-IMP α, a potential alternative pathway for VirE2 import, differences in NLS selectivity) and others will be discussed below.

Systems to Study Import in Plants

The essential breakthrough that permitted the biochemical identification and purification of the factors now known to be involved in nuclear translocation of proteins and other substrates was the development of an in vitro import system utilizing permeabilized animal cells. The method relies on the fact that plasma membranes from animals can be selectively permeabilized with digitonin, a reagent that aggregates to form pores upon binding to cholesterol. This sterol is abundant in the plasma membranes of animal cells but not in other membranes such as the nuclear envelope. The effect of the reagent is to permit the selective depletion of soluble factors from cultured cells while leaving the integrity of the nuclear envelope intact. Nuclear import thus occurs only by authentic facilitated translocation rather than by simple diffusion into damaged nuclei through tears in the nuclear envelope. Import can be directly visualized by microscopy following the addition of fluorescently labeled NLS-containing proteins. Another accepted method to examine import is microinjection into *Xenopus* oocytes or mammalian cells which is analytical and not amenable to fractionation of components. Other approaches that have been reported include the use of purified nuclei or nuclei mixed with *Xenopus* cytosol extracts; however the ease of the permeabilization assay resulted in broad acceptance.

Although digitonin is not the reagent of choice in yeast and plants due to differences in membrane composition, the principle of selective permeabilization has been utilized successfully. A method in yeast was developed in which the plasma membrane of cell wall-less spheroplasts was selectively permeabilized via a simple freeze-thaw technique. As in animal cells, import is dependent upon ATP (which can be converted to GTP for import), temperature, and the presence of exogenous cytosol (from yeast or mammalian cells). Again, alternative methods have been reported.

In plants the development of in vitro import systems was difficult technically due to the fact that plant cells possess a thick cell wall and are highly vacuolated. The first report of in vitro import was an antibody cotranslocation assay in which antibodies to *G-box binding factors* (GBFs) are translocated into the nucleus (presumably by association with the GBFs) of Triton X-100 permeabilized parsley cells. The results indicate that GBFs involved in light-regulated gene expression are imported in response to light and that import is ATP and temperature dependent.

The assay is indirect however relying of protease protection of antibody associated with the nucleus. Two groups reported the development of direct import assays using evacuolated protoplasts from tobacco. The approaches were similar with Merkle et al using Triton X-100 to permeabilize the plasma membrane, whereas Hicks et al used an osmotic shift to achieve permeabilization without detergents. In both cases, direct visualization of fluorescent import substrates indicates specific import that is dependent upon GTP hydrolysis and is specific for proteins containing functional NLSs. Some interesting differences are apparent between import in plants and import in animals and yeast. Import in plants is only partially inhibited on ice compared to an almost complete block in yeast and animals, and import was not blocked by WGA as in animals (perhaps due to the unusual NPC modifications). The most fundamental difference, however, is that import in permeabilized plant cells can occur in the absence of exogenous cytosol. This is not to suggest that cytoplasmic factors are not required in plants as in animals and yeast. However, a significant fraction of specific import factors may be tightly associated with cellular structures such as the cytoskeleton that are not disrupted by the permeabilization techniques used. There is support for this notion.

By direct observation in permeabilized protoplasts and by fractionation of purified nuclei it is clear that in the presence of high concentrations of Triton X-100 significant fractions of At-IMP α remain in the cytoplasm and in association with the nuclear envelope in addition to a soluble pool. Whereas these observations provide an opportunity to investigate potential mechanisms for retention of At-IMP α, unfortunately they limit the utility of the assays for the identification of essential import factors in plants. Alternatives have been used most of which are heterologous systems. As noted Ballas and Cytovsky have used the yeast permeabilized system to demonstrate the function of At-KAPα in the import of VirD2. Other heterologous systems have been used to examine *Agrobacterium* Vir proteins including *Xenopus*, *Drosophila*, mammalian, and yeast cells. Other examples previously cited are the functional characterizations of At-IMP α and rice importins α and β in permeabilized HeLa cells. An alternative heterologous approach is to utilize plant cytosol extracts to support import in animal cells. This approach was examined, and it was found that cytosolic extract from petunia could in fact support import in permeabilized HeLa cells. As in animal cells import is temperature dependent, requires GTP hydrolysis and is blocked by WGA. The final approach that has proven useful is microinjection of import substrates into the stamen hairs of *Tradescantia*. This was valuable in characterizing the nuclear import of VirE2-single stranded DNA complexes. VirE2 facilitates *Agrobacterium* infection by associating with pathogen DNA in the plant cell cytoplasm and assisting in its nuclear import. Import of the complexes was found to be dependent upon GTP hydrolysis and was inhibited by WGA. The inconsistency of WGA inhibition upon injection compared to a lack of inhibition in permeabilized cells in unclear. Perhaps the large mass of the VirE2-protein complexes renders them more susceptible to import inhibition than simple protein substrates.

Regulated Protein Import in Plant Development

Plants lead a sessile life style and thus have evolved sensitive mechanisms to control development and withstand environmental challenges. A large number of gene products are induced in response to light, which is an essential developmental stimulus. Seedlings respond to light by undergoing photomorphogenesis, a process that includes altered morphology (shorter stems, leaf development), the development of chloroplasts and induction of the photosynthetic machinery (greening). Even fully differentiated plants must continually respond to quantitative and qualitative differences in light and other environmental challenges such as temperature fluctuations. For the purposes of illustrating the important roles that nuclear protein import play in such responses, we will discuss several interesting examples in which regulation of import potentiates a response to environmental cues.

Photomorphogenesis

Genetic screens have identified components in light signaling that include photoreceptors and downstream components that couple light signals to gene expression during photomorphogenesis. The understanding of photomorphogenesis including the mechanisms of light-modulated gene expression is major area of plant biology. In *Arabidopsis*, screens for defective or inappropriate responses to light have resulted in mutants that are photomorphogenic (i.e. they possess a light-grown phenotype) in complete darkness and define at least loci known as the COP/DET/FUS loci. One protein that acts as a negative regulator of photomorphogenesis is COP1. Mutations in COP1 result in plants that develop a light-grown phenotype including chloroplast development in complete darkness. When expressed as a GUS fusion protein in *Arabidopsis*, COP1 localizes primarily to the nucleus in the dark in leaves and shoots but is exclusively nuclear localized in the roots. However in the light COP1 partitions between the nucleus and cytoplasm in leaves and shoots. COP1 possesses, in addition to a bipartite NLS, a zinc-binding domain, a coiled-coil domain, and WD-40 repeats. Recent characterization of COP1 domain structure indicates that the amino terminal region containing the zinc-binding domain is essential for basal function, whereas the carboxyl terminal domain is necessary for the repression of photomorphogenesis in the dark. Although the precise mechanism by which COP1 targeting is regulated by light is not fully understood, a potential route by which photomorphogenic repression is achieved is via interaction of COP1 with the *basic leucine-zipper* (bZIP) transcription factor Hy5. Hy5 binds to G-box containing promoters for light regulated genes such as ribulose bisphosphate carboxylase/oxygenase small subunit (RBCS) and *chalcone synthase* (CHS). It has been identified as a suppressor of the *cop1* mutant and has been shown to interact physically with COP1 protein. Several studies indicate that Hy5 is an exclusively nuclear protein that is abundant in the light but degraded in the dark. Furthermore, the degradation is clearly dependent upon interaction with COP1 in the nucleus because a truncated Hy5 protein lacking a COP1-interaction domain is no longer degraded in the dark. Thus, COP1 functions as a repressor of photomorphogenesis by signaling the selective degradation of downstream affectors including Hy5 which participates gene expression essential for development. Recent evidence suggests that degradation of HY5 is mediated by the ubiquitin pathway by interaction with a COP1-containing proteosome and that HY5 degradation can be further modulated by phosphorylation of the COP1 interaction domain.

Expression of HY5 is interestingly itself under negative regulation in the light by the action of a recently discovered calcium-binding protein, SUB1. The story is more complicated as COP1 appears to interact via its coiled-coiled domain with yet another protein (CIP1) that is cytoplasmic and is capable of interacting with the cytoskeleton. One hypothesis is that CIP1 is involved in the retention of COP1 in the cytoplasm in the light essentially excluding it from the nucleus.

The perception of light involves multiple photoreceptors that detect blue light (cryptochrome), UV-B, and red light (phytochrome). The best characterized family of receptors are the phytochromes which are soluble proteins possessing a tetrapyrrole chromophore for activation by red light. After light absorption, the inactive form of phytochrome (Pr) is converted to the active far-red light absorbing form (Pfr) that participates in signal transduction and expression of genes involved in photomorphogenesis. The morphological consequences of red light perception include characteristic hypocotyl shortening and red light dependence of seed germination. There are five genes encoding phytochromes in *Arabidopsis* (phy A through phy E). The best characterized are phy A and phy B each of which detects red light in different ways as developmental cues. Phy A is rapidly degraded in the light and much more abundant in dark-grown plants than phy B. PhyA is responsible for the so-called *very low fluence responses* (VLFR) and for absorption of continuous far-red light known as the high irradiance responses (HIR). Phy B is stable in the light and is responsible for the detection of red light known as the *low fluence responses* (LFR) which are reversible by far-red light.

For the induction of gene expression in response to light signals there must be communication between the soluble photoreceptors (that are for the most part cytoplasmic) and the nucleus. Recent experiments indicate that phy A and phy B are transported from the cytoplasm to the nucleus in response to red light. Rice phy A and tobacco phy B were fused to *green fluorescent protein* (GFP) and overexpressed in tobacco. When adapted to growth in the dark, phy A-GFP and phy B-GFP were not detected. However upon exposure to as little as 5 min of far-red light phy A-GFP (i.e., VLFR) localized to the nucleus. In contrast, upon exposure to red light (i.e., LFR), but not far-red light, Phy B-GFP was found in the nucleus and the localization was reversible upon irradiation by far-red light. Furthermore, the nuclear localization of phy B is dependent upon the presence of the chromophore. The kinetics of the light-dependent relocalization have been examine in detail. These results are consistent with the biological activities of phy A and phy B and indicate that their regulated targeting in response to the quality of the red light is an important step in phytochrome signaling. The mechanism of import inhibition in the dark is hypothesized to involve the masking of putative NLSs within the carboxyl terminus via structural changes that are dependent upon the presence of functional chromophore or perhaps a specific retention of Pr in the cytoplasm in the dark.

The light-regulated nuclear import of several classes of transcription factors has also been described recently and may provide additional pathways for the control of photomorphogenesis. Nuclear translocation of the *common plant regulatory factor* (CPRF) proteins in parsley cells has been shown to be under red light control and is far-red reversible. The CPRFs are bZIP transcription factors that bind to G-box elements found adjacent to many light-regulated genes. Of the three CPRFs that have been examined (CPRF1, CPRF2, CPRF4), CPRF2 was found in the cytoplasm in the dark and relocated to the nucleus in the light. Immunolocalization indicates that phy A HIR and phy B LFR responses are involved implicating CPRFs in phytochrome signaling and provides regulation in addition to light-modulated nuclear import of the receptors. A deletion analysis of CPRF2 reveals two potential domains involved in cytoplasmic retention in the dark. Neither domain has homology to the COP1 retention factor CIP1 nor to a retention domain in the bZIP factor G-box binding factor 1. However, one retention domain of CPRF2 shares 25% identity with an α-helical retention domain from mammalian heat shock factor 2.

Other examples of light-regulated nuclear import are known. Using the previously discussed parsley in vitro antibody cotranslocation assay, Harter et al have found evidence for a cytoplasmic pool of GBF-transcription factors involved in light-regulated gene expression. The GBFs are another class of bZIP transciption factors that bind to G-box elements and participate in light-regulated gene expression. Upon exposure to white light, GBFs were found in the nucleus, presumably due to light stimulated relocalization. More recently, Kircher et al have cloned several CPRFs from parsely and using the antibody co-translocation assay find that parsley CPRF1, CPRF2 and GBF2 are translocated to the nucleus in response to UV light. GBFs have been examined further using three different *Arabidopsis* GBF genes fused to the reporter GUS and examined by transient expression in soybean protoplasts. About 50% of one fusion protein (GUS:GBF2) was found in the nucleus in the dark, whereas this increased to about 80% upon exposure to blue light. Deletion analysis of a different fusion protein

(GUS:GBF1) resulted in an increase in nuclear localization from a maximum of 50% to about 90%. This analysis may have identified a region involved in cytoplasmic retention of GBFs in the dark. One caviat is that GUS:GBF1 itself is not under light control being about 50% nuclear under all conditions tested. Given the importance of light in plant development additional examples of modulated nuclear import will no doubt be identified.

Other Examples of Regulated Import

Plants must continually respond to environmental and pathogen challenges and examples indicating the involvement of regulated import are being discovered. In tomato, the import of several heat shock transcription factors (Hsfs) requires protein-protein interaction. The expression of HsfA1 is constitutive but is accompanied by the expression of several heat shock inducible forms called HsfA2 and HsfB1. HsfA2 has been shown in tomato and tobacco protoplasts to be mostly cytoplasmic upon heat shock, even though related factors such as HsfA1 are translocated to the nucleus under these conditions. If a short region is deleted from the carboxyl terminus HsfA2, the protein is strongly localized to the nucleus. Interestingly, when coexpressed with HsfA1 in tomato protoplasts, HsfA2 is efficiently translocated following heat shock. The cotranslocation is dependent upon the physical interaction of HsfA1 and HsfA2 as demonstrated by coimmunoprecipitation and a two-hybrid assay. The stress induction of HsfA2 and its interaction with constitutively expressed HfA1 to form a transcriptionally active heterodimer provides a mechanism for dynamic changes in the intracellular distribution of HsfA2.

We have already discussed aspects of the *Agrobacterium* system in which the pathogen utilizes endogenous plant import components to assist is the infection process. Nuclear import in plants can also be under viral control. An interesting example occurs in plants infected with the squash leaf curl virus (SqLCV), a geminivirus. The virus encodes two movement proteins, BR1 and BL1, which cooperativey participate in cell-to-cell spread of the virus. BR1 is an NLS-containing protein that shuttles between the nucleus and cytoplasm and binds to single-stranded DNA. BL1 is localized to peripheral regions of cytoplasm and appears to function in the movement of the BR1:viral DNA complexes across the cell wall to adjacent cells. When expressed transiently in tobacco protoplasts BR1 is strongly localized to the nucleus. However when coexpressed with BL1, BR1 is relocalized to the cell periphery via specific interaction between the proteins providing a mechanism for delivery of viral genomes to the cell periphery for cell-to-cell spread. Other examples of viral protein nuclear import and cytoplasmic retention controlling nuclear import in viruses exist.

Besides the specific examples of regulated nuclear import cited above, there is almost certainly broader control of development and environmental responses through modulation of the nuclear import apparatus itself. For example in rice, it is known that light exposure results in the down-regulation of importin α in leaves and dark-grown seedlings. Potentially broad control of import in plants could be imparted through phophorylation, which has been clearly implicated in both overall control of the cell cycle and in the specific regulation of imported proteins in animals and yeast.

Recent Advances in Plant Nuclear Translocation

Several recent advances in our understanding of nuclear protein translocation in plants are worthy of mention as they have a potential impact of the field in general.

Targeting to the NPC

Many of the essential components of the import pathway have been identified animals and yeast, and are beginning to be identified in plants. In addition the NPCs have been the focus of intense investigation in animals and yeast. One fundamental question that has received little or no attention is how proteins in the cytoplasm are targeted to the NPCs for import. The notion that proteins are freely soluble in the cytoplasm where they associate with importins and diffuse to the NPCs for translocation is too simplistic. It is known that organelles and mRNAs can be transported along the cytoskeleton to specific sites. In fact, animal viruses can be targeted to the nucleus along microtubules, and there are strong indications that plant viral movement proteins can associate with the cytoskeleton. Could the cytoskeleton play a role in transporting complexes to the NPC for nuclear import?

Immunolocalization of At-IMP α in tobacco protoplasts is suggestive of cytoskeleton, extending from the nucleus throughout the cytoplasm to the cell periphery. In addition as noted previously, At-IMP α like other importin αs contains hydrophobic armadillo repeats implicated in protein-protein interactions including association with the cytoskeleton Furthermore, At-IMP α cannot be fully depleted from the cytoplasm of permeabilized cells indicating a tight association with intracelluar components. These observations prompted Smith and Raikhe to investigate the role of the cytoskeleton in NPC targeting using double-immunofluorescence and confocal microscopy. Importin α was found to colocalize with microfilaments and microtubules in tobacco protoplasts, whereas depolymerization of cytoskeleton results in loss of the cytoskeleton-like staining pattern. Depolymerization of microtubules results in diffuse cytoplasmic staining. Interestingly, depolymerization of microfilaments results in accumulation of receptor in the nucleus, suggesting that microfilaments may be involved in retention of importin á in the cytoplasm. An examination of At-IMP α association in vitro in a cytoskeleton-binding assay indicates that association with microtubules and microfilaments requires the presence of a functional NLS. The NLS-dependent association of At-IMP α with cytoskeleton may represent a mechanism for the assembly and transport of import complexes to the NPC. Based upon the data, a working model has been proposed in which microfilaments serve as sites for assembly of importin α-NLS protein complexes. Transport to the NPC would likely require the participation of a microtubule motor protein. It is possible that proteins translated from polysomes associated with the cytoskeleton could be assembled into complexes following synthesis. The model is supported by observations of the movement of NLS-containing substrates along neurons toward the nucleus, which is microtubule dependent. It is unclear at this time what role importin β would play in the formation of cytoskeletal import complexes. Although At-IMP α appears to function in import in an importin β-independent manner, this is probably not true for other importin αs. Other connections between importins and the cytoskeleton are becoming apparent. For example, it is now known that importin β can inhibit microtubule assembly in *Xenopus* egg extracts, and it is suggested this serves to suppress aster assembly until interaction with Ran-GTP releases importin β and assemble can proceed during mitosis.

Nuclear Export

Recently the nuclear shuttling protein BR1 from SqLCV protein has been examined for a *nuclear export signal* (NES) that would permit its export from the nucleus as has been found in the viral protein HIV. This signal, like NLSs, is not strictly conserved, but is a leucine-rich hydrophobic sequence of 10 to 13 amino acids. Such as motif was found within BR1, and when its three leucine residues were mutated to alanines, viral pathogenicity was lost indicating the essential nature of these residues. It was further reasoned that only export competent BR1 protein could be relocalized to the cell periphery by association with BL1. In fact, GUS-BR1 fusion proteins when coexpressed in tobacco protoplasts are only relocalized to the cell periphery when the

NES is present. Fusion proteins without the NES remain in the nucleus. Using this assay it was demonstrated that the NES from the *Xenpous* transcription factor IIIA can functionally replace the endogenous NES of BR1 and even restore viral infectivity. A homologue of the NES transporter CRM1 also known as exportin was recently cloned from *Arabidopsis* (AtXPO1) and characterized functionally in a yeast two-hybrid assay. The AtXPO1 interacts with the functional NES from AtRanBP1a as well as the NES from HIV Rev. In another protoplast expression system utilizing an NLS and NES fused to GFP, the NES from HIV Rev functions in export, which is inhibited by leptomycine B as in animal cells. The NES from AtRan BP1a, but not a version in which three leucines were mutated to alanines, also functions in this assay. These are the first examples in plants of characterized NESs and suggest that along with the protein import pathway, there is functional conservation with animals and yeast. Again, the interesting biology will reside in the details of how such pathways are utilized in plants, and the development of a straightforward assay for nuclear export in plants should encourage progress.

Regulation of Nuclear Import and Export of Proteins

The nuclear envelope separates the theatres of two major cellular processes in eukaryotes: transcription takes place in the nucleus whereas proteins are synthesized in the cytoplasm. The localization of these processes in two different compartments of the cell implies that macromolecules must be exchanged very rapidly and efficiently between the nucleus and the cytoplasm in order to ensure proper regulation of signaling and metabolism of a living cell.

All transport processes across the nuclear envelope take place at very large multi-protein complexes called *nuclear pore complexes* (NPCs), which provide the gates between the nucleus and the cytoplasm. It has been known for many years that nucleo-cytoplasmic transport processes of macromolecules are receptor-mediated, but only in the last few years was it revealed that most of the nuclear transport processes depend on importin β-like protein receptors. The genes encoding these receptors constitute a small gene family, and they are named after the member that was first identified at the molecular level. These receptor proteins share limited sequence identity, they can interact with the regulatory GTPase Ran in its GTP-bound form, they are able to interact with nucleoporins, and they shuttle continuously between the nucleus and the cytoplasm. For every class of the many different macromolecular transport cargos, like proteins, mRNA, tRNA, ribosomal subunits, snRNPs, there exists a specific receptor, or a combination of receptors.

Nuclear Import of Proteins

Karyophilic proteins that contain a classical basic nuclear localization signal (NLS), whether monopartite (SV40-like) or bipartite, are imported into the nucleus by the import receptor importin β (also called karyopherin β). However, importin β does not bind directly to these import substrates. In these cases, importin α (also called karyopherin α) serves as an adapter between the cargo protein and the nuclear import receptor itself. Therefore, the first step in the nuclear import of a NLS-containing protein is the specific recognition of the NLS in the cytoplasm by importin α, which constitutes the soluble NLS receptor. Importin β binds co-operatively to the basic amino terminus of importin α called importin β-binding (IBB) domain. This leads to the formation of the triple import complex consisting of importin α, importin β and the NLS-containing protein, which then docks as a single entity to the NPC via importin β and is subsequently translocated into the nucleus. Within the nucleus, the concentration of Ran-GTP is high. Since the binding sites on importin β for Ran-GTP and importin α overlap, Ran-GTP is competing with importin α for the binding to importin β, which leads to the dissociation of the triple import complex and hence to the release of the import cargo in the nucleus. Importin β is recycled to the cytoplasm in complex with Ran-GTP, although there is also evidence that it may leave the nucleus in a Ran-independent manner. After hydrolysis of GTP on Ran in the cytoplasm, Importin β is ready for a new import cycle. Importin α is exported to the cytoplasm by a specific export receptor, termed CAS, which is also a member of the importin β family.

Although most of the proteins that are imported by importin β depend on recognition by importin α, there are also proteins that do not need this adapter. These proteins bind directly to the import receptor importin β and contain a more archaical nuclear localization signal that resembles the IBB domain of importin α. There are also many other proteins that are imported into the nucleus and that contain neither such an import signal nor a classical basic NLS. Nuclear import of these proteins depends on other receptors, which in most cases are also members of the importin β family.

Nuclear Export of Proteins

Nuclear export of proteins was first investigated in the case of the inhibitor (PKI) of the cAMP-dependent protein kinase (PKA) in humans and in the case of the HIV protein Rev. Like in the case of nuclear import, these proteins contain short signals with a low degree of primary sequence conservation, which are permanent and transferable. These *nuclear export signals* (NES) are both sufficient and necessary to label a protein for rapid nuclear export and they show a specific spacing of long-chain, hydrophobic amino acid residues, often leucines. These leucine-rich NESs are specifically recognized in the nucleus by another member of the importin β family of transport receptors, termed exportin 1 (XPO1, also called CRM1). Here, the export receptor XPO1 binds its substrate directly and co-operatively with the GTPase Ran in its GTP-bound form, the concentration of which is high in the nucleus. This triple export complex next interacts with nucleoporins on the nuclear side of the NPC via XPO1 and is subsequently translocated into the cytoplasm. There, the export complex is dissociated by the co-ordinated action of two cytosolic regulatory proteins, the Ran-binding protein 1 (RanBP1) and the GTPase-activating protein for Ran (RanGAP1). This leads to the release of the export cargo into the cytoplasm, and the hydrolysis of GTP on Ran renders this step irreversible. The export receptor XPO1 re-enters the nucleus on its own, due to its ability to interact with nucleoporins, and is ready for a new export cycle. Although this is the best-investigated scenario for the nuclear export of proteins, the leucine-rich NESs are not the only signals that confer nuclear export, as exportin 1 is not the only export receptor.

Regulatory GTPase Ran

The above description of the basic steps of nuclear import and export of proteins already demonstrated that the GTPase Ran plays a central role in the regulation of the directionality of nuclear transport processes that depend upon the nuclear transport receptors of the importin β family. Ran is a remarkable protein since it is the only Ras-like GTPase which is soluble and which shuttles continuously between two cellular compartments. In addition, its GTPase cycle is distributed in a characteristically asymmetric fashion between the nuclear and the cytoplasmic compartments, since the proteins regulating the Ran GTPase cycle show very specific localizations. The guanine nucleotide-exchange factor for Ran, called RCC1 in humans, is bound to chromatin and therefore shows a strictly nuclear localization. In contrast, the GTPase-activating protein for Ran (RanGAP1) and its co-activator RanBP1 are confined to the cytoplasm. As a consequence, the concentration of Ran-GTP is high in the nucleus and low in the cytoplasm.

Based on these characteristics of the Ran GTPase cycle, Ran is able to provide directionality to nuclear transport processes since the importin β-like transport receptors interact with Ran-GTP and their substrates in a specific manner, depending on whether they are import receptors or export receptors. Importins, like importin β, bind to their substrates only in the absence of Ran-GTP, whereas exportins, like XPO1 and CAS, bind to their substrates only in presence of and co-operatively with Ran-GTP. This explains why import complexes form only in the cytoplasm and are dissociated in the nucleus, while export receptors bind to their cargo only in the nucleus and release it in the cytoplasm.

Shuttling of Ran between the nucleus and the cytoplasm is mainly accomplished on one hand by nuclear export of Ran-GTP complexed with an importin β-like transport receptor, as part of an export complex. On the other hand, nuclear import of Ran is ensured by a specific import receptor for Ran-GDP termed NTF2. The latter is an example for receptor-mediated nuclear import which does not depend upon importin β-like transport receptors. Recently, the NTF2-related export protein 1 (NXT1) was identified as a nuclear transport factor that continuously shuttles between the nucleus and the cytoplasm. In contrast to NTF2, NXT1 binds Ran-GTP and regulates both XPO1-dependent and XPO1-independent nuclear export processes. In this way, NXT1 may also contribute to nuclear export of Ran. Hydrolysis of GTP on Ran in the cytoplasm by the regulatory proteins RanBP1 and RanGAP1 is thought to ensure completeness of the Ran GTPase cycle and hence to ensure recycling all factors necessary for nuclear transport processes rather than to directly provide energy to these processes. It has been shown that GTP hydrolysis on Ran is not necessary for translocation across the NPC.

Interestingly, Ran seems to be a multi-functional protein since, in addition to its central role in regulating the directionality of nuclear transport processes, Ran-GTP also plays a major role in regulating spindle-formation during mitosis in mammalian cells. It has been reported that the mitotic role of Ran is largely mediated by importin β, which inhibits spindle formation and sequesters protein factors required for an aster promoting activity. In addition, Gruss et al demonstrated that importin α binds and thereby inactivates a microtubule-associated protein (TPX2) that is required for spindle formation. TPX2 is displaced from importin α by Ran-GTP. Thus, Ran-GTP functions by locally releasing protein cargoes from nuclear transport factors, which serve to regulate spindle formation in mitosis.

Plant Factors and Plant-Specific Features of Nuclear Transport

At least six genes encoding importin α homologues have been described in *Arabidopsis thaliana*. One importin α protein, AtIMP alpha, was shown to be able to bind to three different classes of nuclear import signals that are present in plants, another importin α protein was shown to bind to *Agrobacterium* VirD2. In addition, genes encoding importin β homologues have also been isolated from rice, *Capsicum* and tomato, whereas genes encoding importin β homologues have been characterized only in rice. Recently, the nuclear export receptor XPO1 that specifically binds to leucine-rich NESs has been functionally characterized in *Arabidopsis*.

The *Arabidopsis* protein shares 42-50% identity with its functional human, *S. cerevisiae* and *S. pombe* homologues, it interacts with *Arabidopsis* Ran and with NESs of plant and human proteins including the HIV protein Rev, and is sensitive to the cytotoxin leptomycin B. Export activity within a plant cell was demonstrated in vivo using an assay system with GFP fusion proteins, which also revealed that the Rev NES is fully functional in plants, thereby demonstrating the high conservation of this nuclear export pathway between species phyla and even kingdoms.

Genes encoding the GTPase Ran have been isolated from several plant species, including 5 genes from tobacco and 3 genes from *Arabidopsis*. In addition, three genes encoding the Ran-specific regulator RanBP1 have been isolated from *Arabidopsis*, two of them by a yeast two-hybrid screen with a Ran mutant that is permanently blocked in its GTP-bound form.

Although analysis of plant nuclear transport factors as well as nuclear import and export in plants has confirmed that the basic processes are highly conserved between organisms, there are some plant-specific features. The development of in vitro nuclear import systems for plants has revealed that nuclear import is not inhibited at 4°C, as in animal cells. Also, in contrast to animal nuclear import, the lectin wheat germ agglutinin does not block this process in plants. Interestingly, At-IMPα, one of the plant importin α proteins, was reported to be able to function as nuclear import receptor without binding to importin α. In addition, importin-α has been co-localized with elements of the cytoskeleton in plant cells, suggesting an implication of these structural elements in nuclear import. It is not known to date, whether the latter finding is unique to plants or whether this property of importin α is also shared in other organisms.

Regulation of Nuclear Transport as a Tool to Regulate Signaling

Taken together, a cell has to invest plenty of energy to guarantee continuous and rapid exchange of macromolecules between the nucleus and the cytoplasm. All the proteins involved, from nucleoporins to nuclear transport receptors including their regulatory proteins, must be produced in great numbers. In addition, energy in the form of GTP is consumed during each and every transport cycle. However, the nucleocytoplasmic transport of macromolecules also provides numerous possibilities for the regulation of signal transduction processes. Not only the activity of specific factors but also their localization in a specific compartment may be subject to regulation.

Since the first step of the nuclear import of a NLS-containing protein is its binding to importin α, interference with this recognition step provides a perfect checkpoint for the regulation of the localization of such a protein. If importin α is unable to interact with the NLS of a protein, the protein in question will stay cytosolic. There are many ways to interfere with the binding of importin α to a NLS, like protein modifications such as phosphorylation, interaction of the NLS-containing protein with another protein shielding the NLS from importin α, regulated conformational changes of the NLS-containing protein which may also result in shielding of the NLS, and cytoplasmic anchoring of the NLS-containing protein by interaction with a fixed structure, overriding nuclear import. The fact that such actions which interfere with binding by importin may be eliminated or induced (i.e., switched off and on) very quickly upon a signal makes these mechanisms perfect tools for the regulation of the localization of a protein, such as a transcription factor.

Although the regulation of nuclear import has been investigated much more extensively, the presence of a NES within a protein provides the same potential for regulatory mechanisms with respect to protein localization as does an NLS. If a protein contains both signals, combinations of different mechanisms of the regulation of protein localization become possible. One transport step may be default and the other regulated, or both may be regulated, resulting in a binary switch mechanism. In addition, not only may the localization of a protein as such be regulated, but also the half-life of its localization within the nucleus or the cytoplasm may be actively controlled. This provides an alternative to protein turnover as regulatory mechanism to control the half-life of the localization of a protein in the cytoplasm or in the nucleus.

All these possibilities for regulating protein localization are indeed operational. Together with the regulation of protein activity, they provide a network of mechanisms for the control of signaling. Signal transduction by light provides very illustrative examples for the importance of nucleocytoplasmic partitioning in plant signaling cascades. Research of the last few years has uncovered regulation of the localization of proteins at different levels of light signal transduction pathways.

Nucleocytoplasmic Partitioning in Light Signal Transduction

Plants and light

Plants as immobile organisms have to cope with changing environmental conditions at the place where they grow. Because plants are not able to escape unfavorable conditions they depend upon reliable information about environmental factors like temperature, water supply, and light. One of the most important environmental factors for plants is light. Light not only serves as source of energy for photosynthesis, but also constitutes a morphogenic signal that is perceived by plants to sense changes in the natural environment. Light regulates a wide range of developmental processes and adaptations during the entire life cycle. The onset of seed germination, the developmental switch from skoto- to photomorphogenesis of the young seedling, the detection of neighbors competing for the incident light or the onset of the generative phase and flowering are all driven by light. To perceive the surrounding light conditions, plants have evolved at least three different photoreceptor systems to monitor light quality and quantity ranging from UV to the infrared parts of the spectrum. These are (i) the UV-B receptors characterized by action spectroscopy, (ii) the blue/UV-A receptors cry1, cry2 and phototropin and (iii) the red/far-red reversible phytochromes.

With regard to signal transduction of light, nucleocytoplasmic partitioning has been shown to play a role in the case of at least three classes of molecules. These are the phytochrome photoreceptors themselves, bZIP transcription factors, and the negative regulator of photomorphogenesis COP1.

Phytochrome system

Phytochromes (phy), a group of plant photoreceptors involved in a number of light-dependent processes are the best characterized photoreceptors. In higher plants, phytochromes are encoded by small multigene families. In *Arabidopsis*, five members are known (*phyA* to *phyE*). Phytochromes are synthesized in darkness in their physiological inactive red light-absorbing form (Pr). The inactive Pr-form can be reversibly transformed by absorption of a photon into the physiological active, far-red light-absorbing form (Pfr). Each phytochrome is thought to have a different role in light signaling; PHYA and PHYB have been best characterized. In *Arabidopsis* these two phytochromes are expressed in most cell types. PHYA is most abundant in dark-grown plants, it is light-labile in its Pfr-form and responsible for the non-photoreversible VLFR (very low fluence response) and the HIR (high irradiance response) classes of phytochrome-regulated responses. PHYB is the receptor for the classical red/far-red reversible (LFR) and continuous red light responses and is light-stable.

During the past few years insights into the mechanisms of phytochrome signaling have been substantially increased by genetic and molecular approaches. The identification of components involved in phytochrome-mediated signal transduction and recent studies about the intracellular localization of the photoreceptors implicate a tightly regulated interaction of the nuclear and cytosolic compartments.

Phytochrome-regulated intracellular partitioning of phytochromes

Until recently the dominating view has been that plant photoreceptors are localized in the cytoplasm. Physiological studies in algae, mosses, and ferns showed action dichroism for cloroplast orientation, polarotropism, and phototropism. These observations indicated that the photoreceptors regulating these responses are localized in the cytoplasm in an oriented manner, presumably in association with the plasmalemma or other membrane structures. With regard to phytochromes, this hypothesis was supported by immuncytochemical studies on the Pfr-dependent formation of sequestered areas of phytochrome (SAPs) in the cytoplasm of monocotyledonous seedlings, despite observations of phytochrome-dependent transcription in nuclear run-on experiments. The view of an exclusively cytoplasmic localization of phytochromes was further challenged by Sakamoto and Nagatani. These authors could demonstrate nuclear localization of phyB fragments fused to the GUS (β-glucoronidase) reporter in transgenic plants, pointing to functional NLS sequences in the photoreceptor. Additionally, in this study a substantial increase in the amount of phytochrome in purified nuclei of plant tissues irradiated with red light was observed. More recently, members of the same laboratory could complement a PHYB-deficient *Arabidopsis* mutant by using a protein fusion consisting of full-length PHYB and the in vivo marker protein GFP. The results clearly show a light-dependent nuclear import of PHYB:GFP accompanied by the characteristic formation of speckles of PHYB-GFP inside the nuclei. A more detailed study, also applying PHY-GFP fusion proteins, extended this observations to tobacco plants. The authors could demonstrate nuclear uptake of phytochrome A and phytochrome B, dependent on the respective light requirements for the functionally distinct photoreceptors. Whereas nuclear import of PHYB:GFP requires red light (high amounts of Pfr) and is red/far-red photoreversible (LFR), nuclear uptake of PHYA:GFP can be initiated even by short far-red pulses (VLFR, low amounts of Pfr) and continuous far-red light. Additionally, it was shown that in continuous light the kinetics of Pfr-dependent nuclear import of PHYA is an order of magnitude faster than that of PHYB. Further studies on the light-regulated partitioning of phytochromes have refined and extended these observations. The analysis of nuclear import and speckle formation of the phytochromes of Arabidopsis with supplementary functions revealed for PHYC and PHYE similar kinetics as for PHYB. Interestingly, although PHB and PHYD are closely related genes they showed the largest difference. PHYD:GFP displayed very slow and heterogeneous nuclear import and only one or two nuclear speckles were found after 8 hours of irradiation. The capacity to complement the corresponding photoreceptor mutant in *Arabidopsis* was also demonstrated for a PHYA:GFP fusion protein. The validity of the above-described light requirements of nuclear import for the endogenous protein could also be demonstrated by immunolocalization analysis on the intracellular partitioning of PHYA in pea seedlings. In a reverse approach the localization of functionally impaired versions of phytochrome photoreceptors were analysed which originally were identified by genetic studies. In all cases investigated the respective amino-acid exchanges in PHYA or PHYB lead to aberrant localization patterns.

Taken together, these results show a strong correlation of the light requirements for physiological responses regulated by the respective phytochromes and the intracellular partitioning of these molecules and suggest a dominant role in light signal transduction for photoreceptors imported into the nucleus. This view has been strongly supported by recent findings derived from genetically characterized photo-transduction mutants in *Arabidopsis*.

Nuclear components of the phytochrome signaling pathway

The analysis of mutants defective in PHYA- and/or PHYB-mediated signal transduction has identified a number of proteins involved in the regulatory process leading to the respective physiological responses.

As regards the nuclear import and function of active photoreceptors, of special interest was the identification of an *Arabidopsis* mutant that showed altered PHYB-dependent light signaling (*poc1*). The corresponding gene product, also characterised as PIF3 (phytochrome interacting factor), which was isolated by a two-hybrid approach, is a *basic helix-loop-helix* (bHLH) transcription factor of nuclear localization. In the heterologuous yeast system PIF3 physically interacts preferentially with the Pfr confomers of PHYB and, to a lesser extend, of PHYA. Recently, it was shown in planta that light signals lead in deed to a rapid and transient co-localisation of PIF3 with the Pfr forms of PHYA, PHYB, PHYC and PHYD in small speckles within nuclei of Arabidopsis plants expressing microscopically distinguishable CFP and YFP fusions of the respective protein pairs. Subsequently to this interaction, these early nuclear structures disappear within minutes and the negative regulator PIF3 is degraded with a half-time of about 10-20 minutes. Because the formation of the early and transient type of nuclear speckles of PHYB:GFP is lost in a *pif3* mutant it is tempting to speculate that these structures are involved in marking the bHLH factor for degradation. The identification of another bHLH transcription factor which is a positive regulator in PHYA signaling supports the hypothesis of a direct and distinct effect of phytochromes imported into the nucleus on transcriptional networks regulating photomorphogenesis (*hfr1*). The Pfr forms of PHYA and PHYB can physically bind only to heterodimers of HFR1 and PIF3, but not to HFR1 homodimers. Because of the low abundance of *hfr1* mRNA in plants treated with continuous red light compared to plants irradiated with far-red light it is tempting to speculate whether the abundance of this and related factors could determine the specificity of phytochrome signaling. In this light, it would be of major interest to elucidate the binding specificities of bHLH transcription factors regarding homo- and heterodimerisation on the one hand and the binding of the dimers to promoter elements of individual phytochrome-regulated genes on the other hand.

However, to elucidate the exact function of the various types of PHY-associated nuclear speckles in signal transduction, it is important to identify their components on a molecular level. In our lab, we recently purified PHYB-containing nuclear speckles from 4 week-old Arabidopsis plants and analysed their composition by MALDI-TOF. About 80% of the proteins identified share high homology to proteins shown to be present in the interchromatin granual clusters (ICGs) of animal cells. The precise biological role of ICGs is not known even in animal cells but they are considered to be involved in the storage, modification and recruitment of factors necessary for transcription and splicing. We note, however, that the N-terminal fragment of PHYB fused to GUS:NLS did not show nuclear speckle formation in continuous light, yet it successfully complemented the phenotype of a *phyB* mutant. These data may indicate that nuclear import but not the formation of nuclear complexes may be essential for PHYB signal transduction. Thus we conclude that, given their multiple sizes, their transient and dynamically changing appearance, it remains a challenging task to understand the molecular functions of the light-induced nuclear protein complexes.

Other factors involved in phytochrome signal transduction have also been shown to be nuclear proteins. In the case of PHYA-signaling SPA1, FAR1, FIN219, and EID1 have been characterized as nuclear factors, further underlining the importance of the nuclear compartment in light signaling.

Potential mechanisms of cytosolic retention and nuclear import of phytochromes

In etiolated seedlings and dark adapted plants, where phytochromes exist in their photobiologically inactive Pr forms, PHYB:GFP and PHYA:GFP are localized in the cytoplasm. A mutated version of PHYB:GFP which is not able to bind its chromophore, is confined almost exclusively to the cytoplasm of tobacco irrespective of light conditions. These results indicate that the conformational change of Pr to the physiological active Pfr form is a necessary pre-requisite for the nuclear import of phytochromes. Localization experiments with truncated versions of PHYB:GFP fusion proteins clearly indicate the presence of a functional NLS within the C-terminus of the photoreceptor. The light-independent exclusive nuclear localization of the C-terminal half of PHYB suggests an important role for the N-terminal part of phytochromes in cytosolic retention in darkness. In contrast to tobacco, in Arabidopsis the addition of extra NLSs to PHYB:GFP does lead to a light-independent nuclear import of this modified photoreceptor protein which physiologically results in a hypersensitive PHYB phenotype. Taken together, these data suggest a cytosolic retention mechanism for phytochrome B in its Pr form.

The switch making possible the interaction with the nuclear import machinery could be the release of the photoreceptor from cytosolic retention by the light-dependent conformational change to its Pfr form. So far no information is available on the molecular mechanism of the cytosolic retention of phytochromes, which is therefore a major target of signal transduction research at the present time.

Phytochrome-regulated nuclear import of the bZIP transcription factor CPRF2

In addition to their suggested direct function in the nucleus, photoreceptors, especially PHYA has also been shown to mediate cytosolic events in its Pfr form. Besides acting via trimeric G-proteins, cGMP and calcium/calmodulin on greening and anthocyanin production as revealed by pharmacological studies, PHYA is a phospoprotein and is considered to be a protein kinase. Recently, PKS1 (phytochrome kinase substrate 1), a cytosolic protein identified in a yeast two-hybrid screen, was demonstrated to be phosporylated by PHYA in vitro.

In addition, Pfr-dependent phosporylation events in the cytoplasm could lead to nuclear import of downstream regulatory proteins, which was shown to be the case for a family of basic leucin-zipper motif (bZIP) transcription factors. CPRF proteins (common plant regulatory factors) bind to G-box promoter elements and are thought to play a role in regulating light-responsive genes in parsley, like the gene encoding chalcone synthase. In an initial biochemical study analyzing cellular fractions of parsley cells, a light-regulated and phosphorylation-dependent translocation of G-box binding proteins from the cytosol into the nucleus was demonstrated. Clear evidence for light-initiated nuclear uptake of G-Box binding factors in parsley was provided by using an in vitro nuclear transport system.

Further characterization of several members of the CPRF gene family by immunocytochemistry and transient expression of GFP fusion proteins revealed that only CPRF2 is localized almost exclusively in the cytoplasm in the dark. The cytosolic retention of this transcription factor is released by red light treatment and is at least partially red/far-red light photoreversible. This observation points to the involvement of phytochrome in this light-regulated translocation process. In good agreement with earlier observations, phytochrome-dependent phosporylation within the C-terminus of CPRF2 by a cytosolic serine kinase is proposed to be a pre-requisite for nuclear import. This study as well as localization experiments using truncated versions of CPRF2 point to the cytosolic retention of this transcription factor in a high molecular weight complex in darkness. Transient expression of truncated CPRF2 fused to GFP in parsley protoplasts leads to the conclusion that two structural motifs in the N-terminus, distinct from the NLS-harboring bZIP domain of the factor are necessary to prevent nuclear import in darkness. Additionally, the N-terminal domain can confer cytosolic retention to another nuclear bZIP factor in domain-swap experiments. It is therefore tempting to speculate that phytochrome-dependent phosphorylation of CPRF2 leads to conformational changes within the protein that releases the factor from a cytosolic retention complex. After release, CPRF2 could interact with the nuclear import machinery, translocate into the nucleus, and bind to light-regulated target genes. Very recently, a putative retention protein of an *Arabidopsis* CPRF2 homolog was identified by a screening approach in a heterologuous system, but its function *in planta* has not been corroborated yet.

Blue light photoreceptors and the intracellular partitioning of the transcription factor GBF2

Phototropin and cryptochromes (CRY) belong to a class of blue light-absorbing photoreceptors controlling UV-A/blue light-dependent responses in *Arabidopsis*. Phototropin is a plasmalemma-associated flavoprotein that is thought to mainly sense the direction of the incident light and that mediates phototropism, the orientation of plants and plant organs towards light. Cryptochromes mediate hypocotyl shortening, cotyledon expansion, and anthocyanin production in blue light and play an important role in the entrainment of the circadian clock. Interestingly, the latter function of setting the circadian clock is also committed by conserved cryptochrome homologues in mammals and *Drosophila*.

Cryptochrome 1 (CRY1) was the first member of the long-sought blue-light-absorbing photoreceptors to be identified and cloned from the HY4 mutant of *Arabidopsis*.CRY1 displays a striking similarity to photolyases, but lacks photolyase activity and has a unique C-terminal extension. FADH, the catalytically active chromophore is attached to the N-terminus of the photoreceptor; the second chromophore could be either a pterin or deazaflavin. A second member of the cryptochrome class of blue light photoreceptors, CRY2, has been identified in *Arabidopsis* and is distinguished mainly by a different C-terminal extension.

Microbeam irradiation indicates that fern homologues of higher-plant CRY photoreceptors are present in the cytoplasm, but they are also found to be associated with the nucleus. Recent studies in the fern *Adiantum capillus-veneris* showed that at least two members of the CRY family are localized in the nucleus and that the nucleo-cytoplasmic distribution of these photreceptors is, at most, only slightly influenced by light. With regard to higher plants, transient transformation of *Arabidopsis* CRY1 fused to GFP in plant cells indicate that the receptor is localized in the nucleus in the dark. Localization of CRY1 in light-treated cells has not been reported yet; recently, however, light-dependent cytosolic enrichment of a C-terminal fragment of CRY1 fused to the GUS marker gene was shown in transgenic *Arabidopsis*. It is therefore tempting to speculate about a similar light-dependent intracellular partitioning of the photoreceptor itself.

Analysis of the intracellular partitioning of the second member of the cryptochrome photoreceptor family, CRY2, was performed independently by two groups, either by CRY2:GUS or CRY2:GFP fusion proteins driven by the constitutive 35S promotor. These studies indicate a nuclear localization of the photoreceptor, but if over-expressed marker fusion proteins represent the endogenous situation properly remains to be elucidated. Blue light irradiation of cells that transiently express a fusion protein consisting of CRY2 and the marker protein RFP (red fluorescing protein) led to the formation of a speckled pattern of CRY2:RFP inside the nucleus. This study also provides evidence for physical interaction between phytochrome B and CRY2 as well as co-localization of both proteins in nuclear speckles forming under irradiation conditions combining red and blue light. This finding is remarkable because phytochromes and cryptochromes often regulate similar photomorphogenic responses and interdependencies of both photoreceptor systems are well-known.

Similarly to the observation of a phytochrome-dependent, red light-initiated nuclear import of CPRF2-GFP in parsley, another bZIP protein, GBF2, was demonstrated to be a candidate for nuclear import that depends upon blue light. By comparative analysis of the intracellular partitioning of GUS fusion proteins of several members of the G-box binding factor (GBF) family of *Arabidopsis* it was shown that GBF2, in contrast to the nuclear GBF1 and GBF4, is evenly distributed in the cytoplasm and the nucleus of soybean cells kept in darkness. When these cells were irradiated with blue light, a substantial increase in nuclear GUS:GBF2 was detected. In contrast, irradiation with red light did not change the intracellular distribution of the bZIP protein, pointing to the involvement of blue light photoreceptors in this process. It is not yet known which of the blue light photoreceptors plays a role in the regulated intracellular partitioning of GBF2. Since binding of members of the GBF family to G-Box promoter elements is necessary for the proper expression of light-regulated genes, and since the trans-acting activity of at least one member of the GBF family has been demonstrated (GBF1), regulated nuclear import of GBF2 could represent an important step in blue light signaling.

Photomorphogenic repressor protein constitutive photomorphogenesis 1 (COP1)

COP1 also provides a well-investigated example for the nucleo-cytoplasmic partitioning of a regulatory protein and its implications for signaling in plants. The COP1 protein is localized in the nucleus during seedling germination in darkness and functions as a repressor of photomorphogenesis by suppressing the expression of light-inducible transcripts in the nucleus. Regulation of COP1 activity is negatively controlled by light and involves a light-dependent re-localization of the protein into the cytoplasm. As a result, COP1 levels are drastically reduced in nuclei of hypocotyl cells that are transferred to light as compared to those that were kept in darkness.

The localization of COP1 also has a tissue-specific component, as COP1 levels are constitutively high in the nuclei of root cells, which do not undergo photomorphogenesis. The COP1 protein contains an amino-terminal zinc binding RING finger domain, a carboxy-terminal WD-40 repeat domain, and in between a domain with the potential to form a coiled-coil structure. Detailed analysis of the localization of COP1 fragments in combination with their physiological effects revealed discrete domains that mediate light-responsive nuclear and cytoplasmic localization of the protein: a bipartite NLS within the core domain that mediates nuclear localization, and an amino-terminal domain including parts of the RING domain and parts of the coiled-coil domain that has cytoplasmic retention activity.

In addition, a motif that mediates targeting of the COP1 protein to subnuclear foci has been described, which overlaps the cytosolic retention signal and the putative α-helical coiled-coil domain. How the switch between the cytosolic and the nuclear localization of the COP1 protein is accomplished at the molecular level is under investigation. COP1 acts as a transcriptional regulator by interaction with other regulatory proteins in the nucleus, like with the COP1-interactive protein 7 (CIP7) or with the bZIP transcription factor HY5. HY5 is constitutively nuclear, binds to G-box motifs within promoters of several light-inducible genes, and is necessary for their optimal expression. Recently, it was shown that the nuclear interaction of COP1 and HY5 leads to the degradation of HY5, which is thereby negatively regulated by COP1 at the level of protein stability.

Concluding Remark

In general, the completion of the *Arabidopsis* genome will facilitate the functional characterization of nuclear transport factors in plants. Many of the factors that are known from animals and yeast and have been identified in the *Arabidopsis* genome on the basis of sequence homology have not been investigated to date. As to the role of nucleocytoplasmic partitioning of proteins involved in the signal transduction of light, several interesting questions are waiting to be solved. To begin with the photoreceptors, the molecular mechanisms of the light-dependent re-localization of phytochromes between the nucleus and the cytoplasm are as yet unknown. Whether nuclear import of phytochromes depends upon importin α/β heterodimers or upon different import receptors is another question to be solved.

The mechanism of their cytosolic retention in the dark and their release in the light may be a more challenging problem, since this is supposed to hold the key for the regulation of the nuclear import of phytochromes. In addition, apart from the nature of the speckles formed after the import of phytochromes in the nucleus and from the molecular mechanisms of the nuclear function of phytochromes, a very interesting question is whether phytochromes are degraded in the nucleus or are, at least in part, transported back to the cytoplasm.

As a consequence of the latter hypothesis, at least a portion of the phytochrome pool in a cell would then show light-dependent shuttling between the nucleus and the cytoplasm. Retention mechanisms and nucleocytoplasmic shuttling are also postulated in case of the bZIP transcription factor CPRF2 and in case of the photomorphogenic repressor COP1, to give only two examples. The molecular mechanisms of these processes may differ from the corresponding processes of phytochromes. However, it would be interesting to know if some of the components that confer light-dependent regulation to nucleocytoplasmic partitioning of phytochromes, CPRF2, and COP1 are shared.

7

NUCLEAR IMPORT OF DNA

Nuclear trafficking of macromolecules usually brings to mind the nuclear import of NLS-containing proteins and certain RNAs and the export of NES-containing proteins and mRNAs. One macromolecule whose nuclear import is often overlooked and under-appreciated is exogenously administered DNA. While extra-chromosomal DNA may not be a "normal" species in the cell, its nuclear localization is integral to the life cycles of many pathogens and necessary for the success of transfections in the laboratory and gene therapy in the clinic. Moreover, the movement of DNA from the cytoplasm to the nucleus remains one of the major barriers to efficient gene transfer and expression. Without localization of DNA to the nucleus, no transcription, replication, integration, maintenance, or "*gene therapy*" can take place. Surprisingly, there has been relatively little attention directed toward either discovering or exploiting the mechanisms used by the cell to direct DNA to the nucleus, despite its importance in gene therapy. The discussion that follows will highlight our working knowledge of the mechanisms of DNA nuclear import in both non-viral and viral systems.

NUCLEAR ENVELOPE IS A BARRIER TO GENE DELIVERY

That the nuclear envelope presents a major barrier to gene transfer and viral infections was realized over 20 years ago in seminal experiments by Capecchi and others in which plasmids that had been microinjected into the cytoplasm were found to be virtually incapable of directing gene expression while those injected into the nucleus were highly proficient for gene expression. Using similar microinjection strategies, Graessman demonstrated that when 1000 to 2000 copies of a plasmid were injected into the cytoplasm, less than 3% of the

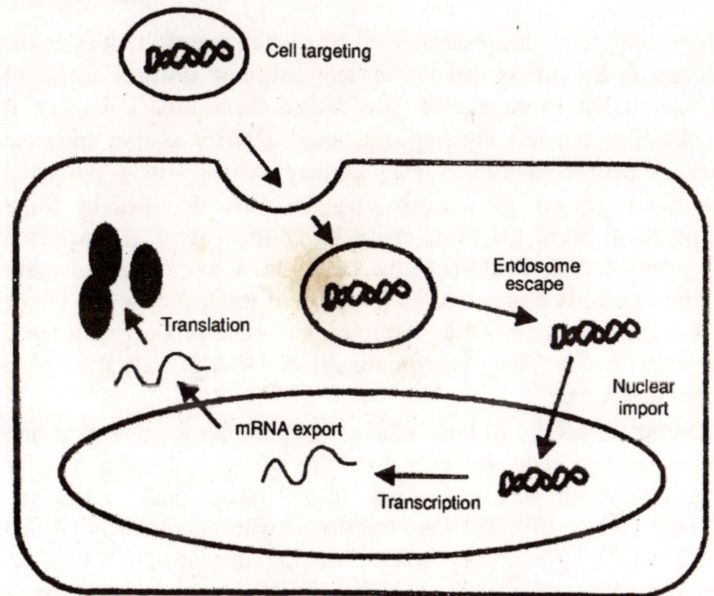

Fig. 7.1. Cellular barriers to gene delivery.

expression was seen as compared to cells injected in the nucleus with the same number of plasmids. Other experiments in a variety of mammalian cell types, as well as in *Xenopus* oocytes has confirmed that plasmids injected into the cytoplasm are much less capable of directing gene expression than those injected into the nucleus. The same is true for many viral genomes: direct nuclear injection, instead of cytoplasmic injection, of the DNA genome from SV40 or a reverse-transcribed retroviral genome resulted in 10 to 100-fold more infectious virus particles in a given time.

During mitosis, the nuclear envelope breaks down, eliminating a major barrier to gene transfer. If plasmids or viral DNA genomes are present in the cytoplasm, they have unencumbered access to the nuclear compartment during this stage of the cell cycle. By contrast, in non-dividing cells, the nuclear envelope provides a substantial barrier to the DNA. Indeed, it has been demonstrated that retroviruses cannot productively infect non-dividing cells due to the fact that their reverse-transcribed viral genomes (rtDNA) cannot traverse the nuclear envelope to gain access to the nucleus. However, when the cells undergo mitosis, rtDNA can localize to the nucleus, integrate, and lead to new rounds of replication. Similarly, it is greatly appreciated by researchers the world over that non-dividing cells (growth-arrested cells, confluent cells, primary cells, etc.) are exceedingly difficult to transfect. Again, the main reason for this is that the nuclear envelope restricts access of exogenous DNA to the site of transcription. In at least a number of the cells of an actively dividing population, DNA is able to enter the nucleus and express itself. Using primary human airway epithelial cells, it was shown that actively dividing cells, identified by BrdU incorporation, were ten times more likely to express a transferred gene product than BrdU-negative cells. In a more recent study, synchronized cells transfected by a variety of techniques in the G2 or G2-M stage expressed 50- to 300-fold more gene product than those transfected in G1. Thus, nuclear import of DNA is crucial to gene transfer reactions.

During transfections, large numbers of plasmids are delivered into the cytoplasm of cells, but only a fraction of these plasmids make it to the nucleus for gene expression. In one recent study, HeLa and CV1 cells (1-2 × 10^5 cells) were transfected with fluorescently labeled plasmids (1.25 μg) using liposomes, and the amount of DNA in the cytoplasm and nucleus was quantified by flow cytometry. Within 2 hours, approximately 2000 copies of plasmid were found to be intracellular in either cell type, and by 24 hours, the number had increased only slightly. These numbers are in close agreement with previous studies. However, when the amount of DNA in the nucleus was measured, more than 60% of the plasmids were in the nuclei of HeLa cells, versus 30% for CV1 cells. One interpretation of these results is that different cell types may be more or less efficient at nuclear targeting of DNA than others. Alternatively, differences in cell division rates between the cells or differences in rates of cytoplasmic degradation of plasmids could also account for the differences. Nonetheless, relatively few plasmids enter the nuclei of dividing cells, let alone those of non-dividing cells. In studies from our laboratory, it took approximately 30 to 100-times more plasmid injected into the cytoplasm of dividing cells to observe the same level of gene expression as compared to nuclear-injected DNAs, again illustrating that nuclear import is a relatively inefficient process, even when the nuclear envelope breaks down.

Nuclear Import of DNAs in Non-Dividing Cells

Based on the data above, it could appear that it is necessary to use dividing cells for successful transfections and certain viral infections (e.g., retroviruses). The question then arises, does DNA ever enter the nuclei of non-dividing cells? Based on a number of criteria and experimental data, the answer is yes. Indeed, a number of viruses, including HIV, SV40, adenovirus, herpes simplex virus, and hepatitis B virus and bacterial pathogens such as *Agrobacterium tumefaciens* have developed very efficient ways to infect non-dividing cells. Furthermore, plasmids have been successfully delivered to a variety of tissues in animals and humans that are made up largely of cells that either do not divide or do so very slowly. Studies over the past 5 to 10 years have begun to elucidate some of the pathways used by viruses and plasmids to enter the nuclei of non-dividing cells. The common theme of these various mechanisms is that they all employ proteins complexed to the DNA for nuclear transport. The next section will highlight some of the better studied non-viral, viral, and bacterial systems.

Plasmid Nuclear Import

Nuclear import of plasmids in the absence of cell division was first demonstrated by Jon Wolff and colleagues. Plasmids were microinjected into the cytoplasm of cultured myotubes and nuclear localization was followed either directly using biotin-labeled plasmids, or indirectly by gene expression from the plasmid. They showed that plasmids were able to localize to the nucleus in these quiescent cells in a dose-dependent manner in an energy-requiring process. Co-injection of the lectin WGA that blocks signal-mediated transport through the NPC inhibited gene expression, leading the authors to conclude that the DNA entered the nuclei through the NPC. However, the import they detected was not efficient. Using biotin or gold-labeled plasmids, they were unable to detect nuclear DNA after cytoplasmic injection until 20 hours post-injection. Again, this illustrates the inefficiency of DNA nuclear uptake.

In a similar set of experiments, but using epithelial cells, our laboratory also showed that plasmids were able to enter the nuclei of non-dividing cells using an NPC-mediated pathway. Synchronized, confluent (i.e., contact-inhibited), or aphidicolin-arrested cells, as well as asynchronous populations have all been used to demonstrate that plasmids can enter the nuclei in the absence of cell division. To confirm that the cells did not divide over the course of our experiments, fluorescent markers that were too large to translocate across the NPC were coinjected with the DNA; if the marker localized to the nucleus, it indicated that the cells had either divided during the course of the experiment or that the DNA had been injected into the nucleus accidentally. Our initial experiments were performed in African green monkey kidney epithelial cells using protein-free, purified SV40 DNA (5,243 bp) which was detected post-injection by in situ hybridization. The advantage of this approach is that it is direct and does not rely on the transcriptional activity or post-transcriptional processes that must occur in order to detect the expression of a gene product as an indicator of DNA nuclear import. When SV40 DNA (about 5,000 copies per cell) was injected in to the cytoplasm, the DNA could be detected diffusely within the cytoplasm immediately following injection. Within 2 to 4 hours of injection, much of the in situ signal was localized in the perinuclear region, perhaps suggesting that the DNA was accumulating at the nuclear envelope awaiting import. Finally, by 6 to 8 hours post-injection, the majority of the DNA was localized to the nucleus. Once in the nucleus, the DNA accumulates in distinct regions of the nucleus that co-localize with proteins involved in transcription and splicing (e.g., SC-35), indicating that the DNA is functional for transcription. This time course of SV40 DNA nuclear accumulation is in agreement with that observed by Nakanishi et al, who followed nuclear import by measuring large T-antigen expression from the cytoplasmically injected SV40 DNA or virions.

Nuclear transport of SV40 DNA was shown to utilize the NPC for its translocation based on the inhibitory effects of the lectin WGA or an antibody (mAb414) that has been shown to block the central pore of the NPC. Energy depletion also inhibited nuclear localization of the DNA as did low temperatures. These results demonstrate that plasmids enter the nucleus using the same nuclear pore complexes as do NLS-containing proteins and small RNAs. It was also demonstrated that nuclear import of the DNA was not affected by inhibition of translation using either puromycin or cycloheximide. However, when RNA polymerase II-mediated transcription was inhibited using several different drugs, nuclear import of SV40 DNA was halted. A similar link between transcription and nuclear transport has also been observed for the import of the A1 hnRNA binding protein. The A1 protein shuttles into and out of the nucleus using the importin homologue transportin. When RNA polymerase II activity is inhibited, A1 is not re-imported into the nuclei of heterokaryons, while other nuclear proteins are transported. Thus, transcription is required for nuclear import of A1 (as well as plasmids), although how remains unclear. One possibility for the transcription-dependent plasmid nuclear import is that proteins similar to A1 in their dependence on active transcription for nuclear import, bind to plasmids in the cytoplasm and facilitate their nuclear uptake. However, at present, this remains to be determined.

Although protein-free SV40 DNA is readily taken up by nuclei of non-dividing cells, many other plasmids are not. When plasmids including pBR322, pUC19, and pGL3 basic are injected into the cytoplasm, they do not localize to the nucleus within 8 to 12 hours of injection. However, these plasmids do gain access to the nuclei when the cells divide. The simplest interpretation of these results is that plasmid nuclear import is sequence-specific and that the SV40 genome contains a sequence that can mediate nuclear uptake. Indeed, when as little as 50 bp of the SV40 enhancer region is cloned into any of the other bacterial plasmids that normally remain cytoplasmic, they are targeted to the nucleus with the same kinetics as the entire 5.2 Kbp SV40 genome. Furthermore, the intranuclear localization of the plasmids are the same: they remain absent from nucleoli and have a punctate staining pattern in the nucleoplasm.

The SV40 enhancer contains binding sites for a number of well-defined, general transcription factors. Among others, AP1 (fos and jun), AP2, AP3, NF-κB, Oct-1, TEF-I, and TEF-II have been shown to bind to this 72 bp region of DNA. Like other nuclear proteins, these transcription factors are synthesized in the cytoplasm and contain NLSs for their nuclear import. Under normal circumstances in the cell, once translated, these proteins are transported into the nucleus where they recognize their binding sites within chromosomal promoters and enhancers and bind to the DNA to carry out their roles in regulation. However, if exogenous DNA is present in the cytoplasm (as in the case with transfections or gene therapy), the transcription factors can bind to their binding sites on this DNA to form a protein-DNA complex. One or more of the transcription factors bound to the DNA may have an NLS that is exposed at the surface of the complex such that it can interact with the importin machinery for nuclear import. Thus, the cytoplasmic plasmid containing the SV40 enhancer will be transported into the nucleus by a "*piggy-back*" mechanism, similar to the mechanism of the HIV preintegration complex.

One caveat for the above model for sequence-specific DNA nuclear import is that at least one or more of the transcription factors that bind must possess an NLS that is away from the DNA binding domain and free to interact with the importin proteins. Thus, zinc-finger transcription factors such as SP1 would not be able to mediate DNA nuclear import because the NLS is embedded within the DNA–binding zinc-finger domain; once bound to the DNA, the NLS would be buried at the protein-DNA interface. Similarly, the NLS of the c-jun component of the AP1 transcription factor would also be unable to bind to the importins because its NLS is immediately adjacent to the DNA binding region of the protein. By contrast, NF-κB could potentially be the adapter protein between DNA and the import machinery because its DNA-binding domain and NLS are structurally separated.

Based on the above model for plasmid nuclear import, it could follow that any eukaryotic promoter or enhancer sequence could act as a scaffold for transcription factor binding and as a nuclear targeting sequence. However, this does not appear to be the case. Several very strong viral promoters have been tested for their ability to promote nuclear import of DNA, and found not to have import activity. The immediate early promoter and enhancer from CMV (CMV_{iep}), the Rous sarcoma virus LTR, the Moloney murine leukemia virus LTR, and the herpes simplex virus *thymidylate kinase* (TK) promoter all are incapable of causing a plasmid to localize to the nucleus in the absence of cell division. Although each of these promoters and enhancers contain multiple binding sites for a variety of transcription factors, they do not function as DNA nuclear targeting sequences. The simplest explanation would be that the SV40 enhancer binds to one or more proteins that none of the others bind to and that these unique proteins are responsible for import. The only identified transcription factor that binds to the SV40 enhancer and not the other sequences is AP2. However, AP2 has been shown to bind to another promoter that is not targeted to the nuclei of all cells, so it is not the responsible protein. The more likely explanation for why the SV40 sequence is unique is that it is the overall organization and structure of the transcription factor-DNA complex that is important and that more than one transcription factor is needed for nuclear localization. To support this hypothesis, our laboratory has examined the nuclear import activity of plasmids containing mutations that disrupt most of the individual binding sites within the 72 bp enhancer repeat and found that disruption of any individual binding site kills import activity. Thus, it is likely that the three-dimensional structure and organization is necessary for DNA nuclear localization.

Several other sequences have been identified that act as DNA nuclear targeting sequences. The common feature of these sequences is that, unlike the SV40 enhancer which acts in all cell types, they act only in specific cell types. Based on our model for plasmid nuclear localization, it is possible that binding of cell-specific transcription factors to DNA could result in cell-specific nuclear import in those cells in which the transcription factors are expressed; in cells that do not express the proteins, no DNA-protein complexes could be formed and no nuclear import would occur. Indeed, the *smooth muscle gamma actin* (SMGA) promoter, which binds to a set of transcription factors found only in smooth muscle cells is transported into the nuclei of smooth muscle cells, but not the nuclei of endothelial cells, fibroblasts, or epithelial cells. The minimal sequence of the SMGA promoter necessary for nuclear transport encompassed the first ~400 bp of the promoter from –404 to +25. This region contains binding sites for both general (C/EBP and AP2) and smooth muscle specific transcription factors (SRF and Nkx3). When a plasmid containing the 400 bp SMGA promoter was injected into the cytoplasm of SRF-expressing stable transfectants of non-smooth muscle cells (CV1 cells), some of the plasmids were able to localize to the nuclei of the cells, whereas they remained completely cytoplasmic in the parent CV1 cells. However, the level of nuclear import was less than in smooth muscle cells, suggesting that additional transcription factors other than SRF alone are needed for maximal nuclear import of the DNA.

More recently, it has been reported that the incorporation of NF-κB binding sites alone in a plasmid can increase the nuclear localization of the plasmid in HeLa cells in a regulated fashion. In this study, a series of 5 consensus binding sites for NF-κB were cloned into the pGL3-control plasmid and the plasmid was transfected or microinjected into the cytoplasm of cells. Nuclear import was followed indirectly by gene expression or directly by labeling the DNA with a fluorescently-labeled *peptide nucleic acid* (PNA). In transfection studies, the NF-κB site-containing plasmid expressed about the same as the parent plasmid under normal circumstances, but expressed gene product much more robustly in the presence of the NF-κB activator TNF-α. Further, more of the NF-κB plasmid appeared to localize to the nucleus in the presence of TNF-α than did the parent plasmid. Because TNF-α is known to promote the nuclear translocation of NF-κB, the authors' conclusion was that the activated transcription factor bound to its sites on the plasmid and facilitated transport, as in our model. However, it should be pointed out that the plasmids used in this study also contained the SV40 enhancer. Consequently, it is unclear whether inclusion of NF-κB binding sites alone would support nuclear import of the DNA.

Nuclear Import of Plasmids in Cell-Free Systems

To characterize the mechanisms of DNA nuclear import in more detail, and to identify the proteins involved, several groups have utilized digitonin-permeabilized cells. Wolff and colleagues demonstrated that linear fragments of labeled double-stranded DNA can localize to the nuclei of permeabilized cells in a reaction that is inhibited by agents that block the NPC and by energy depletion. DNA nuclear import was also saturable, but was not competed by excess NLS-containing proteins, suggesting that the DNA is entering the nucleus through a pathway distinct from that of classic NLS-containing proteins. In contrast to their, and other's, findings that in microinjected cells 5 to 15 Kbp plasmids can be imported into the nucleus, they found that nuclear uptake of the DNA in permeabilized cells was size-dependent. DNA fragments less than 1000 bp in length were able to accumulate in the nuclei, but longer fragments remained excluded from the nuclei. Furthermore, DNA nuclear import was inhibited by the addition of cytoplasmic extracts. This is in contrast to what has been seen for the nuclear import of NLS-containing proteins and U snRNAs. Perhaps the reasons for these differences was that the DNA fragments used were highly conjugated to fluorescent tags. Thus, the properties of the DNA may be vastly different from native DNA. By contrast to the results with linear fragments of DNA, when the Wolff group used plasmids, they found that nuclear uptake in permeabilized cells was dependent on cytoplasmic extracts only when the DNA was covalently cross-linked to a number of NLS peptides. However, these NLS-conjugated plasmids failed to localize to the nuclei of microinjected cells as determined by fluorescence microscopy, but did show modest increases in gene expression compared to unmodified plasmids when cytoplasmically microinjected into the cells.

In a separate set of experiments, using permeabilized cells and intact plasmids that were fluorescently tagged with a triplex-forming PNA, our laboratory demonstrated that plasmid nuclear uptake was time-, energy, and temperature-dependent and utilized the NPC for translocation. Furthermore, nuclear entry of the plasmids, as well as NLS-containing proteins was absolutely dependent on the addition of cytoplasmic extracts. When using purified importinα, importinβ, and Ran, plasmids were excluded from the nuclei unless nuclear extract was also provided. The probable reason for this is that the nuclear extracts are providing a source of transcription factors. Because importinα, importinβ, and Ran do not bind to DNA directly, DNA-binding, NLS-containing proteins must also be added to act as adapters between the DNA and the importins. As seen in microinjected cells, DNA nuclear entry was sequence specific: a 4.2 kbp plasmid lacking the SV40 enhancer was excluded from the nuclei while an isogenic plasmid containing the SV40 sequence localized to the nucleus efficiently. Plasmids up to 14 kbp were efficiently imported into the nuclei within 4 hours, although larger plasmids were not tested. Thus, like most other nuclear localizing macromolecules, plasmids were imported into the nuclei using the same mechanisms.

Nuclear import of plasmids has also been studied in synthetic nuclei reconstituted from phage λ DNA and interphase extracts from *Xenopus laevis*. Nuclear import of NLS-containing proteins has been extensively characterized using this system. To study DNA nuclear uptake, 48.5 kbp linear λ DNA was labeled along its length with Cy3 and added to reconstituted nuclei. In its protein-free, extended state, naked double stranded λ DNA is ~2 nm in diameter and about 15 μm in length. However, once added to cellular extracts, the DNA would become covered with proteins and condensed to take on a different hydrodynamic shape. Like NLS-mediated uptake, import of λ DNA was inhibited by WGA, suggesting that the DNA enters the nucleus via the NPC. However, although ATP depletion and low temperatures effectively inhibited BSA-NLS import, the treatments had no effect on DNA nuclear uptake. Because of this it is likely that this linear DNA is entering the nuclei through a pathway different from that seen in intact cells. To study how the DNA moved through the NPC, a 3 μm polystyrene sphere was attached to one end of the DNA and the DNA was then added to reconstituted nuclei. Within several minutes, laser tweezers were used to immobilize DNA-bead complexes adjacent to nuclei and the bead was pulled away from the nucleus. When some of the beads were pulled, the nuclei followed, indicating that the DNA was bound to the nucleus. In other cases, the bead could be pulled a defined distance from the nucleus (although less than the total persistence length of the DNA) before it recoiled toward the envelope. Measurements were taken at 2 second intervals to determine the kinetics of uptake. Import of the DNA occurred with an initial rate of 28 nm/sec, with or without ATP present. Further, the presence of an NLS at the end of the DNA distal to the bead (i.e., at the NPC-binding end) caused no increase in nuclear import rates. These results suggest that such linear DNA enters the nucleus by a mechanism distinct from that of NLS-mediated proteins. Because energy is not required, it is unlikely that molecular motors are involved, but rather a model is favored whereby the DNA is "ratcheted" into the nucleus in a serpentine fashion. Similar models have been postulated for the nuclear import of ssDNA (e.g., T-DNA from *Agrobacterium tumefaciens*) or peptide translocation into the endoplasmic reticulum. Such a mechanism may also be involved in plasmid nuclear import for the translocation of the bulk of the DNA; the NLS-transcription factor complex may be necessary for initial localization and binding to the NPC, but the remainder of the DNA could be pulled in by a similar mechanism.

Alternative Pathways for Plasmid Nuclear Uptake

It is well documented that two of the strongest promoters for in vivo expression are the CMV immediate early promoter and enhancer and the RSV LTR promoter. When plasmids carrying these promoters are injected into the post-mitotic myotubes in skeletal muscle from rodents or primates, robust gene expression is observed, even as quickly as 15 minutes post-injection. Because these cells are non-dividing, the DNA is somehow gaining access to the intact nucleus, although how remains to be determined. In Dowty's experiments, a plasmid using the RSV LTR to drive luciferase expression (i.e., no SV40 sequence) was efficiently imported into the nuclei of myotubes by a pathway using the NPC. A common thread to all of the above studies is that they employed skeletal muscle. One possible explanation is that skeletal muscle represents a tissue unlike any others in the body. Another finding that suggests that this at least partially could be the case is that skeletal muscle is known to express plasmid-encoded genes for extended periods of time, in some cases over the lifetime of the experimental animal, whereas all other cells in the body tend to express plasmid-encoded genes on the order of days to weeks. While differences in patterns of gene expression are routinely seen in muscle versus other tissues, this may not be the sole reason. Another alternative is that plasmid nuclear import may not be absolutely sequence-dependent, but rather, the presence of certain sequences such as the SV40 enhancer increase the rate of nuclear targeting. As such, promoters like the CMV_{iep}, which we have shown does not localize to the nucleus within 8-12 hours, may drive nuclear import at reduced rates. Simple kinetics could then account for the gene expression that is seen within minutes of plasmid injections into muscle: the more plasmid delivered to the cytoplasm, the more likely at least some will be driven into the nucleus. By microinjecting single myotubes in vivo, it has been shown that 10^6 plasmids (lacking SV40 sequences) are needed for gene expression. Once in the cytoplasm, any plasmid will be quickly coated with proteins. It is more than possible that at least some of these DNA-binding proteins will contain NLSs, creating a complex of DNA weakly bound to a few nuclear-localizing proteins. When compared to the SV40 sequence which binds multiple transcription factors with exposed NLSs, import of the other DNAs could appear much slower, and in the time frame of our experiments, not at all.

Several experiments from Peter Traub's group have suggested that there may exist another pathway for DNA nuclear import. The intermediate filament protein vimentin is a cytoplasmic protein, but when bound to single- or double-stranded oligonucleotides, it migrates to the nucleus. Similarly, when supercoiled plasmids, regardless of sequence, were bound to vimentin, through the cryptic DNA-binding activity of this protein, they caused the rapid movement of the protein into the nucleus. Although the experiment was not performed, these results suggest that supercoiled DNAs may have the ability to rapidly migrate into the nucleus. However, in our hands, this is not the case. Thus, these results suggest that the nuclear import of supercoiled DNA/vimentin complexes may be due to the complex rather than either of the parts.

Viral Nuclear Import

A number of DNA and RNA viruses must target their genomes to the nucleus in order to complete their life cycles. As for plasmid nuclear import, in all cases, the viral genome is targeted to the nucleus through the interactions of the nucleic acid with one or more viral proteins. To summarize, strategies for the nuclear import of viral genomes can be categorized into three groups: (i) targeting of the viral capsid to the nuclear envelope and "*dumping*" of the DNA into the nucleus, (ii) the use viral capsid proteins to target the DNA to the nucleus, and (iii) the use of core proteins to target the DNA into the nucleus.

Herpes simplex virus (HSV) enters the cell through the interaction of a group of envelope glycoproteins with proteins on the cell surface. Once internalized, the viral capsid, surrounded by a set of tegument proteins, is targeted toward the nucleus via the cytoskeleton. Fluorescently-labeled HSV capsids have been shown to target to the NPC using the microtubule-associated motor protein, dynein; depolymerization of the microtubule network abolished intracellular movement of the capsids, but not cell entry. Once the capsid reaches the nuclear envelope, the core "*docks*" and uncoats at the NPC. Experiments using isolated rat liver nuclei have shown that HSV capsids can bind to the NPC only when cytoplasmic extracts or importin β

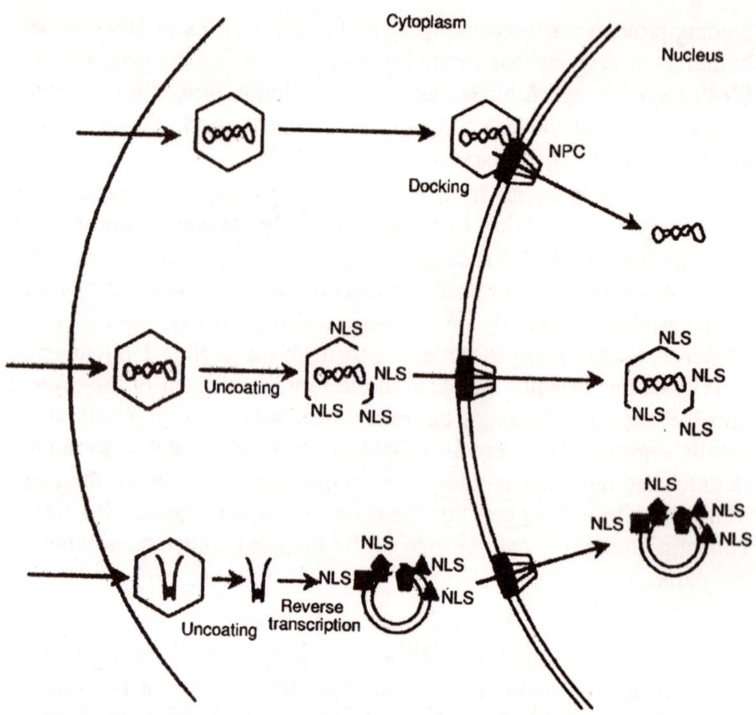

Fig. 7.2. Major nuclear import strategies used by viruses.

and RAN and an energy source are provided. This implies that NLS-containing proteins within the tegument and/or capsid are also required for NPC targeting. Immediately following NPC binding, capsid-associated DNA becomes DNaseI-sensitive, mimicking the in vivo uncoating at the NPC. Somewhat similar to HSV, adenovirus enters the cell, although by escape from endosomes, and also targets to the nucleus using the cytoskeleton. Based on studies using fluorescently-labeled viruses and real-time video microscopy, microtubules have also been implicated in the nuclear targeting of adenovirus, although other studies have not confirmed this. Like HSV, the nuclear import machinery is needed for nuclear entry of adenoviral DNA, but hsc70 appears also required, perhaps to aid in the disassembly of the capsid before nuclear entry of the genome. Hepatitis B virus also targets its capsid to the NPC in a process dependent on phosphorylation of the core protein and use of its NLS.

Other viruses, including SV40 and baculovirus, target their genomes into the nucleus using capsid proteins that accompany the DNA into the nucleus. The best studied of these is SV40. When SV40 virions were microinjected into the cytoplasm of SV40 permissive cells, the injected virion proteins localized to the nucleus and were immediately followed by large T-antigen expression, indicating that the viral genome had also entered the nucleus. Virion (or more likely virion protein-DNA complex) entry into the nucleus and subsequent T-antigen expression could be blocked by co-injection of WGA or antibodies against NPC proteins, or by energy depletion, all treatments known to inhibit translocation through the NPC.

The NPC also appears to be used for the nuclear targeting of incoming SV40 virions during an infection since cytoplasmic microinjection of anti-Vp3 antibodies blocked T-antigen expression during cell surface-mediated infection. This study also showed electron dense particles, interpreted to be intact or partially dissociated virion particles, in transit through the NPC. However, whether these were indeed intact viral particles was not convincingly proven. While it has been largely assumed that uncoating of the papovaviruses occurs exclusively in the nucleus, there is a great amount of evidence demonstrating that uncoating can occur in the cytoplasm. For example, the DNA in over 50% of infecting virions recovered from cytoplasm were nuclease-sensitive within 30 minutes of infection, indicating that they had at least partially uncoated. Thus, it is more probable that the virus partially uncoats in the cytoplasm and that the viral DNA remains associated with many of the capsid proteins and then enters the nucleus as a protein-DNA complex. The SV40 capsid consists of 360 copies of the major capsid protein and 72 copies of the minor two proteins. All three viral structural proteins contain well defined *nuclear localization signals* (NLSs) that direct the proteins to the nucleus. Furthermore, all three proteins bind DNA. Consequently, even if the virus uncoats losing 90% of its capsid proteins, the genome will still be complexed with almost 50 NLS-containing proteins. Since the incoming viral genome is already coated with a large number of NLS-containing proteins, nuclear import is much more rapid than that seen when purified SV40 DNA is injected.

A third common theme used by viruses to target their genomes to the nucleus is best exemplified by HIV. Like oncoretroviruses, HIV is internalized into the cytoplasm and the virus uncoats. The incoming viral RNA genome is then reverse transcribed in the cytoplasm into a double stranded DNA molecule of approximately 14 kbp. A set of viral proteins that are carried within the virus particle remain associated with the viral genome during reverse transcription. This complex is referred to as the preintegration complex, and is made up of at least 4 proteins: nucleocapsid (NC), matrix (MA), integrase (IN), and Vpr. Both matrix and integrase possess NLSs that have been proposed to play a role in the nuclear import of HIV. Vpr also plays a role in nuclear import, although its role is less well defined. When the NLS of either matrix or integrase is mutated or Vpr is deleted, the infectivity of the resulting virus is greatly diminished. While the matrix protein appears to function by donating its NLS for DNA nuclear import, experiments suggest that Vpr may both promote association of importinα with the matrix NLS and act directly as an importinβ-like protein to facilitate nuclear import of the complex. Thus, the model for HIV preintegration complex nuclear targeting is that the reverse-transcribed viral DNA is transported into the nucleus through the interactions of its associated NLS-containing proteins, a mechanism similar to that seen for plasmids and other viral genomes.

Nuclear Import of Single-Stranded DNA

All of the preceding discussion has dealt with the nuclear import of double-stranded DNA, either in its linear or circular forms. Single-stranded DNAs are also transported to the nuclei of cells by very similar mechanisms. The most extensively studied ssDNA in terms of nuclear localization is that of the plant pathogen *Agrobacterium tumefaciens*. *A. tumefaciens* transforms host cells by transferring a segment of a single-stranded DNA, referred to as the T-DNA (transferred DNA), derived from its tumor-inducing (Ti) plasmid. Nuclear import of T-DNA complexes into host cell nuclei is very fast and efficient, suggesting an active nuclear transport process. Furthermore, T-DNA import has been shown to be inhibited in presence of competitor molecules such as BSA-NLS or agents that block the NPC. Like viruses, the T-DNA transfer process utilizes specialized bacterial virulence proteins provided by the pathogen for nuclear transport.

It has been shown that for effective T-DNA transfer, virulence proteins are not only necessary for the assembly of the nucleoprotein complex, but also for its subsequent nuclear import in the host cell. Briefly, the virulence proteins VirD1 and VirD2 are shown to be required for the excision of the T-DNA from its adjacent sequences within the Ti plasmid. Subsequently, virulence protein E2 (VirE2) is recruited to the complex by its single-stranded DNA-binding properties and is thought to exert its influence on T-DNA transfer efficiency by coating and compacting the excised DNA sequence necessary both to preserve the T-DNA integrity and to facilitate its nuclear import. All three proteins contain well-defined NLSs. Subsequent to transfer of the T-DNA complex into the host cell, VirD2 through its nuclear localization signal (NLS) is thought to guide the entire complex in to the nucleus. However, for import of short stretches of single stranded DNA, VirD2 alone has been shown to be sufficient.

Agrobacterium-mediated T-DNA transfer to plant cells is the only known example for interkingdom DNA transfer and is widely used for plant transformation. However, the interest in this field has broadened by a recent report showing that reconstituted complexes consisting of the bacterial virulence proteins VirD2, VirE2, and single-stranded DNA can mediate nuclear import of nucleic acids in permeabilized mammalian cells in vitro and in *Xenopus* oocytes. Furthermore, in mammalian cells, a VirD2 mutant lacking its C-terminal nuclear localization signal was deficient in import of the ssDNA-protein complexes into nuclei and import of DNA-protein complexes was dependent on importinα, Ran, and an energy source, suggesting a classical importin-mediated nuclear import pathway.

Nuclear Import of Oligonucleotides

It is well established that, like double-stranded DNA, oligonucleotide polymers can be used to specifically express, or even more importantly, modulate the expression of a gene. Short stretches of single stranded DNA and RNA molecules have been shown to exert complementary antisense effects or enzymatic activities to modulate and in some cases to completely abolish expression of selective host genes. The ability to specifically target a single gene has brought antisense oligonucleotides to the fore front of rational drug design with great potential for research as well as clinical applications.

Antisense oligonucleotides are synthetic polymers of oligonucleotides usually with builtin structural modifications, such as a phosphorothioate backbone, to modulate their stability, cellular uptake, and/or toxicity. Antisense oligonucleotides are generally designed to bind through Watson-Crick base pairing to a unique complementary RNA sequence and this hybridization selectively interferes with nuclear processing, transport and/or cytoplasmic translation of the target molecule as well as its degradation in an RNase H-dependent manner. Although the intended site of action in the majority of cases is cytoplasmic, it is important to emphasize that in many cell types and using different delivery methods, oligonucleotides have been shown to localize to the nucleus and in some cases this nuclear localization has been shown to correlate with an increase in antisense activity.

Any oligonucleotide polymer used to specifically express or modulate the expression of a gene, must first reach the cytosol to be able to exert its biological effect. A variety of studies have shown that oligonucleotide uptake by cells is time-, temperature-, energy-, concentration-, cell type- and size-dependent, and can be affected by chemical modifications. Kinetic analysis of oligonucleotide binding and uptake has generally pointed to a bimodal process where receptor mediated uptake predominates only at low micromolar concentrations whereas fluid-phase endocytosis prevails at higher concentrations. Abundant evidence has also been presented demonstrating the existence of cell surface receptors that bind oligonucleotides. Taken together these data also support the notion that a receptor-mediated but passive uptake, referred to as adsorptive endocytosis, may be the prevailing mechanism of oligonucleotide uptake. Cationic liposomes have also been used extensively for the successful delivery of oligonucleotides to cells. However, the success rate for liposome mediated antisense oligonucleotide delivery varies depending on the cell type, particular liposome formulation used, and cell culture conditions utilized. Nevertheless, using a cationic liposome assisted delivery and the C6 glioma cell line as an in vitro model, Islam and colleagues have shown a 10-12 fold increase in cellular uptake of both biotin-labeled and radiolabeled oligonucleotides. Furthermore, in the absence of liposomes, internalized oligonucleotides were shown to be sequestered mostly within endosomal and lysosomal vesicles. In the presence of cationic lipids, distribution of internalized oligonucleotides showed an enhanced penetration of the cytosol and increased accumulation within the nucleus consistent with an increased endosomal release. Finally, electroporation has also been used to attain high intracellular concentrations of oligonucleotide in a large proportion of viable cells. Even more interestingly, following electroporation transfected oligonucleotides rapidly localized to the nucleus and remained predominantly nuclear for at least 2 days. Taken together, these results demonstrate that the route of cellular uptake can have profound effects on the intracellular distribution of nucleic acid.

More and more compelling evidence implies that antisense DNA may exert their effects in the nucleus. For example, uptake studies with ^{35}S-labeled phosphorothioate oligonucleotides have shown that following cellular uptake labeled oligonucleotides were found in significant amounts in the nucleus. In contrast to other studies, nuclear uptake was not blocked by WGA, suggesting a diffusion driven process, rather than an importin-mediated one. However, using oligonucleotides conjugated to a photoactivatable, radiolabeled crosslinker, it has been shown that, even following uptake, internalized oligonucleotides remain largely associated with proteins. Although several candidate proteins were detected, internalized oligonucleotides were predominantly found to be associated with a 75 kD protein that appears to be membrane-associated. Therefore, it has been suggested that the majority of intracellular oligonucleotides remain associated in vesicles with the same protein to which they bind on the cell surface and only a small percentage of non-protein-bound cytosolic oligonucleotide can be detected. Additionally, the majority of free cytosolic oligonucleotides readily accumulated in the nuclei where they were shown to associate with a new set of nuclear oligonucleotide binding proteins.

More recent evidence suggests that phosphorothioate oligonucleotides utilize the classic NPC-mediated pathway for nuclear localization. Further, at least a fraction of the nuclear pool of oligonucleotides are in dynamic balance with the cytoplasmic pool and are able to continuously shuttle between the nucleus and the cytoplasm while maintaining their antisense activity. The shuttling of phosphorothioate oligonucleotides was shown to be an active transport process and sensitive to treatment with WGA indicating an NPC-mediated translocation process. Nucleocytoplasmic shuttling was only moderately affected by disruption of the Ran/RCC1 system, suggesting a transport process similar to that of U snRNA, tRNA and mRNA export. However, oligonucleotides without a phosphorothioate backbone chemistry do not share these characteristics and are only weakly restricted in their migration upon chilling, ATP depletion and WGA treatment. Therefore, whether a particular oligonucleotide is going to be subject to active nuclear transport or rather moved by diffusion may indeed depend on its affinity for protein binding partners.

In another study of nuclear targeting and retention of oligonucleotides, using high resolution imaging and fluorescently tagged phosphorothioate oligonucleotides, Lorenz et al have shown that under conditions expected to display optimal antisense activity, phosphorothioate oligonucleotides predominantly localize to the cell nucleus (but not to the nucleoli). Under these conditions, nuclear oligonucleotides accumulated in a number of bright spherical foci referred to as "*phosphorothioate bodies*". Furthermore, these nuclear foci formed in both transfected as well as microinjected cells, suggesting that formation of nuclear phosphorothioate bodies was independent of the oligodeoxynucleotide delivery method used. Also, formation of these nuclear foci did not correlate with antisense activity or the primary sequence of oligonucleotide used. Ultrastructurally, these nuclear foci have been shown to correspond to electron-dense structures of 150-300 nm diameter and resemble nuclear bodies that were described to occur at a lower frequency in non-treated control cells. These foci developed in living cells within minutes after the introduction of the oligonucleotides and they were relatively stable entities that underwent noticeable reorganization only during mitosis. These findings support the notion that, following uptake and endosomal release, phosphorothioate oligonucleotides associate with a new set of nuclear proteins including the nuclear matrix.

Nuclear Import of *Agrobacterium* T-DNA

Agrobacterium-mediated genetic transformation is a process by which genetic material is transported from the bacterium into the host nucleus, where it stably integrates. The transferred DNA (T-DNA) is escorted, by two bacterial proteins, as a single-stranded DNA-protein complex (a T-complex), which mediate its transport to the host nucleus. The large size and mass of this DNA-protein complex raise questions as to the molecular machinery and mechanism by which the T-complex passes the nuclear pore barrier. Recent studies have revealed the important role of specific host proteins in interacting with and guiding the T-complex through the nuclear pore, and to its point of integration.

Agrobacterium-mediated genetic transformation of plant cells is a unique and complicated process by which genetic material is transported from the bacterium into the host nucleus, where it stably integrates. Although in nature, *Agrobacterium* transforms mainly dicotyledonous plants, under controlled culture conditions it has been shown to possess a broader host range which includes monocot plants, yeast and other fungi, and even human cells. In modern plant breeding, *Agrobacterium* is widely used for plant genetic engineering. The molecular basis for the transformation process has therefore been the subject of numerous studies over the past several decades.

Agrobacterium tumefaciens is a gram-negative bacterium which is the causative agent of crown-gall disease in many dicotyledonous plant species. The disease symptoms result from the transfer, integration and expression of a specific DNA fragment (known as transferred DNA or T-DNA) from the bacterial tumor-inducing (Ti) plasmid into the plant cell genome. The machinery needed for the generation of transferred T-DNA and for its transport into the host cell is encoded by a serie of chromosomal (*chv*) and Ti-plasmid-encoded virulence (*vir*) genes. Interestingly, although the bacterial proteins possess many of the functions needed for the transformation process, host-plant factors also play a crucial role in the T-DNA nuclear import and integration. In this section, we discuss the role of both *Agrobacterium* and host proteins during *Agrobacterium* T-DNA long voyage in the host cell, from its cytoplasmic point of entry to the nucleus.

Genetic Transformation Process

The T-DNA fraction is a specific DNA segment located on the *Agrobacterium* Ti plasmid. The T-DNA itself is not sequence-specific and is defined exclusively by its left and right borders which are two 25-bp direct repeats. Since the T-DNA does not contain any specific sequences coding for its processing, translocation into the host cell cytoplasm, nuclear import or integration any sequences placed between its borders will also be transferred and integrated into the host genome. Moreover, the borders are the only *cis*-required elements needed to define and process the T-DNA, and T-DNA molecules can therefore technically be located on other genetic elements within the *Agrobacterium* cells, rather than on the Ti plasmid itself. Indeed, the practice of using a small, low-copy binary plasmid, carrying a well-defined T-DNA region, in conjunction with large Ti plasmids which do not contain their original T-DNA (also known as "*disarmed*" Ti plasmids), is the molecular basis for the use of *Agrobacterium* as a vector for plant genetic engineering.

The transformation process begins with the induction of the *Agrobacterium* VirA-VirG sensory machinery and subsequently the virulence (Vir) proteins (a2) by host-specific small phenolic signal molecules. The VirD1-VirD2 protein complex acts as an endonuclease which nicks both borders on the noncoding strand of the T-DNA region.Through a strand-replacement process (a3), a single-stranded (ss) T-DNA molecule is released from the Ti plasmid with a single VirD2 molecule which remains attached to its 5'-end. The T-strand-VirD2 complex (also called "*immature T-complex*") is then coated with numerous VirE2 molecules to form the transported form of the T-DNA (known as a "*mature T-complex*", or simply, "*T-complex*"). The T-strand is then transported into the host cytoplasm either as an immature (b1) or mature (b2) T-complex through a *vir*B/*vir*D4-encoded channel, travels through the host cytoplasm (b4) and is finally imported into the host nucleus. Once inside the nucleus, the T-complex is stripped of its bacterial chaperones and is randomly integrated into the host genome through an unknown mechanism.

Although much is known about the mechanism by which the Vir proteins mediate the production of the T-strand and the mature T-complex, little is known about the mechanism governing the T-complex transport into the host cell, its voyage through the host cytoplasm and its integration. Recent studies have begun filling this gap by producing new information about the T-complex nuclear import mechanism, especially about the role that specific host factors, interacting with the T-strand bacterial chaperones, the VirD2 and VirE2, play during the transformation process.

T-Complex Export to Plant Cells

The T-strand is a ssDNA molecule which is a copy of the coding strand of the T-DNA region. Molecular evidences suggest that the T-strand may associate with two *vir* gene products, Vir and VirE2. Following T-strand production, VirD1 is released while VirD2 remains covalently attached to the 5'-end of the border nick sites; thus, VirD2 associates with the T-strand throughout its journey to the point of integration in the host nucleus. VirE2, a ssDNA-binding protein, is presumed to coat the T-strand along its length. As do most ssDNA-binding proteins, VirE2 binds ssDNA cooperatively and without sequence specificity, consistent with the nonsequence-specific nature of the T-DNA itself. Mutational analysis of VirE2 revealed that the amino-terminal part of the protein is important for its binding cooperatively, while its carboxy-terminal portion is essential for ssDNA binding. Importantly, the carboxy-terminal part of VirE2 also contains an RPR motif, which likely functions as a signal for protein export from *Agrobacterium* into plant cells through the VirB/VirD4 channel.

The length of the T-DNA is virtually unlimited, as it has been shown that even very large molecules can be efficiently transported and integrated into the host genome. For example, using modified Ti plasmids, with the right border removed, the entire copy of the bacterial Ti plasmid was transported and integrated into the host genome. Using specifically designed binary vectors engineered to carry large-size inserts, even 150-kb-long T-DNA fragments are transported and integrated into tomato plant cells. In the latter experiment, extra copies of the *virA* and *virE2* loci were necessary to mediate the transport of such large molecules, supporting the role of the VirE2 molecule as an essential chaperone of the T-complex.

It is still unclear whether the T-complex matures in the bacterial or the host cell, although there is little doubt that VirE2 is associated with the T-strand in the host cell cytoplasm. Potentially, the coupling between VirE2 synthesis and T-strand generation, as well as their physical proximity in the bacterial cell, could result in an immediate interaction and the production of mature T-complex within that cell. The strong cooperative interaction between VirE2 and ssDNA, and the suggestion that both are transported through the same channe further support the notion that mature T-complexes are assembled early in the transformation process, within the bacterial cell. Indeed, T-strands and VirE2 are co-precipitated by anti-VirE2 antibodies from extracts of *vir*-induced *Agrobacterium* cells. Nevertheless, other experimental evidence suggests that the immature T-complexes and VirE2 molecules are separately transported into the host cell cytoplasm, where they assemble to form mature T-complexes. First, co-transformation of tobacco plant cells with an *Agrobacterium* strain containing T-DNA but lacking VirE2 and a second strain lacking T-DNA but containing VirE2 restored infectivity of these individually noninfectious bacteria. Second, transgenic plants over-expressing the VirE2 protein could restore the infectivity of a VirE2-deleted *Agrobacterium* mutant. These observations suggest that immature T-complexes and VirE2 molecules are independently exported from the bacterial cells. Indeed, studies have shown that VirE2 export from *Agrobacterium* can be inhibited without affecting T-strand export.Furthermore, VirE1, another protein product of the *virE* locus, was suggested to act as a VirE2 chaperone, preventing it from binding the ssDNA and facilitating its export into plant cells. Recent data, however, indicate that VirE1 is neither exported into the plant cell nor it is required for VirE2 export by the VirB/VirD4 transport system, limiting the VirE1 role to stabilization of VirE2 prior to its translocation into the host cell.

Mature T-complex is most likely to be the structure imported into the host cell nucleus. Several lines of experimental evidence support this assumption. First, infectivity of an avirulent VirE2 mutant *Agrobacterium* strain could be restored when inoculated onto VirE2-expressing transgenic plants. Second, plant cells infected with the VirE2 *Agrobacterium* mutant exhibited lower accumulation of T-strand molecules than those infected with wild-type *Agrobacterium*. Third, VirE2 association with ssDNA molecules in vitro was shown to protect them from exonucleolytic degradation. Thus, T-strands are most probably associated with VirE2—which provides them with the necessary protection within the host cell cytoplasm—and imported into the nucleus as a mature T-complex.

Molecular Structure of the Mature T-complex

Elucidation of three-dimensional structure and computational analysis of the T-complex may assist in understanding the mechanism of its nuclear import and subsequent integration. Complexes, formed in vitro by interaction between purified VirE2 and the bacteriophage M13 ssDNA, were examined by scanning transmission electron microscopy, followed by mass analysis. These analyses revealed that the VirE2-ssDNA association produces rigid and coiled filaments that are 12.6 nm-wide, with a density of 58 kDa/nm, and that each turn of the filament coil contains an average of 3.4 molecules of VirE2 and bases of ssDNA. Based on these parameters, a 22-kb T-strand of the wild-type nopaline-specific *Agrobacterium* is calculated to associate with molecules of VirE2, and a 150-Kbp T-strand derived from the longest artificial T-DNA used to transform plants is predicted to bind VirE2 molecules. The length of the mobilized T-strand, when coated with VirE2 molecules, is estimated to range between 40 nm and 80 microns for T-DNA regions between 20 Kb and 150 Kb.

Both the wide outer diameter (12.6 nm) and the extended length of the mature T-complex suggest an active mechanism for its nuclear import. Indeed, the T-DNA outer diameter exceeds the orifice of the nuclear pore diffusion channels, but it is easily compatible with the size-exclusion limit of the nuclear pore, which reaches nm during the process of active nuclear uptake. Thus, packaging of the T-strand into the T-complex may be essential for nuclear import because, in the absence of VirE2, free T-strand molecules would likely fold into significantly larger structures unable to penetrate the nuclear pore. For example, in vitro (i.e., in a simple solvent), consideration of long free ssDNA as a polymeric random coil, rather than an extended thread, suggests a diameter of nearly 300 nm because the typical size of a polymeric random coil is approximately the geometric mean of the extended length and the persistence length, which, of course, is much larger than the entire nuclear pore complex.

Nuclear import of VirE2-ssDNA complexes may also be facilitated by their apparent rigidity, which would prevent their folding into random polymeric coils with large diameters. On the other hand, it remains unclear how relatively rigid T-complexes, whose length exceeds the entire diameter of a typical plant cell, move through the host cell cytoplasm and fit into its nucleus without extensive folding or supercoiling. Thus, some restructuring and bending may be required, especially for the nuclear import of long T-DNA complexes.

T-Complex Nuclear Import

T-complexes are polar molecules and their nuclear import is thought to occur in a polar fashion. The VirD2, attached to the 5'-end of the T-strand, is presumed to act as the nuclear import pilot, guiding it not only to the nuclear pore but also to the point of integration. Various studies, using either direct immunolocalization or translational fusions with such reporters as β-glucuronidase (GUS) and GFP, have shown that VirD2 molecules specifically accumulate in the plant cell nucleus. Sequence analysis of the VirD2 protein reveals that it contains two *nuclear localization signals* (NLSs) located at the carboxyand amino-terminal ends. The VirD2 carboxy-terminal bipartite NLS is the only sequence that appears to function during *Agrobacterium* infection, as it was shown that T-DNA expression and tumorigenicity are reduced in plants infected with *Agrobacterium* carrying carboxy-terminal NLS-deletion mutants of VirD2. The reduced infection efficiency, rather than complete blockage suggests that other components of the T-complex participate in the nuclear import process. Indeed, VirE2, the T-strand coating protein, was also found to accumulate in plant cell nucleus. Sequence analysis of the VirE2 protein revealed a single NLS, located in the middle of the molecule. Mutations within the central region of the VirE2 sequence decreased *Agrobacterium* tumorigenicity but did not affect the ssDNA-binding activity or stability of the protein.

Several experimental approaches have been used to demonstrate the direct function of VirE2 in T-DNA nuclear import. Using microinjection of in vitro-formed T-complexes, the ability of VirE2 to direct fluorescently labeled ssDNA into the plant cell nucleus was studied. First, microinjection of fluorescent single or double stranded (ds) DNA alone resulted in clear cytoplasmic fluorescence with no evident nuclear staining. This clearly demonstrated the inability of DNA alone to enter the host nucleus under these experimental conditions. Second, microinjection of in-vitro-formed VirE2-ssDNA complexes, but not of VirE2-dsDNA, resulted in efficient nuclear accumulation of the labeled DNA, thus indicating the requirement of ssDNA-VirE2 complex formation. Third, nuclear import of VirE2-ssDNA was blocked by nuclear import-specific inhibitors such as wheat germ agglutinin and nonhydrolyzable analogs of GTP, indicating that VirE2-mediated nuclear import of ssDNA is an active process which requires ATP. In a different approach, a mutated *Agrobacterium* strain, lacking the entire VirE2 as well as the specific carboxy terminal NLS of VirD2 was used to study the function of VirE2 during the transformation process. This mutant produced tumors on VirE2-expressing transgenic plants, but not on wild-type tobacco plants. These observations demonstrated that VirE2 alone, when expressed in the host cells, can mediate the import of transferred T-strands into the nucleus, in the absence of NLS from any other known T-DNA-associated protein.

A certain functional redundancy between VirE2 and VirD2 in the nuclear import of T-complex may exist. Nevertheless, it is possible that the combined action of VirE2 and VirD2 is required for the polar translocation of T-complex into the nucleus in a manner which may be important for the later integration step. Various studies suggest that T-DNA integration is indeed a polar process, although it is still unclear whether it begins at the 5'- or 3'-end of the T-strand molecule. One possibility is that the T-complex polar structure dictates the polarity of its nuclear import and integration. In such a scenario, functional variations between VirD2, which is covalently bound to the 5'-end of the T-strand, and VirE2, which covers the entire T-strand, may determine the polarity of the transport and integration.

Several approaches have been utilized to investigate whether VirD2 and VirE2 may perform different but complementary functions during nuclear import of the T-complexes. In the first approach, using a heterologous living nonplant cell system which lacks nuclear transport machinery of the *Agrobacterium* natural host cells, differences between VirD2 and VirE2 nuclear import could be discerned. VirD2 localized in the nucleus of *Xenopus* oocytes, *Drosophila* embryos, human kidney and HeLa cells, and yeast cells. In addition, VirD2 nuclear accumulation in animal cells was blocked by known inhibitors of nuclear import, thus demonstrating the functionality of its evolutionarily-conserved NLS. VirE2, on the other hand, did not target to the nuclei of *Xenopus* oocytes, *Drosophila* embryos, HeLa cells or yeast cells, suggesting that its nuclear import may be plant-specific.

In another approach, permeabilized human cells were used to study the differences between VirD2 and VirE2 nuclear import. In this system, both VirD2 and VirE2 accumulated in the nuclei of permeabilized HeLa cells. Nevertheless, while VirD2 alone, when covalently bound to fluorescently labeled ssDNA, promoted T-strand nuclear import, VirE2 failed to do so. These observations suggest that during transformation, both VirD2 and VirE2 are required for T-complex nuclear import. Indeed, a recent study showed that both VirE2 and

VirD2 are required for efficient import of the T-DNA complex into plant cell nuclei. Using this approach, in vitro-formed VirE2-VirD2-ssDNA complexes were tested for their import into plant nuclei in vitro. Whereas VirD2 was sufficient for the import of short ssDNA, VirE2, in combination with VirD2, was essential for the import of long ssDNA molecules. Interestingly, RecA, a protein that can bind ssDNA and form thin DNA-protein complexes that are only 5.2-9.3 nm-wide, is capable of replacing VirE2 during nuclear import of long T-DNAs, but not during earlier events of T-DNA transfer to plant cells. It was thus suggested that VirD2 and VirE2 perform complementary functions in T-complex nuclear import. While VirD2 initially directs the T-complex into the nuclear pore, VirE2 may shape it in a transferable form and assist translocation of the entire T-complex into the host cell nucleus. Collectively, functional differences between VirD2 and VirE2 suggest that (i) in plant cells, VirE2 and VirD2 employ different cellular factors for their nuclear entry, and (ii) animal cells lack the subset of factors that recognize VirE2 and help its nuclear uptake in plant cells.

Host Cell Proteins that Interact with VirD2 and VirE2

It is well established that both VirD2 and VirE2 are required for the efficient nuclear import of the *Agrobacterium* T-strand. During this process, VirD2 and VirE2 must function together. Their import into the host cell nucleus probably occurs *via* different cellular mechanisms, and thus VirD2 and VirE2 are expected to utilize different cellular factors for their nuclear import. In this scenario, the single VirD2 molecule, attached to the 5'-end of the T-strand, and the more abundant VirE2, which covers the remainder of the T-strand molecule, do not compete directly for the same host factors. Such physical and functional arrangements would enable VirD2 and its host interactors to initiate nuclear import of the T-complex, while VirE2 and its interactors likely drive the import process to completion. Several host factors have been identified in past years that interact with *Agrobacterium* VirD2 in a yeast two-hybrid protein-protein interaction assay.

One set of VirD2-interacting host proteins are members of a large cyclophilin family of peptidyl-prolyl *cis-trans* isomerases (PPIases), which are highly conserved in plants, animals, and prokaryotes. Specifically, VirD2 was found to interact with the *Arabidopsis* cyclophilin DIP1 and three more isoforms of *Arabidopsis* cyclophilins, Roc1, Roc4, and CypA, in both the yeast two-hybrid system and in vitro. Specific inhibition of the VirD2-CypA interaction in vitro was accomplished by adding cyclosporin A, which is known to bind cyclophilins and block their PPIase activity. Cyclosporin A also inhibited *Agrobacterium*-mediated genetic transformation of *Arabidopsis* roots and tobacco cell suspension cultures. The exact biological role of cyclophilins during *Agrobacterium* infection is still unclear. They have been proposed to maintain the proper conformation of VirD2 within the host cell cytoplasm and/or nucleus during T-DNA nuclear import and/or integration. Indeed, in addition to their enzymatic activity, cyclophilins may act as molecular chaperones that aid in protein folding in animal cells. Two additional *Arabidopsis* proteins, DIP2 and DIP3, which bind VirD2 in the two-hybrid assay, have been reported, but no information as to their identity or biological role were provided. More recently, two more members of the cyclophilin family, Roc2 and Roc3, have been reported to bind VirD2. However, their biological function during *Agrobacterium* transformation has not been examined.

Another host cellular factor that binds VirD2 is the tomato DIG3 protein. This protein, a type 2C serine/threonine protein phosphatase (PP2C), was found to interact specifically with the VirD2 NLS region. An *Arabidopsis abi1* mutant, knocked out in a PP2C homolog, exhibited higher sensitivity to *Agrobacterium*-mediated genetic transformation, than did wild-type plants. In addition, over-expression of DIG3 in tobacco protoplasts specifically inhibited nuclear import of a GUS-VirD2 NLS fusion protein. It was thus suggested that phosphorylation of the VirD2 NLS region may potentiate VirD2 nuclear import, while its dephosphorylation by DIG3 PP2C would negatively regulate this process.

A third type of VirD2-interacting host protein is a member of the growing karyopherin α family, known to mediate nuclear import of many NLS-containing proteins. Specifically, VirD2 was found to interact with AtKAPα, both in vivo and in the yeast two-hybrid system. AtKAPα was found to possess the classical features typical of karyopherin a proteins: it contains eight contiguous repeats of the "arm" motif and four amino-terminal clusters of basic amino acids. In animal and yeast cells, the "arm" motifs are thought to recognize the imported cargo through its NLSs while the amino terminal basic domain is thought to interact with the karyopherin β proteins. Indeed, AtKAPα interaction with its cargo, the *Agrobacterium* VirD2, required the presence of the VirD2 carboxy-terminal NLS. Functionally, AtKAPα showed high homology to the yeast karyopherin α, Srp1p, and could complement this gene function in a temperature-sensitive *srp1-31* yeast mutant. In addition, the yeast Srp1p could functionally promote nuclear import of fluorescently labeled VirD2 in permeabilized yeast cells. Using this yeast-derived nuclear import assay, it was also shown that VirD2 nuclear import is dependent upon its NLS. First, VirD2 lacking its NLS failed to localize in the cell nucleus and remained cytoplasmic, and second, a synthetic peptide corresponding to the VirD2 NLS blocked the nuclear import of fluorescently labeled VirD2, probably due to competition with VirD2 for interaction with Srp1p/AtKAPα. In a recent report, VirD2 was found to interact with three more members of the *Arabidopsis* karyopherin α family, which differed from AtKAPα, and in another report an *Arabidopsis* mutant, knocked out in one of its karyopherin α genes, was resistant to *Agrobacterium* infection. It was thus proposed that, during *Agrobacterium* infection, karyopherin α proteins, which naturally function as part of the nuclear import machinery, directly promote nuclear import of VirD2 and its cognate T-strand.

Recently, two additional host proteins, CAK2Ms and TBP, have been reported to interact with VirD2 in vivo and were suggested to function in the last step of the transformation process, T-DNA integration. CAK2M was identified by its ability to phosphorylate transiently expressed VirD2 in alfalfa protoplasts. CAK2M, a conserved plant ortholog of cyclin-dependent kinase-activating kinases, was also found to interact and phosphorylate the C-terminal regulatory domain of RNA polymerase II largest subunit. The latter protein is known for its ability to recruit TATA box-binding proteins (TBPs), not only for the regulation of transcription, but also for the control of transcription-coupled repeat (TCR), which ensures preferential and effective removal of DNA lesions from transcribed genes. Interestingly, VirD2 was also found to tightly associate with TBP in cells that has undergone *Agrobacterium*-mediated transformation. However, it is still unknown whether CAK2M recruits VirD2 to TBP, and whether phosphorylation of VirD2 by CAK2M regulates VirD2 association with the 5'-end of the T-strand. Nevertheless, these observations show that VirD2 is recognized by widely conserved nuclear factors in eukaryotic cells and raises the possibility that CAK2M plays a function in the intra-nuclear voyage of the T-complex to the point of integration while TBP may function as a mediator between VirD2 and the host DNA repair machinery during the integration process.

Since VirD2 and VirE2 utilize different cellular routes for their nuclear import, they most probably interact with different host factors. Indeed, unlike VirD2, VirE2 did not interact with AtKAPα in the yeast two-hybrid assay, but was found to interact with another *Arabidopsis* protein, VIP1. VIP1 contains a conserved stretch of basic amino acids (basic domain) abutting a heptad leucine repeat (leucine zipper), two structural features characteristic of the basic-zipper (b-ZIP) proteins which are known to localize to the cell nucleus. Indeed, VIP1 was capable of promoting the nuclear import of GFP-tagged VirE2 in mammalian cells and, using a genetic assay for nuclear import and export, VIP1 expression was shown to promote nuclear import of VirE2 in yeast cells. Down-regulation of VIP1 in plant cells, using antisense transgenic plants, blocked the nuclear uptake of GUS-tagged VirE2, but not of GUS-VirD2, thus demonstrating the specific role of VIP1 in the nuclear import of VirE2 molecules in plant cells. Moreover, VIP1 antisense transgenic plants showed resistance to both transient and stable genetic transformation by *Agrobacterium*, indicating that they were blocked at the nuclear import stage of the transformation process.

VIP1 is a nuclear protein which has been found to interact with *Arabidopsis* karyopherin α in the yeast two-hybrid assay. Moreover, in yeast cells, VIP1 nuclear import depended on the presence of the cellular Srp1 protein, indicating that VIP1 is imported into the cell nucleus via the karyopherin α-dependent pathway. It was thus suggested that VIP1 interacts with VirE2 in the cell cytoplasm and carries it into the cell nucleus by a "*piggy-back*" mechanism, serving as a molecular adaptor between VirE2 and the host cell karyopherin α.

VIP1 may represent the cellular factor involved in the plant-specific nuclear uptake of VirE2, because it induces VirE2 nuclear import in nonplant systems, and because no animal or yeast homologs of VIP1 have been found in protein databases. Moreover, the low cellular levels of VIP1 found in various plant tissues suggest that, in nature, *Agrobacterium*-mediated transformation may not occur at its maximal possible efficiency. In fact, inoculation of various plant tissues of most *Agrobacterium*-susceptible plant species results in the transformation of an extremely low number of cells, even with very dense *Agrobacterium* inoculum. Indeed, over-expression of VIP1 in transgenic plants resulted not only in higher susceptibility to *Agrobacterium* infection, but also in faster nuclear import of the T-DNA.

Because, as a b-ZIP protein, VIP1 may be involved in transcription, associating with the chromosomal DNA either directly or through other components of transcription complexes, and because VirD2 was also found to interact with members of a transcriptional complex, VIP1 and VirD2 interactors may function in the intranuclear transport of the T-complex. Thus, VIP1 may perform a dual function: facilitating nuclear targeting of VirE2 and playing a role in the intranuclear transport of VirE2 and its cognate T-strand to the site of integration. A similar dual role in nuclear and intranuclear transport has been suggested for the yeast Kap114p protein, of which functions are to import the TBP into the cell nucleus, and to target it to the promoters of genes to be transcribed. Besides VIP1, three other VirE2-interacting proteins have been reported, but not identified or characterized.

VirE3, a Bacterial Substitute for the Host Protein VIP1

Recent studies have shown that, in addition to the *Agrobacterium* immature T-complex and its VirE2 chaperone, two other bacterial proteins, VirF and VirE3, are exported from the bacterium to the host cell. VirF is an F-box protein that probably participates in the targeted proteolysis of as yet unidentified substrates.The function of VirE3 is just beginning to emerge. Recent studies suggest that VirE3 may act as a substitute for the cellular protein VIP1 in facilitating the nuclear import of VirE2. First, VirE3, like VIP1, targets to the nucleus of plant, mammalian and yeast cells. VirE3 was found to interact with AtKAPα in the yeast two-hybrid system, suggesting its nuclear import via the karyopherin α pathway. Importantly, VirE3 not only interacted with VirE2 in the yeast two-hybrid system, but also mediated its nuclear import in mammalian cells, thus mimicking the VIP1 function in this heterologous nuclear import system. Indeed, VirE3 complemented the VIP1 function in the nuclear import of VirE2 *in planta*, as was evidenced by its ability to direct GUS-tagged VirE2 into the cell nucleus in VIP1 antisense plants. It was thus suggested that, because VIP1 is not an abundant cellular protein and may represent one of the limiting factors for transformation, *Agrobacterium* may have evolved to produce VirE3 and export it into the host cell to at least partially complement the cellular function of VIP1, necessary for the infection.

Model for T-DNA Nuclear Import and Intranuclear Transport

Recent years have brought great progress in our understanding of the biological activities of VirD2 and VirE2, as well as in the identification of some of their cellular and bacterial interactors. A multitude of functional data has allowed developing a model for nuclear import and intranuclear transport of the *Agrobacterium* T-complex. The transport begins with the assembly or the entry of the mature T-complex into the host cell cytoplasm. If the mature T-complexes form already within the bacterial cells, they are directly transported into the plant cell (b2). If, however, mature T-complexes are formed only in the host cell cytoplasm, VirE2-VirE1 complexes, as well as immature T-complexes, are transported independently into the host cell, possibly through the same VirB/VirD4 channe (b1). In the latter case, before entering the cytoplasm, VirE2 must dissociate from VirE1 and attach to the VirD2-T-strand molecules to form mature T-complexes (b3). In parallel to the mature T-complex, two other *Agrobacterium* proteins are transported into the host cell cytoplasm through the same VirB/ VirD4 channel, - VirE3 (b4) and VirF, of which VirE3 most likely functions in the T-complex nuclear import, and VirF probably acts in the cell nucleus to uncoat the T-complex prior to its integration.

Once inside the cytoplasm, the T-strand, shaped and protected by its chaperones VirD2 and VirE2, begins the journey to the host cell nucleus and its resident genome. During this voyage, VirE2, cooperatively coats the T-strand, shapes it into a coiled filament and protects it from cellular nucleases. VirD2, on the other hand, acts as the T-complex pilot and guides it to the nuclear pore. Because of their very large size and rigid coiled shape, T-complexes cannot move through the cytoplasm in a simple Brownian motion, let alone passively diffuse through the nuclear pore. It is unknown whether or not the T-complex undergoes structural changes as it travels through the cytoplasm, but this coiled "telephone cord"-like complex may stretch, thus reducing its outer diameter and facilitating the import process, once it arrives to the nuclear pore.

During transport, the T-complex bacterial chaperones, VirD2 and VirE2, presumably interact specifically with their respective cellular factors, as well as with the *Agrobacterium* VirE3 protein. First, cellular cyclophilins may bind VirD2 (c1) to maintain its active conformation or perform some other, as yet unknown, functions as the T-complex moves in the cytoplasm, and/or during its nuclear import. Since cyclophilin binding does not involve the NLS region of VirD2 it may occur concurrently with VirD2 interaction with AtKAPα (c2) (and/or other member of this protein family), which interacts directly with the carboxy-terminal NLS signal of VirD2 and mediates its nuclear import. In animal and yeast cells, karyopherin á usually functions in a heterodimer with karyopherin β1. In this complex, the α subunit recognizes the NLS signal of the transported protein molecule, while the β subunit mediates docking of the entire NLS-karyopherin α/β complex at the nuclear pore and its interaction with Ran GTPase. It is still unclear whether karyopherin α1 is also involved in the nuclear import of VirD2 (c2) since, in a heterologous mammalian in vitro system, *Arabidopsis* karyopherin α has been shown to function alone, independently of karyopherin α1. Nevertheless, since *Arabidopsis* karyopherin α carries the karyopherin α-binding motif, but does not contain sequences known to be required for binding to the nuclear pore or Ran, it is highly likely that an as yet undiscovered karyopherin β1 is involved in VirD2 nuclear import. Indeed, in rice, karyopherin β1 has been isolated and shown to interact with karyopherin α, and an *Arabidopsis* mutant, knocked-out in a putative karyopherin β gene, is resistant to *Agrobacterium* transformation. Regardless of involvement of karyopherin β in VirD2 nuclear import, binding of AtKAPα to VirD2 at the 5'-end of the T-strand may orient the entire T-complex, initiating its directional nuclear import.

Since VirE2 does not bind directly to the host karyopherin α, but rather interacts with the host factor VIP1 which in turn binds to karyopherin α (c3), nuclear import of VirE2 and most of the T-complex molecule, may occur by a "*piggy-back*" mechanism. In this scenario, VIP1, which is a nuclear protein, interacts both with VirE2 and with the host karyopherin α. Thus, nuclear import of VIP1 leads to nuclear import of its interacting VirE2 and, by implication, of the T-complex. Although the involvement of other, as yet unidentified, plant factors, such as an hypothetical VirE2-specific karyopherin α or β, in VirE2 nuclear import cannot be excluded, the ability of the *Agrobacterium* VirE3 protein to replace VIP1 (b4) further supports the crucial need for an adaptor molecule between VirE2 and the host cell nuclear import machinery during the T-complex nuclear import; because VIP1 is not an abundant cellular protein, its function can be complemented, at least partially, by VirE3.

The combined action of AtKAPα and VIP1/VirE3 in the nuclear import of VirD2 and VirE2, respectively, further supports the notion of the T-complex polar translocation, since VirD2 and VirE2 do not directly interact with and compete for the same cellular proteins in their nuclear import. Polar translocation may be a common feature in the nuclear transport of many naturally occurring nucleic acid-protein complexes. For example, nuclear export of a 75S premessenger ribonucleoprotein particle in *Chironomus tentans* initiates exclusively at the 5'-end of the RNA.

Once inside the nucleus, perhaps even before the entire t-complex molecule has completely traversed the nuclear pore, the VirD2 NLS region may become dephosphorylated by PP2C (d1); this VirD2 dephosphorylation has been proposed to regulate its nuclear import. Within the cell nucleus, VirD2 may also interact with CAK2M and TBP (d2). Because both CAK2M and TBP are members of the plant RNA transcription machinery, their interactions with VirD2 may further guide the entire T-complex into the site of integration in the host chromosome. Similar to VirD2, VirE2 also associates with a putative member of plant transcriptional complexes, VIP1 (d3). In addition to facilitating VirE2 nuclear import, VIP1 may also function in the intranuclear transport of the T-complex, leading it to chromosomal regions where the host DNA is more exposed and, thus, more suitable for T-DNA integration. Here again, the combined, and noncompetitive action of VirD2 and VirE2, through their interaction with different host factors, may represent the molecular basis for the polar nature of T-DNA integration.

Future Directions

In this chapter, we describe the significant progress achieved in our understanding of nucleic acid transport during the *Agrobacterium*-host interaction. Major advances have been made in identifying host factors and their role in the nuclear import of VirD2, VirE2 and the T-complex. Further developments in this field will most likely come from the identification of additional cellular participants and regulatory components of the transport pathways. For example, plant molecular motors potentially involved in T-complex shuttling through the cell cytoplasm to the nuclear pore still need to be identified, and the mechanism by which the long T-complex molecule moves through the nuclear pore awaits better characterization. Perhaps the best way to achieve these goals is to combine biochemical, molecular, genetic, bio-physical and cell biological techniques. For example, identifying and characterizing plant mutants with altered susceptibility to *Agrobacterium* infection revealed the importance of two karyopherin proteins, as well as the possible involvement of the host cytoskeleton in the transformation process. In addition, bio-physical studies, using reconstituted frog nuclei, revealed the role of dynein-like motors in the microtubule-based movement of in vitro-formed T-complexes toward the cell nucleus. The importance of understanding the molecular mechanisms governing T-complex nuclear and intranuclear transport is difficult to overestimate. This fundamental knowledge will have a profound effect on our understanding of the general cellular mechanisms by which nucleic acids and proteins are imported into the nucleus and help us to design new strategies for the production of agronomically important plants resistant to *Agrobacterium*, and enable the development of improved genetic engineering procedures for the efficient nuclear delivery and integration of foreign genes.

INDEX

A
A. cordata, 138, 186
A. glutinosa, 186
A. rhizogenes, 65
A. rubra, 186
A. subcordata, 138
A. tumefaciens, 65, 288
A. woerlitzensis, 179
Abies, 188
Acca, 138
Acer, 187, 191
Acer pseudoplatanus, 219
Acer rubrum, 80
Achimenes, 90
Actinidia chinensis, 83, 137
Adiantum capillus-veneris, 128, 270
Aesculus, 179
Aesculus hippocastanum, 97, 104
Aggressors, 63
Agrobacterium, 239, 243, 244, 246, 251, 255, 262, 289, 292, 293, 294, 295, 296, 297, 298, 299, 300, 301, 302, 303, 304, 305, 306
Agrobacterium rhizogenes, 66
Agrobacterium tumefaciens, 65, 277, 284, 288, 292
Albizia lebbek, 95
Albizia richardiana, 210
Allium carinatum, 104
Allium cepa, 146
Alnus, 138
Alnus crispa, 186
Alstroemeria, 87
Amaranthus paniculatus, 172
Amelanchier, 149
Amelanchier spicata, 80
Amorphophallus rivieri, 172
Androgenesis, 49
Annona squamosa, 186
Anoectangium thomsonii, 168
Anthurium, 90, 93
Anthurium scherzerianum, 92
Anti-auxin, 32
Antirrhinum, 141
Antirrhinum majus, 132
Apical meristems, 17
Arabidopsis, 210, 242, 244, 246, 247, 248, 252, 253, 254, 258, 262, 264, 265, 266, 267, 270, 271, 273, 299, 300, 301, 302, 304, 305
Arabidopsis thaliana, 143, 262
Armoracia, 90

Aronia melancarpa, 226
Artocarpus heterophyllus, 209
Asclepias rotundifolia, 86
Asexual, 70
Asparagus, 81, 82, 123, 217, 235
Asparagus cladodes, 40
Atropa belladonna, 34, 97, 168, 229
Auromonas elodea, 232
Autogamous, 71
Autotrophic, 78
Axenic, 72

B

Basic helix-loop-helix, 267
Basic leucine-zipper, 252
Begonia, 171, 218, 225
Begonia franconis, 28, 131, 169
Begonia hiemalis, 153
Belt and braces, 171
Beta vulgaris, 138
Betula, 149
Betula nigra, 80
Bougainvillea, 226
Brassica campestris, 212
Brassica napus, 97, 178, 212, 216
Brassica nigra, 124, 137, 184, 193
Brassica oleracea, 138, 172
Bravo, 129
Bryophyllum, 184
Bryophyllum tubiflorum, 146

C

C. crenata, 187
Callus, 94
Callus colonies, 41
Camellia sasanqua, 220
Canna indica, 187
Capsicum, 262
Capsicum annum, 186
Cardamine pratensis, 136, 144
Carex flacca, 169
Carica, 94
Carica papaya, 97, 128, 175
Casein hydrolysates, 136
Castanea, 186, 208, 217
Castanea sativa, 150
Catasetum, 34
Catasetum trulla, 34
Catharanthus roseus, 118, 143, 172
Caulogenesis, 48
Cell suspensions, 94
Cephalotaxus, 218
Cercis canadensis, 149
Chalcone synthase, 252
Chamaecyparis obtusa, 139
Chelate, 162
Chenopodium rubrum, 36, 224
Chironomus tentans, 305
Chrysanthemum, 93, 129, 171, 192
Chrysanthemum morifolium, 87
Cichorium, 90
Cichorium intybus, 92
Citrus, 32, 41, 86, 90, 94, 95, 111, 112, 113, 124, 178, 180, 184, 196, 215
Citrus aurantifolia, 97
Citrus limon, 113
Citrus sinensis, 178
Cleavage, 107
Cleome iberidella, 131
Coconut water, 179
Cocos nucifera, 179
Coffea arabica, 130
Coleus blumei, 138
Colocasia esculenta, 96
Common plant regulatory factor, 254
Conditioning, 24
Convolvulus, 33, 90, 171
Convolvulus arvensis, 133
Coronilla varia, 140
Correa decumbens, 226

Corydalis yanhusuo, 211
Cucumis melo, 223
Cucumis sativus, 196
Cuscuta reflexa, 97
Cydonia, 149
Cymbidium, 82, 98, 194
Cynara scolymus, 231
Cytokinin, 32

D

Dahlia, 117, 144
Daphne odorata, 84
Datura, 32, 49, 56, 63
Datura innoxia, 48, 97, 216
Datura stramonium, 179, 182
Daucus carota, 52, 97, 134, 157, 179, 180, 188, 191, 223
De novo, 17, 106
Delphinium, 144, 191
Dendrobium, 129, 191
Dendrobium fibriatum, 149
Di-haploids, 13
Digitalis obscura, 202, 206
Dioscorea, 110, 232
Dioscorea alata, 205
Dioscorea deltoides, 222
Disanthus cercidifolius, 223
Disarmed, 293
Docks, 285
Doritaenopsis, 178
Drosophila, 243, 244, 251, 270, 298
Dumping, 285

E

Echinopsis, 203
Eleusine coracana, 93
Embryo sac, 49
Embryogenesis, 48
Embryoid, 48
Engineering, 65
Epidendrum obrienianum, 34
Epiphyllum, 90
Erwina herbicola, 68
Eucalyptus, 83, 155, 158
Eucalyptus ficifolia, 84
Eucalyptus globulus, 175
Eucalyptus gunnii, 192
Euchema, 235
Euphorbia fulgens, 174
Euphorbia longan, 210
Euphorbia peplus, 86
Explants, 9, 20
Extra vitrum, 77, 195, 203

F

Fast-growing calluses, 105
Feeder effect, 24
Feijoa, 138
Forsythia intermedia, 80
Fraxinus, 189, 204
Fraxinus pennsylvanica, 173
Freesia, 103
Friable, 35
Fuchsia hybrida, 210
Funaria, 148

G

G-box binding factors, 250
Gelidium, 230
Gelled, 6
Gene, 65
Gene engineering, 65
Gene therapy, 274
Gerbera, 90, 153
Gerbera jamesonii, 49, 171
Germinated, 29
Geum, 233
Gibberellins, 8
Ginkgo, 48
Ginkgo biloba, 179
Gladiolus, 110, 211, 212
Glehnia littoralis, 177

Gloxinia, 83
Glutamate dehydrogenase, 126
Glycine max, 139, 153, 174, 186, 209
Glycyrrhiza echinata, 177
Gossypium, 127
Green fluorescent protein, 253
Gymnocladus dioicus, 80, 85
Gynogenesis, 50

H

Habituation, 9
Haplopappus gracilis, 127, 133, 155, 161
Haworthia, 173
Helianthus, 190
Helianthus annuus, 157
Helianthus tuberosus, 193
Hemerocallis, 144, 191
Hevea, 226
Hevea brasiliensis, 153, 211
Homozygous, 71
Hordecale, 56
Hordeum distichum, 35
Hosta, 80, 90, 95
Humulus, 82
Hyoscyamus, 49
Hyperhydric, 86, 121

I

Immature T-complex, 293
In ovulo embryo culture, 31
In vitro, 1, 131
Indica, 188
Intensity, 77
Ipomoea, 143, 219
Ipomoea batatas, 210
Iris, 144
Isatis tinctoria, 34

J

Japonica, 188
Juglans regia, 179

K

Kalanchoe, 184
Kalanchoe blossfeldiana, 81

L

L. austricum, 56
Lapageria rosea, 223
Larix deciduas, 157
Lemna perpusilla, 168
Less active, 105
Leucospermum, 236
Lilium, 90
Lilium auratum, 205, 227
Linum, 177
Linum perenne, 56
Linum usitatissimum, 97
Lolium multiflorum, 104
Low fluence responses, 253
Lupinus alba, 82

M

Macleaya cordata, 40
Macro-methods, 71
Macro-propagation, 71
Magnolia, 127
Magnolia soulangiana, 225
Major plant nutrients, 114
Malus, 171, 188, 233
Malus domestica, 80, 233
Mature T-complex, 293
Medicago sativa, 95, 126, 147
Megaspore, 49
Meristemoids, 91
Meristems, 19
Mesembryanthemum crystallimum, 146
Metroxylon sagu, 235
Microplants, 19
Micropropagation, 19, 72
Mitosis, 70
Mixed cultures, 80
Mixotrophic, 78

Morphogenesis, 17
Morphogenetic, 48
Morphogenic, 48
Morphology, 18
Multiple shoots, 88
Musa accuminata, 188
Mutarotases, 242

N

N. petunoides, 33
N. tabacum, 33, 126
Nasturtium officinale, 34
Nautilocalyx leaf, 226
Nemesia strumosa, 188
Nerine bowdenii, 196
Nicotiana, 31, 33, 49, 63, 93, 195, 206, 216, 226
Nicotiana glutinosa, 203
Nicotiana rustica, 82
Nicotiana sylvestris, 43
Nicotiana tabacum, 52, 120, 133, 136, 172, 188, 207
Nipple flasks, 23
Node culture, 87
Nodules, 91
Nuclear export signal, 257, 260
Nuclear localization signals, 239, 287, 297
Nuclear pore complex, 240, 258
Nurse cells, 39
Nurse cultures, 142
Nursed, 39

O

Ocimum, 169
Oncidium varicosum, 34
Organized, 91
Organogenesis, 17
Organogenetic, 48
Organogenic, 48
Oryza sativa, 132
Ostericum koreanum, 228

P

P. echinata, 171
P. syringae, 69
Panax ginseng, 211
Panicum miliaceum, 180
Papaver somniferum, 56
Paphiopedilum, 187
Paphiopedilum ciliolare, 178
Passage, 26
Passiflora, 229
Paul's Scarlet, 125, 126
Pelargonium, 31, 46, 101, 103, 226
Pennisetum purpureum, 97
Peptide nucleic acid, 281
Petunia, 50, 52, 63, 143, 188, 229
Petunia hybrida, 188
Phaseolus, 186
Phosphorothioate bodies, 291
Physcomitrium pyriforme, 207
Physiology, 18
Phytochromes, 264
Picea, 190
Picea abies, 195
Picea glauca, 133
Pictacia, 149
Piggy-back, 247, 279, 302, 305
Pinus, 159, 202
Pinus coulteri, 195
Pinus pinaster, 121
Pinus radiata, 234
Pinus strobus, 129, 171, 233
Pinus taeda, 161, 174, 182
Pistacia, 79, 149
Pisum sativum, 92
Plantago ovata, 235
Plantlets, 29, 75
Plasmodesmata, 209
Poinsettia, 223
Polyethylene glycol, 44, 108
Poncirus, 90

Populus, 149
Post-zygotic, 30
Pre-zygotic, 30
Primary callus, 35
Prostanthera striatifolia, 226
Protocorm, 95
Protocorm-like bodies, 96
Prunus, 150, 152, 160, 161
Prunus amygdalus, 137
Prunus avium, 150
Prunus glandulosa, 138
Prunus tenella, 149
Pseudomonas elodea, 232
Pseudomonas syringae, 68
Pseudotsuga, 202
Pseudotsuga menziesii, 137, 174, 179
Psidium guajava, 149
Pteridium, 207
Pyrus, 141

R

R. xanthina, 196
Radicula aquatica, 81
Ranunculus sceleratus, 97
Raphanus sativus, 208
Recurrent somatic embryogenesis, 97
Reduced, 119
Relative humidity, 237
Repetitive embryogenesis, 97
Reseda luteoli, 126
Rex aquifolium, 97
Rhizobium, 33, 118
Rhizogenesis, 48
Rhododendron, 93
Ribes nigrum, 226
Robinia, 91
Robinia pseudoacacia, 33
Rosa, 84, 126, 130, 194, 223
Rosa hybrida, 149
Rosa persica, 196
Rumex acetosa, 217
Rumex acetosella, 210

S

S. cerevisiae, 262
S. khasianum, 30
S. pombe, 262
Saccharum officinarum, 210
Saintpaulia, 90
Saintpaulia leaf, 89
Santalum, 226
Schizosaccharomyces pombe, 248
Secondary shoot primordia, 93
Self-fertilizing, 67
Sellowiana, 138
Semi-organized, 91
Senecio hybridus, 129
Senescence, 32
Sequoia, 149
Sequoiadendron giganteum, 149
Sexual, 70
Shamouti, 105, 106
Shoot tip culture, 79
Silene alba, 136, 144
Sinapis alba, 208
Sinensis, 138
Sinningia, 90
Smooth muscle gamma actin, 281
Solanaceae, 52
Solanum, 30, 49, 63
Solanum curtilobum, 92
Solanum melongena, 30, 33
Somaclonal variants, 20
Somatic embryo, 48
Somatic embryogenesis, 19
Sphingomonas paucimobilis, 232
Sporophytic, 49
Stationary phase, 10
Stellaria media, 81
Stevia, 93
Stevia rebaudiana, 92

Streptocarpus, 90
Suction pressure, 197
Super Yellow, 129
Supradomes, 93

T

T-complex, 293
Taraxacum, 90
Taxus, 174
Test tubes, 17
Theobroma cacao, 97, 104
Threonine deaminase, 138
Thuja orientalis, 97
Thymidylate kinase, 280
Tipping, 79
Tissue culture, 19
Todea barbara, 96
Torenia, 139, 149
Torenia fournieri, 131, 139
Totipotency, 2, 20
Tradescantia, 251
Trifolium pratense, 158
Trifolium repens, 97
Trigonella, 137
Triticale, 184
Triticum, 56
Triticum aestivum, 140
Tropaeolum majus, 82
Trueness to type, 83
Turgor pressure, 197

U

Ulmus, 149
Undefined, 115, 170

V

V. riparia, 173
Vanilla planifolia, 34
Very low fluence responses, 253
Vicia faba, 224
Vicia hajastana, 25, 184
Vigna aconitifolia, 210
Vitamin, 172
Vitis, 155, 161, 226
Vitis hybrids, 97
Vitis vinifera, 97, 173

W

Wheat germ agglutinin, 241
Wolffia microscopica, 168

X

Xenopus, 240, 243, 244, 249, 251, 257, 275, 289, 298
Xenopus laevis, 283

Y

Yeast extract, 176

Z

Zamia integrifolia, 97
Zamia latifolia, 139
Zea mays, 88, 140, 179, 210
Zoysia japonica, 174, 175